NSK

H

Vegetable Crops

Dennis R. Decoteau

The Pennsylvania State University

Prentice Hall
Upper Saddle River, NJ 07458

Library of Congress Cataloging-in-Publication Data

Decoteau, Dennis R.
    Vegetable crops/Dennis R. Decoteau.
        p. cm.
    Includes bibliographical references.
    ISBN 0–13–956996–0
    1. Vegetables. 2. Truck farming. I. Title

SB321 .D393 2000
635—dc21   99–045819

Publisher: *Charles E. Stewart, Jr.*
Associate Editor: *Kate Linsner*
Managing Editor: *Mary Carnis*
Production Liaison: *Eileen O'Sullivan*
Production Editor: *Lori Harvey, Carlisle Publishers Services*
Director of Manufacturing & Production: *Bruce Johnson*
Manufacturing Manager: *Ed O'Dougherty*
Marketing Manager: *Ben Leonard*
Cover Design: *Miguel Ortiz*
Interior Design: *Carlisle Communications, Ltd.*
Formatting/Page Make-up: *Carlisle Communications, Ltd.*
Printer/Binder: *Courier Westford*

Printed in the United States of America

10 9 8 7 6 5 4 3 2 1

ISBN 0-13-956996-0

Prentice-Hall International (UK) Limited, *London*
Prentice-Hall of Australia Pty. Limited, *Sydney*
Prentice-Hall Canada Inc., *Toronto*
Prentice-Hall Hispanoamericana, S.A., *Mexico*
Prentice-Hall of India Private Limited, *New Delhi*
Prentice-Hall of Japan, Inc., *Tokyo*
Pearson Education Asia Pte. Ltd., *Singapore*
Editora Prentice-Hall do Brasil, Ltda., *Rio de Janeiro*

To Chris and Megan

Contents

Contents

vii

▬ PART 3   SPECIFIC VEGETABLE CROP CHARACTERISTICS AND GROWING PRACTICES   187

Contents   xxiii

Preface

Experimenting in science and/or dabbling in art, feeling good about growing and caring for a living organism, having access to more nutritious food, and supporting a large industry that employs many people and on which the nation is partially dependent for its well being—all of these experiences and others encompass the interest and passion involved in the growing of vegetables.

The commercial vegetable industry is a highly sophisticated enterprise that contributes to a safe and plentiful food supply. A safe and plentiful food supply is one of the nation's most important national security issues and vegetable crops certainly play an important role in this issue. Allied industries, besides those directly involved with growing such as supplying fertilizers, chemicals, and equipment, also contribute to the economy of the nation.

The growing of vegetable crops in the home garden provides unlimited enjoyment and satisfaction and fulfills many of our daily needs for nutrition and exercise. Gardening is often mentioned as the number-one pastime in the United States. Vegetable gardens obviously play an important role in the positive view of gardening by the public. Health and human nutrition are becoming more important topics in society, and the health benefits of vegetables are being cited more often by consumers as an important factor in determining what foods they eat.

Vegetable Crops provides an overview of the principles and practices of growing vegetables crops. While the emphasis is on commercial production, much of the information should also be useful to the backyard and weekend gardener. While the topic of vegetable production is large, this book attempts to highlight important concepts in a condensed but informative manner.

Key Features

Vegetable Crops is divided into three main parts. Part I presents a general overview of the topic of vegetable crops with discussions on the history of vegetables, the vegetable industry, classifying vegetables, and how environmental factors affect crop growth. Part II highlights general practices that may be used in the production of most vegetable crops including discussions on preparing the field, choosing cultivars, planting, watering, extending the growing season, handling pests, and harvesting and marketing. It also includes discussions on the uses and nutritional benefits of eating vegetables. Part III presents botanical and production information for specific vegetable crops. These crops are grouped for discussion in a pragmatic, convenient, and generally accepted manner according to characteristics of family grouping and crop use.

Target Audience and Uses of the Book

This book is designed for use in vegetable crop production-type courses that may be taught at universities, junior colleges, or community colleges. It also should serve as a reference book for vegetable growers, researchers and Cooperative Extension Service personnel, home gardeners, high school vocational/technical schoolteachers, and others who desire information on vegetables. It is envisioned that the actual uses of this book will be as varied as the list of topics discussed.

While the primary intended audience for this book are students enrolled in a one-semester Vegetable Crops course, the actual design of this book was based on a two-course sequence that was taught for many years at Clemson University. Parts I and II contain many of the general topics discussed in the first course and Part III is the specific crop information that was typically the basis of the second three-credit course. In later years the two courses were combined into one four-credit course with some of each course serving as the basis of the combined course.

Production information presented for crops discussed in this book was primarily gleamed from various Cooperative Extension publications, research findings, and professional experience. Successful practices used for growing specific vegetable crops often vary from location to location and even farm to farm making it nearly impossible to fairly represent in one book all practices that may be successfully used for producing a crop or to define practices for all potential locations and environments. While specific crop production information (i.e., cultivars, fertilizer rates, and plant spacing) is presented (especially within the chapters of Part III), this information is for discussion purposes and is not intended to serve as a substitute for crop recommendations from the local State Cooperative Extension Service or grower experience. It is important that Cooperative Extension publications and personnel be consulted for specific location and up-to-date recommendations for a crop of interest before any planting is considered.

Acknowledgments

My colleagues at The Pennsylvania State University and Clemson University are gratefully thanked for their support and friendship through the years. My thanks go to the many students that I taught while I was on the faculty at Clemson University for providing me with the energy and experiences to learn and gain confidence in my teaching abilities. I also would like to acknowledge Dr. Roy Ogle, Professor Emeritus at Clemson University, from whom, upon his retirement in 1987, I inherited the Vegetable Crops teaching program at Clemson University. Dr. Ogle graciously provided me with many teaching resources (including some of the pictures and slides used in my classes and included this book) so that I could carry on the tradition that he helped establish of providing a quality education for students of Vegetable Crops at Clemson University. The following reviewers provided valuable feedback and constructive ideas during manuscript preparation: Steve J. Bell, Merced College; Ann Emmsley, Maui Community College; and William S. Sakais, University of Hawaii.

Finally, I would like to thank all the editors and professionals at Prentice Hall for helping me pull this together and keeping me on track.

About the Author

Dr. Dennis R. Decoteau is a Professor and the Head of the Department of Horticulture at The Pennsylvania State University. He received his B.S. degree in Environmental Studies from the University of Maine at Fort Kent and his M.S. and Ph.D. degrees in Plant Science/Physiology from the University of Massachusetts. Dr. Decoteau has won numerous teaching and research awards, including the Outstanding Teacher Award in Horticulture from Clemson University, the Outstanding Research Paper on Teaching from the American Society for Horticultural Sciences, the L. M. Ware Distinguished Research Award from the Southern Region American Society for Horticultural Science, and the IV Congreso International De Neuvas Technologias Agricolas Award. He has taught several undergraduate and graduate courses, including Vegetable Crops. He has written extensively in the scientific press and for the general public.

OVERVIEW OF VEGETABLE CROPS

chapter
1

History of Vegetable Crops

History of Mankind

The human race came into existence about 2 million years ago, but *Homo sapiens* (thinking man) has only existed for some 60,000 years. It is believed that our earliest ancestors were food-gatherers that spent most of their time hunting animals, catching fish, and collecting edible plants. Ancient cave art often depicted important plants that were used for food. Plants also appear to have been used by primitive man for uses other than food, since flowers for ornamental purposes were included in some of the burials of 60,000 years ago.

About 10,000 years ago, man changed from primarily gathering food to producing food. Eventually, through cultivation of plants and domestication of animals, humans were able to produce sufficient food to free them from the constant search for their next meal. With the development of a dependable food source, humans began to live together and villages and towns came into existence.

Food production became an activity that all members of the family could participate in. Members would assist, in some way, in the planting, weeding, and harvesting of their food plants. As the food supply was expanded to feed larger populations and even to provide surpluses, trade and occupational specialization occurred and jobs

evolved that provided man with resources, such as money or materials, as payment for his efforts, which he didn't need to develop for himself.

It is suspected that humans were in the Americas as early as 15,000 years ago. The first Americans probably crossed the Bering Strait from Asia during the most recent Ice Age. It is also suspected that the first Americans followed large game during the game's migration and the humans settled where large numbers of game were present.

History of Agriculture and Horticulture

Evidence indicates that agriculture and horticulture originated during the Neolithic Age (ca. 8000 B.C.) in the semiarid mountainous regions near the river valleys of Mesopotamia, in present-day Iraq. About that time, men and women began collecting grain plants and keeping domesticated animals such as sheep and dogs. Wheat and barley seeds dated circa 6750 B.C. have been found in evacuations in Iraq. Early centers of horticultural development included the Near East, Southeast Asia, Meso-America, and possibly Western Sudan.

The further domestication of plants and animals paralleled the development of villages, towns, and cities along the Nile circa 3500 B.C., in present-day Egypt. The Egyptians are also credited with developing technologies for food storage, such as pickling and drying, and the agricultural technologies of drainage, irrigation, and land preparation.

The Romans, circa 500 B.C. to A.D. 500, developed efficient agricultural production systems that became the cornerstone of their strength. They effectively copied or borrowed anything they thought was beneficial for their society from the people that they conquered. Pliny the Elder (A.D. 23–79) produced a major work describing Roman agriculture and the use of crop varieties, or cultivars, and legume rotation.

During the collapse of the Roman Empire (ca. A.D. 500) and before the Renaissance, monasteries became important reservoirs for horticultural and agricultural skills perfected by the Romans. Monasteries often had cultivated fields of grain and vegetables as well as orchards. The monks in the monasteries maintained important collections of herbal and medicinal plants. Many of the plants kept in cultivation in the monasteries would become important later during the Renaissance.

Since many vegetables had their earliest uses as medicinal, herbals are some of the earliest and most important manuals about vegetables. The origin of these early plant books is traced to the Greek interest in cataloging and describing plants. As originators of the study of botany, the Greeks produced writings that listed common plants and often supplied their medicinal usage. For much of the Middle Ages, there was little distinction between medicine and botany, as plants were used to cure ills.

The greatest authority of medicinal plants up to the Renaissance was the Greek physician Pedanius Dioscorides (ca. A.D. 20). Dioscorides was the most authoritative writer in botany for 16 centuries and his work *De Materia Medica* is considered one of the most important herbals of all time. He traveled widely and studied plants wherever he went. For over 1,500 years his work was considered the final authority on the pharmaceutical uses of herbs and plants. His work described roughly 500 plants and 1,000 different medications.

The development of the printing press resulted in a greater demand for herbals, and the first half of the sixteenth century saw many of them translated into common languages such as French, German, Italian, and English. The Latin and Greek versions were printed and reprinted as well, and served as references alongside the popular contemporary versions.

By the second half of the sixteenth century botany had established itself as a science. Not only were the drawings of the plants more accurate, but with the exploration of new worlds many new plants were discovered and cataloged. It was fashionable during that period for the aristocracy throughout Europe to maintain large gardens full of exotic plants, and to have extensive collections of herbals in their libraries.

During the Renaissance, John Gerard (1545–1612) wrote the *Herball*, which was first published in 1597 and became the best known of the English herbals. Gerard was trained as a barber-surgeon, steadily rising in reputation in his profession while living in London and working as superintendent of Lord Burghley's London and Hertfordshire gardens. He also maintained a large garden of his own and eventually became herbarist to King James I.

The First Cultivated Plants

The tribes in the Old World that engaged in hunting and in gathering wild edible plants made attempts to domesticate dogs, goats, and possibly sheep as early as 9000 B.C. Cultivated plants originated in more than one ancient center of origin with centers of origin of major cultivated crops in both the Old and New Worlds (Table 1.1). A plant's center of origin is the geographical area where a species is believed to have evolved through natural selection from its ancestors. This is also the plant's center of diversity in which a pool of genes exists for use by plant breeders in vegetable crop improvement programs.

Earliest evidence of plant cultivation in the Americas appears to be about 10,000 years ago in South America, where potatoes and a variety of other crops were domesticated. Agriculture probably began in Mexico circa 5000 B.C. Early cultivated plants included maize, avocado, squash, and chili peppers. Research also indicates that agriculture began very early in the Andean valleys of Peru. It seems probable that agriculture originated independently in at least two regions in the New World—Mexico and Peru—and the earliest site may have been Peru.

Maize (corn) appears to have been introduced into North America from South America about 1,700 years ago and was a relatively minor crop until about 800 years ago. Different varieties of maize with different cob shapes and size appeared in different regions. Europeans recorded the existence of short- and long-season corn varieties, but it is not known when or how Native Americans developed these varieties. Also about this time, the common bean was introduced.

Domestication

All of the modern crops that we grow and eat today had their earliest beginnings as wild plants. Because of genetic variation, as a result of sexual reproduction, some wild plants may have developed characteristics that attracted humans to select and eat them instead of other available plants. As a result of eating the plants, seeds and other plant parts were dispersed. This was the beginning step in the process of domestication.

It has been suggested that human latrines may have been testing grounds for the first crop breeders, since these may have been primary spots where seeds of ingested

TABLE 1.1   *Centers of origin of important cultivated plants and commercially important vegetable crops that originated from these centers*

Old World

I. China—The mountains and adjacent lowlands of central and western China represent the earliest and largest centers of origin of cultivated plants and world agriculture. Important vegetable crops that originated from this center include radish, eggplant, several legumes, Brassicas, onions, and various cucurbits.

IIA. Indian (India and Burma)—Important vegetable crops that originated from this center include many legumes and gourds.

IIB. Indo-Malayan (Indochina, Malaysia, Java, Borneo, Sumatra, and Philippines)

III. Central Asia—Important vegetable crops that originated from this center include garden pea, broad bean, carrot, radish, garlic, spinach, and mustard.

IV. Near Eastern (Asia Minor, Iran, Transcaucasia, and Turkmenistan Hoghlands)—Important vegetable crops that originated from this center include cabbage, lettuce, and muskmelon.

V. Mediterranean—Important vegetable crops that originated from this center include beet, parsley, leek, chive, celery, parsnip, and rhubarb.

VI. Abyssinian (Ethiopia and Somaliland)—Important vegetable crops that originated from this center include okra and garden cress.

New World

VII. South Mexican and Central America (South Mexico, Guatemala, El Salvador, Honduras, Nicaragua, and Costa Rica)—Important vegetable crops that originated from this center include corn, common bean, lima bean, sweet potato, pepper, and cherry tomato.

VIIIA. South American (Peru, Ecuador, and Bolivia)—Important vegetable crops that originated from this center include potato, tomato, and pumpkin.

VIIIB. Chiloé (South Chile)—An important vegetable crop that originated from this center is the white potato.

VIIIC. Brazilian–Paraguayan

Adapted from Acquaah, G. 1999. *Horticulture: Principles and Practices.* Upper Saddle River, NJ: Prentice Hall; and Janick, J. 1979. *Horticultural Science.* 3d ed. San Francisco: W. H. Freeman. Used with permission.

fruits would have been deposited. These were also areas where the soils had greater nutrient concentrations than in nonlatrine areas.

Garbage dumps where food scraps were scattered may have also played a role in early plant domestication and growing. Spoiled or rotten fruit would have been placed in these areas and the seeds from these fruits could have germinated. Also, seeds of consumed fruits may have been deposited into these areas and provided another source for future plant generations.

During the 10,000 or 11,000 years that have passed since the beginning of plant domestication, the plants that humans selected as useful to them have undergone profound changes. Artificial selection of plants during the ages differs considerably from natural selection. During natural selection the plants have properties that preadapt them to a wide variety of environmental conditions, hopefully ensuring continuation of the species. Artificial selection breaks down these stabilized systems, creating gene combinations that possibly could not survive in the wild.

The Importance of Dispersal in Plant Domestication

Seed dispersal is one of the more important events to occur for early plant domestication. Dispersal may occur by the plants or by their seeds being blown about in the wind or floating in the water. Some plants are carried by animals by enclosing the seeds within tasty fruit and advertising the fruit ripeness by color or smell. An excellent example of this is the muskmelon.

Other plants produce fruit that is customarily eaten by a particular animal. Strawberries are often eaten by birds, acorns by squirrels, and mangoes by bats. In each of these cases, the animal eating the fruit effectively disperses the seeds.

Other methods of seed dispersal are unintentional. Many wild plants have specialized mechanisms that scatter seeds and generally make them unavailable to humans. An example of this is peas. Pea seeds must be removed from the pod to germinate. Most wild peas have pods that explode (split open) when the seeds within the pod are mature. During pod collection by humans it was primarily only the pods that didn't explode that were collected, and this nonexploding trait was passed on during domestication as a result.

For survival of plant species it was advantageous that seeds produced by the plant had a scattered germination rate (i.e., they didn't germinate at one time). This resulted in plants not having all their seeds germinating at a time when possibly the climate or some other factor may not have been conducive for early plant growth and development. Thick seed coats contributed to seeds undergoing a scattered germination rate. Many ecologically advantageous traits such as thick seed coats have been reduced or removed during domestication to improve the efficiency of seedling establishment and plant growth in large-scale field plantings.

Why Did Some Plants Take Longer to Domesticate?

During the development of agriculture and plant domestication some plants were domesticated earlier or easier than others. Around 10,000 years ago, the earliest domesticated plants appear to have been the Near Eastern crops of cereals and legumes such as wheat, barley, and peas. These crops may have been domesticated earlier because they came from wild ancestors that had many characteristics that were advantageous for the process of domestication. Some of these advantages included seeds that were edible in the wild and could be readily stored, and plants that were easily grown from sowing grew quickly and were self-pollinated. Overall, few genetic changes were required of these crops to go from wild plants to domesticated plants.

Fruit and nut domestication probably began circa 4000 B.C. The fruit and nut crops typically are not harvested until 3 to 5 years after planting. In order to grow these types of crops, people needed to be committed to settled life and could no longer be seminomadic. Fruit and nut crops are often grown by cuttings and this had the advantage that the progeny or descendants of the original plants were identical to the original plant. Some fruit trees cannot be grown from cuttings and it was found to be a waste of effort to grow them from seed. Instead, grafting techniques had to be discovered, perfected, and contributed before suitable domestication could occur.

Criteria Used by Early Plant Gatherers during Domestication

Early plant gatherers applied self-serving criteria to decide which plants to gather. Of utmost importance was size. As a result of continually choosing the largest fruit of the wild plants through the years, many of the crops that underwent domestication have bigger fruits than their wild ancestors.

Another criterion used by plant gatherers was taste. Many wild seeds are bitter tasting, yet their fruits are sweet and tasty. This was important so that seed would not be chewed up and eaten but instead would be expelled. Other criteria for gathering were fleshy or seedless fruit, oily fruits and seeds, and plants with fiber for clothes, and so on.

Adaptive Plant Structures in Vegetable Crops

There is no one plant structure that is consumed as the edible plant part across all the vegetable crops. Instead, the marketable plant part for a particular vegetable crop could be a stem, leaves, root, or flower. As a result it is necessary to understand the basic purposes of the specialized plant structures in the life of an average plant.

Stems

Stems are the supporters and producers of leaves and flowers. A primary function of the stem is to distribute the leaves in space so that the leaves can intercept sunlight for photosynthesis. Stems can also store food and provide for the movement of water and nutrients to and from the leaves. An example of a crop in which the stem is used as the edible plant part is asparagus.

In most vegetable crops the stems are above ground and are called aerial stems. These aboveground stems may be rigid (erect) or flexible and form climbing stems or vines. The stem is made up of distinct areas where leaves and/or buds are attached. These areas are called nodes. The area between two successive nodes is called an internode. Stems usually possess a terminal bud at the tip and axillary buds in the axils of each leaf. The axillary buds may produce stems or flowers.

Some stems undergo modifications and appear as thorns, prickles, or tendrils. Some stems also occur as stolons, producing at intervals erect stems and adventitious roots as a modification of the horizontal or prostrate stems. The rhizomes of some plants such as potato may be thick and fleshy or partially thickened and called tubers. Some plants are propagated by short, stout, erect underground stems in which food is stored. Such a stem is broader than it is long and is known as a corm. Bulbs such as onions have layers of thickened leaves (bud scales) that surround a short, erect stem.

Leaves

On the surface of stems are a number of structures. The most prominent structures are the leaves. The leaves are attached at nodes and may be broad or flat, narrow and elongated, or needlelike.

Leaves are appendages of the stem that have the primary purpose of carrying on photosynthesis. However, some leaves are not photosynthetic, and even photosynthetic leaves have other functions such as storage of food and water, reproduction, root formation, climbing, protection, or flower formation. The leaves of many economical vegetables (spinach, onion, and cabbage) are useful because they store water, salts, food, materials, and vitamins.

Roots

Most roots function to anchor plants and to absorb water and nutrients from the soil. In some cases they also serve as food storage reservoirs. There are two basic types of roots—fibrous and taproots. In the taproot system, the primary root is enlarged and the secondary roots are few and slender. The type of root system and size varies with different species of plants and also with environmental factors such as moisture and soil composition. The enlarged taproots of carrots and beets and the enlarged lateral roots of sweet potatoes are examples of root plant parts that are consumed as the edible portion.

Flowers

Flowers provide plants with the necessary apparatus to carry on sexual reproduction. Sexual reproduction is the foundation for evolution and crop improvement through genetic changes. There are several vegetables in which immature flowers or flower parts are the edible and marketable plant parts. These include broccoli, cauliflower, and Brussels sprouts. Edible vegetable flowers, such as squash flowers, are also becoming more popular.

Vegetable Crop Improvement Programs

Improvement to vegetable crops has been a continuous process since humans began collecting and then growing food. In the process of primitive man selecting the plants he desired and eventually consumed, he passed on these chosen plants and the characteristics that made them be selected to the next set of plants that grew. His criteria for selection may have been fruit size, color, taste, fast rate of growth, resistance to diseases or insects, or simply the lack of toxins to humans. This process of selection was effective, as many of our contemporary cultivated vegetables no longer resemble their primitive ancestors.

Plant breeding as a science is the systematic improvement of plants (Figure 1.1). Plant genetics is the study of the mechanisms of heredity of plant traits and is the

FIGURE 1.1   Small plot field evaluations are used to determine the potential usefulness of new cultivars from plant breeding programs for production in local areas.

underlying science of plant breeding. Genetics as a science owes a tremendous amount of its success to the study of vegetable crops, since Gregor Mendel (1822–1884) did the pioneering work on genetic inheritance in 1865 using garden peas. Today's vegetable breeding and genetic programs continue to utilize the basic concepts of genetics and controlled plant crosses. Crossing is the transfer of genes (in pollen) from one plant to another.

Some plants are self-pollinated, which means that the flowers of the plant use their own pollen for pollination. Many annual vegetables (peas, lima beans, and chick peas) are self-pollinated. Seeds from these plants produce uniform plants very much like the parents. Mass selection and pedigree selection are breeding methods for self-pollinated crops. Mass selection involves selecting the best plants based on appearance (phenotype). Pedigree selection is conducted after creating variability by controlled crossing of two parents.

Cross-pollination is the process of transferring pollen grains from one plant and depositing them on the stigma of the flower of a different one. Seeds from plants that cross-pollinate (corn, cabbage, and spinach) produce hybrid (nonuniform) plants. Hybrid vigor or heterosis refers to the increase in vigor shown by certain crosses as compared to that of either parent. Many cross-pollinated horticultural crops, such as apple, are also vegetatively propagated. Mass selection, recurrent selection, and other methods of breeding may be applied to cross-pollinated species.

Many of today's vegetable crop improvement programs also utilize the modern concepts of molecular genetics, tissue culture, and genetic engineering (Figure 1.2). Genetic engineering is the artificial manipulation or transfer of genes from one organism to another. Genes may be directly transferred to plants using either crown gall bacterium (*Agrobacterium tumifaciens*) as a vector or by bombarding genes into plants using a particle gun. The tomato has been successfully genetically engineered for increased fruit quality. The commercially available genetically engineered tomato cultivar "Flavr Savr" was developed by Calgene (a biotechnology company)

FIGURE 1.2   Tissue culture and genetic engineering are common tools used in modern vegetable breeding programs.

to vine ripen before harvest. These tomatoes have better flavor than the traditional green-harvested tomatoes that are fumigated in storage to enhance ripening. Transgenic tomato plants have also been developed that are resistant to herbicides such as glyphosate.

Review Questions

1. What role did the evolution of ancient man going from primarily food-gatherer to food producer have in the development of villages and towns?
2. What was the role of monasteries and herbals in contributing to our contemporary knowledge of plant uses?
3. What are some of the methods of seed dispersal that are used by plants?
4. What were some of the criteria used by early plant gatherers in selecting plants?
5. Why did cereal and legume crops undergo domestication easier and earlier than fruit and nut crops?
6. What is plant breeding?

Selected References

Acquaah, G. 1999. *Horticulture: Principles and practices.* Upper Saddle River, NJ: Prentice Hall.

Janick, J. 1979. History's ancient roots. *HortScience* 14:299–313.

Janick, J. 1979. *Horticultural science.* 3d ed. San Francisco: W. H. Freeman.

Norstog, K., and R. W. Long. 1976. *Plant biology.* Philadelphia: W. B. Saunders.

Selected Internet Sites

www.aces.uiuc.edu/~sare/history.htm/   A Brief History of Agriculture, University of Illinois College of A.C.E.S.

www.agr.ca/backe.htm/   Agriculture and Agri-Food Canada: History, Agriculture, and Agri-Food Canada.

www.usda.gov/history2/back.htm   History of American Agriculture 1776–1990, USDA Economic Research Service.

chapter

2

Understanding the Vegetable Industry

History

The share of consumer expenditures for food in the United States is the lowest in the world (approximately 12% of the disposable income for an average consumer). This is due in no small part to efficient and progressive agricultural industries that have developed through the years. These industries have evolved from the primitive earliest food production systems developed by the colonists and settlers for survival to the highly sophisticated systems of today's producers that supply a wide variety and relatively consistent supply of food for sale nationally and internationally.

Colonial Period

American colonists were largely self-supporting, growing vegetables for their own use. Early settlers endured a rough pioneer life while adapting to new environments. Small family farms predominated, except for some relatively large plantations that were developed in the southern coastal areas.

The settlers used simple tools such as wagons, plows, harrows, axes, rakes, scythes, forks, and shovels. All seed sowing was done by hand. Most of the early settlers lived near forested areas, which provided wood for housing, fencing, and fuel.

In 1819, Jethro Tull helped revolutionize farming by developing and patenting an iron plow with interchangeable parts. By the 1820s and 1830s farmers, blacksmiths, and other innovators introduced various modifications to the plow that provided a sharper and stronger cutting edge with smoother surfaces so the soil did not stick to either the plowshare or the moldboard. John Deere in partnership with Leonard Andrus began manufacturing steel plows in 1837.

Most early settlers planted a mixture of crops both for home consumption and for market. Field corn was often a mainstay because it gave high reliable yields. Diversification was also needed to supply the various food components for the home table, and vegetables were a mainstay of these plantings.

Manufacturing was making its way into the cities and into agriculture by the 1850s. The 2-horse straddle row cultivator was patented in 1856. The change from hand power to horses characterized the first American agricultural revolution.

Post–Civil War

Many products that were developed prior to or during the Civil War increased the productivity of each laborer in harvesting, planting, and cultivating fields. In 1800, approximately 75% of the population was directly engaged in agricultural production. By 1850, it was less than 60%, and by 1900 less than 40% of the population was engaged in agricultural production.

During these times the number of farms began to decline and farm size increased. As both farming and manufacturing became more productive, they provided people with a broader range of career options in such fields as medicine, law, science, government, and entertainment. The increased productivity of farmworkers led to surpluses of agricultural products and thus lower prices. This affected the livelihood of many farmworkers and the supply of available labor at times exceeded the demands, resulting in unemployment. By the 1890s, agriculture had become increasingly more mechanized.

Industrial expansion, which began about 1895, established concentrated population centers or cities. Agricultural improvements made more food available to support a larger nonagricultural population. These large population centers became largely dependent on special producers for their food supply, and, as a result, commercial production of vegetables developed near population centers.

During the 1910s, big, open-geared tractors came into use in areas of extensive farming. In 1926, a successful light tractor was developed and during the 1930s the all-purpose, rubber-tired tractor came into wide use. By the 1930s, 58% of all farms had cars, 34% had telephones, and 13% had electricity.

Pre–World War II

For much of the early 1900s, American agricultural policy was guided by the philosophy that society would best be served by traditional family-size, owner-operated farms. These family farms relied heavily on local labor, supplies, and consumers to sustain their business.

Post–World War II

After 1945, the rapid transition to volume marketing systems began. The improvement in technology in food handling systems (including refrigeration) developed during wartime, and the changing economic structure of American agriculture, spurred by the highway expansion of the 1950s, favored those growers who could supply the market with a large volume over a prolonged period of time. Small farms were inefficient and many either failed or enlarged to meet new challenges. Growers enlarged through purchases of additional land or through production/marketing cooperatives. They maintained competitiveness by adopting new technology or by stressing high quality in vegetable production and handling. During these times, much less variety was available than today in terms of number, form, and quality of crop products.

The change from horses to tractors after 1945 and the adoption of a group of technological practices characterized the second American agricultural revolution. By 1954, the number of tractors on farms exceeded the number of horses and mules for the first time. Also by 1954, 71% of all farms had cars, 49% had telephones, and 93% had electricity.

1970s to the Present

The 1970s and 1980s were highlighted by a vegetable crop industry that responded to an increasingly diverse population and the products that this population demanded. During the 1970s, no-tillage agriculture was popularized, a trend that continued into the 1980s as more farmers used no-till or low-till methods to curb soil erosion. By 1975, 90% of all farms had phones, 98% had electricity.

In the 1980s, targeted marketing replaced mass marketing. Development of new products occurred at a rapid pace during the 1980s and continued during the 1990s. During the 1990s, more farmers began to use low-input sustainable techniques to decrease chemical applications.

U.S. Agriculture and Crop Production

In 1996, the total U.S. farm cash receipts for agriculture was in excess of $209 billion (Table 2.1). This was about a 16% increase in farm cash receipts from 1992. While cash receipts increased during this time period, government payments decreased by about 20%. This is indicative of the prevalent governmental philosophy during the 1980s and 1990s toward reduced funding for government programs.

Crop cash receipts accounted for in excess of $109 billion, an increase of 28% from 1992. Of this amount, $14.3 billion was due to vegetables and melons. The cash receipts for vegetables and melons increased by 21% from 1992.

Components of the Vegetable Industry

The commercial vegetable industry can be broken down into four broad categories: fresh market, processing, forcing, and niche markets. In addition, vegetables are often the mainstay of many home gardens.

TABLE 2.1 *U.S. farm cash receipts, 1992 and 1996*

Category	1992	1996
	million dollars	
Livestock and products, total	85,637	92,914
Meat animals	47,748	44,382
Dairy products	19,736	22,834
Poultry and eggs	15,524	22,326
Other	2,629	3,371
Crops, total	85,744	109,425
Feed crops	20,099	28,114
Oil-bearing crops	13,286	17,756
Vegetables and melons	11,851	14,349
Fruits and tree nuts	10,179	11,714
Food grains	8,467	11,550
Cotton (lint and seed)	5,192	7,461
Tobacco	2,958	2,796
Other	13,712	15,686
Government payments	9,169	7,286
Total U.S. farm cash receipts	180,550	209,625

Source: USDA–NASS.

Commercial Production

Fresh Market Commercial fresh market vegetables can be produced on either truck farms or market gardens. The differences between truck farms and market gardens are where they are located, the number of different types of crops grown, the relative acreage of each crop grown, and how and where the crops are marketed. In addition, small acreage vegetable growers may direct market their crops through subscription farming, pick-your-own operations, or farmer's markets (also called "curb markets").

Truck Farms Truck farms (Figure 2.1) are often located near transportation systems or highways. They deal with only one or two crops on a substantial acreage for distant marketing, often by truck. The term *troque* means to barter and this is a typical way that truck crops are marketed to buyers. Truck farmers almost always sell to distant markets, because local markets can't handle the volume of selected crops that they produce. Truck farms typically are located in rural areas where land is inexpensive.

Truck farms can be any size operation but they are usually large-acreage enterprises (e.g., a typical truck farm may be 200 acres of tomatoes and 50 acres of watermelons). Truck farms are concentrated in geographical areas and historically were located along the coast. Traditionally, the southern United States had many truck farms because this region could grow crops during times when other areas of the United States could not.

FIGURE 2.1   A typical truck farm grows one or two crops, such as tomatoes, on a substantial acreage for distant marketing.

FIGURE 2.2   Plants for sale at a retail market.

Market Gardens   Market gardens are located near population centers and supply a wide variety of home or locally grown produce or plants (Figure 2.2). A roadside vegetable stand is an example of a market garden.

Market gardens are usually small businesses (6 acres or less), often found on the outskirts of cities. They grow vegetables for local markets and often use intensive agriculture with high rates of fertilizer and access to irrigation (often city water). Market gardens frequently grow multiple crops per year on the same land and use plastic mulch and row covers to extend the marketing season. Market gardens concentrate on high-profit crops and seek to produce high quality. Their success partially depends on demonstrating that the quality of the crops they produce is better than crops that are shipped in.

Other Types of Direct Marketing   While market gardens are one of the more common types of direct marketing for vegetables, there are other creative marketing methods (e.g., subscription farming, pick-your-own operations, and farmer's markets) that vegetable growers can participate in. These marketing methods vary in the amount of labor and capital the grower must provide and location of the marketplace.

Subscription farming involves the grower contracting with the consumer to produce and deliver a specific vegetable. This approach is good for growers who hope to profit from specific special attributes of their vegetables such as products with ethnic appeal, organically grown produce, or gourmet items.

Pick-your-own or U-pick operations (Figure 2.3) require the least grower labor and capital for market facilities. The customers perform the harvesting. This method works well for some commodities and in some locations, but not for all crops or for all growers. Growers who have pick-your-own operations must be able to effectively deal with the public, and must be willing to accept a certain amount of unintentional damage caused by customers.

Farmer's markets (also called "curb markets") are similar to roadside markets, but the retailing function is moved closer to the customer. This enables the grower to offset the potential disadvantage of production location. Staffing needs are simple to plan because operations occur only during specific hours (and on specific days). One disadvantage with farmer's markets is the need to predict sales in advance so that enough vegetables can be harvested and prepared each day. At a farmer's market, growers cannot replenish stock by quickly harvesting additional produce, and rain during the market may result in unsold product that may not be able to be sold on the next market day.

FIGURE 2.3   A pick-your-own eggplant field.

Value of Production and Trends   The value of 25 major fresh market vegetables and melons for 1997 was $7.9 billion (Table 2.2). Tomatoes and head lettuce had the greatest production values, each in excess of $1 billion. Crop production (by weight) for the 25 major vegetables and melons increased each year from 1995 to 1997.

Since 1976, the value of the 14 major fresh market vegetables increased by 106% to a value in excess of $204 million in 1996 (Table 2.3). The greatest increases in crop values during the two decades prior to 1996 were in broccoli (775%), cauliflower (380%), spinach (378%), and cucumbers (278%).

California is the leading U.S. fresh market vegetable production state, accounting for 44% of the area harvested, 49% of the production, and 53% of the value in 1997 (Table 2.4). Other major fresh market states include Florida, Arizona, Georgia, and Texas.

TABLE 2.2   *Principal vegetables for fresh market: production and value by crop, 1995–1997*

Crop	Production			Value		
	1995	1996	1997	1995	1996	1997
	1,000 cwt			1,000 dollars		
Artichokes	819	890	865	61,965	65,416	67,620
Asparagus[1]	2,024	1,989	1,979	177,170	156,623	181,224
Beans, lima	165	136	150	5,280	4,216	4,950
Beans, snap	4,441	3,711	3,790	162,260	154,952	156,377
Broccoli[1]	15,815	15,453	17,315	443,304	409,167	495,515
Brussels sprouts	561	684	630	14,390	20,120	26,800
Cabbage	22,994	24,531	27,395	260,644	244,814	279,291
Cantaloupes	19,278	22,119	23,556	350,698	400,795	417,859
Carrots	23,478	26,760	33,599	394,356	355,829	441,193
Cauliflower[1]	6,528	6,644	6,483	216,548	213,983	197,956
Celery	18,830	18,861	18,062	306,828	199,398	273,445
Corn, sweet	21,399	22,717	22,587	389,288	384,445	398,279
Cucumbers	10,079	9,845	10,957	166,333	186,325	185,194
Eggplant	633	677	738	16,225	18,146	17,558
Escarole/endive	532	612	556	14,642	13,377	13,123
Garlic	4,620	6,125	5,550	140,700	196,333	261,519
Honeydews	4,332	4,737	5,795	89,193	80,405	109,394
Lettuce						
Head	62,349	65,852	68,542	1,463,348	970,798	1,187,830
Leaf	8,924	9,154	9,241	309,477	245,536	244,841
Romaine	8,530	8,651	9,279	215,026	163,132	170,954
Onions	64,182	61,369	63,883	633,692	581,571	648,437
Peppers	14,431	16,953	16,773	452,786	474,801	502,595
Spinach	1,776	1,654	1,903	56,458	48,029	58,682
Tomatoes	34,535	34,564	37,809	891,343	966,679	1,246,843
Watermelons	40,444	44,135	40,734	357,062	275,684	309,230
25 vegetables and melons	391,699	408,823	428,171	7,589,016	6,830,574	7,896,709

[1]Includes processing total for dual usage crops.

Source: USDA–NASS.

TABLE 2.3   *Value of fresh market production of major vegetables: 1976, 1986,*
and 1996

Crop	1976	1986	1996
		value ($1,000)	
Asparagus	35,025	97,941	103,996
Broccoli	42,495	184,665	371,894
Carrots	99,785	213,141	347,164
Cauliflower	42,381	170,020	203,563
Celery	133,336	211,065	197,753
Sweet corn	121,812	209,318	376,798
Head lettuce	462,262	699,273	975,541
Onions	217,202	439,239	580,324
Tomatoes	416,562	788,424	879,318
Cabbage	110,405	—	245,155
Spinach	12,224	—	58,387
Cucumbers	49,404	—	186,939
Irish potato			
Sweet potato	99,054	134,436	204,658

Source: USDA–ERS, Vegetable Yearbook, 1998.

TABLE 2.4   *Leading fresh market vegetable states in 1997*

	Area harvested		Production		Value	
Rank	State	Percent of total	State	Percent of total	State	Percent of total
1	CA	43.7	CA	48.7	CA	53.4
2	FL	10.0	FL	10.3	FL	15.1
3	GA	6.5	AZ	7.7	AZ	6.6
4	AZ	6.4	GA	5.3	GA	3.8
5	TX	4.4	TX	4.0	TX	3.1

Source: USDA–NASS.

The U.S. demand for fresh vegetables (excluding potatoes) increased from 110.4 pounds per person in 1980 to 142.6 pounds per person in 1989. Demand remained relatively constant through 1993, but has since increased to 153.1 pounds per person in 1997. The growth is suspected to be partially driven by increases in per capita income, and in the quality and variety of product being offered to the consumer.

Processing   Processing vegetables are raw products supplied by growers through contracts that specify some of the production techniques, price per ton at a given level of quality, and standards for acceptance of harvest. Growers involved with producing vegetables for processing deal with a low margin of profit and, as a result, must grow large acreages of these crops and use mechanical harvesters (Figure 2.4)

FIGURE 2.4   Harvesting a field of processing tomatoes with a mechanical harvester.

to generate acceptable profits. Processors often supply assistance with the harvest and technical support during the growing of the crop. The cultivars that are used reflect specific quality components dictated by the type of processing. The four major processing crops are sweet corn, peas, snap beans, and tomatoes. Vegetables are processed by freezing, dehydration, and canning. The major processing states are Wisconsin, California, and Minnesota.

The value of 13 major processing vegetables for 1997 was $1.5 billion (Table 2.5). This was slightly down from values in 1995 and 1996. This decrease in value was also observed in total production by weight. Processing tomatoes ($605 million) and sweet corn ($247 million) had the greatest production values of the major processing crops.

Since 1976, the value of the 13 major processing vegetables increased by 48% to a value in excess of $1.5 million in 1996 (Table 2.6). Most of these crops increased in crop values during the two decades prior to 1996, except for canned beets, cabbage, and broccoli. While green peas generally increased in value, there was a decrease in value in canning peas that was more than offset by an increase in the value of frozen peas.

California is the leading U.S. processing vegetable production state, accounting for 21% of the area harvested, 59% of the production, and 44% of the value in 1997 (Table 2.7). Other major processing states include Wisconsin, Washington, Minnesota, and Oregon.

Vegetable Forcing   Vegetable forcing has traditionally been the greenhouse industry. It is the principal part of the vegetable industry that is responsible for growing vegetables out of season. Traditional greenhouse vegetable crops are tomatoes, cucumbers, lettuce, and the production of transplants. Most of the glass greenhouses in the United States have been replaced with greenhouses constructed of double-layer polyethylene (Figure 2.5), or rigid coextruded polycarbonate and acrylic corrugated sheets. Greenhouse production typically uses a nutrient film technique.

TABLE 2.5 *Principal vegetables for processing: production and value by crop, 1995–1997*

Crop	Production			Value		
	1995	1996	1997	1995	1996	1997
	tons			1,000 dollars		
Beans, lima	70,850	72,710	76,280	31,589	33,105	35,006
Beans, snap	695,450	773,560	733,000	120,992	138,103	129,753
Beets	150,000	125,870	122,180	8,814	8,092	8,136
Cabbage	173,870	141,920	183,670	7,549	6,029	8,299
Carrots	585,550	566,580	551,450	46,443	38,098	37,447
Corn, sweet	3,324,150	3,296,330	3,323,540	251,156	258,840	247,839
Cucumbers	610,460	560,670	619,090	135,803	139,330	146,043
Peas, green	492,590	413,960	475,940	131,762	117,596	136,996
Spinach	155,820	165,850	146,940	15,970	17,105	15,729
Tomatoes	11,286,040	11,408,740	9,972,650	713,544	711,121	605,350
10 crops	17,544,840	17,526,190	16,204,740	1,463,666	1,467,419	1,370,598
Asparagus	46,180	43,780	38,920	52,999	53,143	47,571
Broccoli	98,590	63,250	57,590	38,018	24,501	24,371
Cauliflower	41,550	27,640	28,550	18,758	13,174	13,652
3 crops	186,320	134,670	125,060	109,775	90,818	85,594
13 crops	17,731,160	17,660,860	16,329,800	1,573,441	1,558,237	1,456,192

Source: USDA–NASS.

Development of winter production areas in the South and in Central America during the last couple of decades has reduced the overall amount of greenhouse vegetables grown in the United States, but some sections of the greenhouse vegetable industry have shown a healthy growth pattern in recent years. This growth in greenhouse production is predominately limited to several states including Colorado, Florida, Pennsylvania, California, and Arizona (Table 2.8). Overall greenhouse production in the United States is relatively minor compared to European countries (Table 2.9).

Niche Markets   Niche markets for vegetables are those markets that specialize in nontraditional crops. Common niche market vegetables include specialty crops such as unusual or exotic vegetables (Figure 2.6), organically grown vegetables, herbs, and spices. Larger ethnic populations and growth in their cultural expression have increased the demand for product diversity. Also, a broader portion of the population is experimenting with foods once considered ethnic or regional.

Most nontraditional crops often do not have established markets or a visible marketing infrastructure. Therefore the marketing of niche crops generally requires more planning, research, and personal commitment than for traditional crops. Production and marketing research is required for determining whether a niche crop is grown. Crop selection is based on existing market opportunities, potential for market growth, and an assessment of how accessible the market is to an individual producer. Often filling a market need is more expedient and profitable for a vegetable grower than trying to create a market for a new commodity.

TABLE 2.6 *Value of processing for canning and freezing of major vegetables: 1976, 1986, and 1996*

Crop	1976	1986	1996
	value ($1,000)		
Snap beans			
Canning	64,452	62,903	89,912
Freezing	17,443	34,359	44,994
Total	81,895	97,262	134,906
Sweet corn			
Canning	70,619	82,998	139,708
Freezing	37,084	67,823	119,132
Total	107,703	150,821	258,840
Spinach			
Canning	5,719	—	8,416
Freezing	4,850	—	8,689
Total	10,569	—	17,105
Asparagus	23,119	39,028	52,705
Broccoli	25,709	55,074	24,501
Cauliflower	9,333	21,843	13,174
Green peas			
Canning	64,370	45,322	60,147
Freezing	37,612	48,459	57,349
Total	101,982	93,781	117,496
Tomatoes	375,407	472,927	723,914
Cucumbers	79,751	113,400	142,784
Beets, canning	8,383	—	8,092
Carrots	18,638	23,335	39,914
Cabbage	7,166	—	6,029
Lima beans	25,426	—	30,115
Total	1,016,535	1,067,471	1,508,234

Source: National Agricultural Service, USDA.

TABLE 2.7 *Leading processing vegetable states in 1997[1]*

Rank	Area harvested		Production		Value	
	State	Percent of total	State	Percent of total	State	Percent of total
1	CA	20.9	CA	59.1	CA	43.7
2	WI	17.3	WI	7.7	WI	9.8
3	MN	14.8	WA	6.9	WA	7.9
4	WA	11.2	MN	6.0	MN	7.8
5	OR	6.9	OR	3.8	OR	5.4

[1]Lima beans, snap beans, beets, cabbage, carrots, sweet corn, cucumbers for pickles, peas, spinach, and tomatoes.

Source: USDA–NASS.

FIGURE 2.5 A polyethylene-covered greenhouse.

TABLE 2.8 *Greenhouse vegetable production in the United States*

State	Production (# acres)	Major crops
Colorado	69	Tomatoes
Florida	44	Cucumbers, peppers, tomatoes
Pennsylvania	54	Tomatoes, cucumbers
New York	35	Tomatoes
Ohio	35	Tomatoes, other crops
California	28	Cucumbers, tomatoes
Arizona	25	Tomatoes
Mississippi	15	Tomatoes
North Carolina	10	Tomatoes

Modified from Snyder, R. G. 1996. Greenhouse vegetables—introduction and U.S. industry overview. *Proc. Natl. Ag. Plastics Congress* 26:247–252. Used with permission.

TABLE 2.9 *Greenhouse vegetable production by country*

Country	Production (# acres)
Spain	30,000
Holland	11,400
England/Wales	3,000
Canada	710
USA	450

Source: Snyder, R. G. 1996. Greenhouse vegetables—introduction and U.S. industry overview. *Proc. Natl. Ag. Plastics Congress* 26:247–252. Used with permission.

FIGURE 2.6   Vegetable crops such as luffa gourds can be grown for specialty markets.

Another section of the niche markets for vegetables is growing cultivars of traditional crops primarily for their eating characteristics rather than for yield or shipping attributes. These cultivars often have superior taste and are generally distributed through upscale restaurants, farmer's markets, and specialized produce retailers. An example of this is vine-ripened tomatoes. Also, so-called heirloom cultivars, cultivars that were grown generations ago, are now being grown and marketed as specialty items.

Specialty, niche market vegetable crops are grown profitably by careful and intelligent growers. Proper environmental control is critical to the success of specialty crops and often requires additional expenses in the growing of these crops. The market for these commodities has to be sufficiently strong to support a high price. Many of the suppliers of specialty vegetables are well established, and gaining access to the market is often a tremendous obstacle to overcome; however, market gains can be extremely profitable.

Domestic and import production of the 12 main specialty vegetables increased 258% and 797%, respectively, from 1981 to 1996 (Table 2.10). The total amount of production (domestic and import) of these specialty vegetable crops increased from 4.5 to 21 million cwt (hundredweight) during this time period.

Imports and Exports   In order to satisfy the needs of the U.S. consumer, selected vegetables are imported into the United States from other countries. In excess of $1.4 billion of the major vegetables was imported in 1996 (Table 2.11). Mexico was the greatest supplier of imported vegetables to the United States, providing $1.2 billion. Canada provides the second-most amount of imported vegetables, followed by the Netherlands.

TABLE 2.10   *Specialty vegetables, fresh market: U.S. domestic and imported shipments*

Crop	1981	1986	1991	1996
	(1,000 cwt)			
Chinese cabbage				
Domestic	137	149	110	161
Import	0	14	31	45
Total	137	163	141	206
Escarole/endive				
Domestic	440	366	202	185
Import	20	66	145	130
Total	460	432	347	315
Garlic				
Domestic	—	—	—	—
Import	193	362	415	463
Total	193	362	415	463
Greens				
Domestic	974	1,293	1,058	1,288
Import	0	40	164	247
Total	974	1,333	1,222	1,535
Romaine lettuce				
Domestic	746	2,287	4,754	6,945
Import	0	11	0	54
Total	746	2,298	4,754	6,999
Other lettuce[1]				
Domestic	418	2,234	3,647	3,354
Import	42	76	51	131
Total	460	2,310	3,698	3,485
Misc. herbs[2]				
Domestic	0	54	96	139
Import	0	1	181	399
Total	0	55	277	538
Misc. oriental[3]				
Domestic	0	187	367	285
Import	0	7	74	240
Total	0	194	441	525
Parsley				
Domestic	151	184	160	78
Import	0	26	28	74
Total	151	210	188	152
Southern/snow peas				
Domestic	14	42	20	7
Import	26	147	52	244
Total	40	189	72	251
Chile peppers				
Domestic	32	412	414	414
Import	358	685	1,129	1,781
Total	390	1,097	1,543	2,195
Misc. vegetables[4]				
Domestic	699	70	120	92
Import	179	352	436	649
Total	878	422	556	741
Total				
Domestic	3,612	7,279	10,948	12,948
Import	899	1,947	5,230	8,064
Total	4,511	9,226	16,178	21,012

[1]Includes Boston, Bibb, and red and green leaf lettuce.

[2]Includes anise, basil, chives, cilantro, cipolinos, dill, dry shallot, mint, parsley root, thyme, and watercress.

[3]Includes bean sprouts, bok choy, dikon, gobo, and lobah.

[4]Alfalfa sprouts, cardoon, domestic celeriac (celery root), chicory root, Jerusalem artichoke, oyster plant (salsify), radicchio, and tomatillos.

Source: USDA–ERS, Vegetable Yearbook, 1998.

TABLE 2.11   *Selected fresh vegetables: U.S. import value from selected countries and the world, 1996*

Item	Canada	Mexico	Chile	Netherlands	Other	World
			$ 1,000			
Asparagus	29	33,710	2,347	0	23,605	59,691
Broccoli	498	6,020	0	0	0	6,518
Cabbage	6,330	2,025	0	2	22	8,379
Carrots	16,955	5,935	0	0	209	23,099
Cauliflower	2,369	648	0	2	0	3,019
Celery	968	4,432	0	0	392	5,792
Sweet corn	351	4,116	0	0	2	4,469
Cucumbers	6,326	118,311	0	519	2,679	127,835
Eggplant	0	16,581	0	37	388	17,006
Garlic	131	19,804	1,673	0	5,492	27,100
Lettuce	4,051	3,644	8	31	703	8,437
Onions	8,928	126,281	2,523	1,185	10,008	148,925
Okra	0	7,462	0	0	204	7,666
Green peas	15	6,060	0	3	4,390	10,468
Peppers, bell	11,428	111,485	4	43,022	4,742	170,681
Peppers, chile	55	45,594	0	143	475	46,267
Squash	142	80,771	0	4	1,192	82,109
Tomatoes	37,408	580,349	29	42,646	12,036	672,468
Total	95,984	1,173,228	6,584	87,594	66,539	1,429,929

Source: USDA–ERS, Vegetable Yearbook, 1998.

Exports provide additional markets for the U.S. vegetable growers. Of the major vegetables, the amount exported has increased steadily since 1976 (Table 2.12). The leading vegetable crops that were exported in 1996 were the lettuces (640 million pounds) and onions (581 million pounds). The opening of markets in the Pacific Rim are also expanding the opportunities for vegetable growers. For example, Japan opened their markets to U.S. fresh tomato growers for the first time in 1997.

Home Gardens

Home garden vegetables are grown in small quantities in sections of the property of many homeowners. Home gardening has been designated as the number one outdoor leisure activity.

Trends in gardeners' ages are changing. More baby boomers (30- to 50-year-olds) are participating today than in the past. In previous years people gardened to save money. Today many people garden for fresher tasting vegetables, better quality food, better nutrition, and improved health.

Current Fresh Vegetable Consumption Trends

Spending on fresh produce by consumers has been growing steadily. From 1986 to 1992, spending on fruits and vegetables increased from $194 per person to $254. During this time period the sale of dairy products and nonalcoholic beverages decreased.

TABLE 2.12   *Selected U.S. exports of fresh market vegetables, 1976, 1986, and 1996*

Crop	1976	1986	1996
		Amount pounds (1,000)	
Asparagus	10,436	17,603	31,695
Broccoli	—	119,524	278,600
Carrots	69,285	160,629	210,099
Cauliflower	—	78,463	232,764
Celery	122,046	213,910	257,581
Spinach	—	25,677	28,166
Cucumbers	52,600	61,820	70,687
Peppers	39,489	85,526	133,322
Garlic	5,146	8,900	19,976
Cabbage	56,405	64,747	96,897
Sweet corn	—	54,186	90,971
All lettuce	360,802	553,601	640,805
All onions	326,582	261,113	581,387
Tomatoes	212,374	288,053	295,441
Snap beans	14,994	30,138	44,660
Cantaloupe	54,492	105,793	126,834
Watermelons	84,298	58,227	255,260
Other melons	—	47,772	109,941

Source: USDA–ERS, Vegetable Yearbook, 1998.

The top three fresh vegetables bought weekly by consumers are regular tomatoes (44%), vine-ripened tomatoes (41%), and sweet corn (40%) (Table 2.13). Three other fresh vegetable items are purchased at least weekly by about one-third of consumers. These items include Iceberg lettuce (38%), cucumbers (35%), and Roma or Italian tomatoes (32%).

Shoppers are taking advantage of new fresh vegetable items and including them in their diets. Salads and salad mixes are some of the more popular new items being purchased. Following salads and salad mixes are fresh-peeled carrots.

The number of consumers who purchase fresh vegetable items is increasing. Shoppers bought an average of 28.8 different types of vegetables in 1997. Latest per capita consumption of fresh vegetables based on farm fresh weight was 173.6 pounds. Ninety-five percent of consumers said they bought one or more types of lettuce. About 91% of consumers said they had bought Iceberg lettuce, followed by green or red leaf lettuce at 49%, Romaine at 44%, and other specialty lettuces such as Bibb or Boston at 35%. The next-most common item purchased was potatoes by 96% of consumers. Russet potatoes were the type of potato most purchased at 80%. Ninety-three percent of consumers purchased one or more types of tomatoes.

Fresh-cut and convenience-oriented vegetables have shown the most growth in overall purchases among consumers. Salads and salad mixes climbed from 50% of consumer purchases in 1995 to 65% in 1997. Fresh-cut carrots also increased 15 percentage points from 1995 to 1997 in 63% of the individuals surveyed.

TABLE 2.13   *Weekly vegetable purchases. Households that report purchasing the following vegetables at least once a week when in season and available, 1997*

Vegetable	% purchased
Tomato, regular	44
Tomato, vine-ripened	41
Sweet corn	40
Lettuce, Iceberg	38
Cucumbers	35
Tomatoes, Roma or Italian	32
Lettuce, leaf	28
Salads, salad mixes	26
Broccoli, florets	26
Broccoli, whole	25
Beans, yellow/wax	25
Lettuce, Romaine	25

Source: The Packer. 1997. Fresh Trends: A profile of the fresh produce consumer. Reprinted by permission of *The Packer.*

Nutrition/National 5 A Day Program

According to a recent report from the U.S. Surgeon General, the three most important personal habits that influence health are smoking, alcohol consumption, and diet. For the two out of three adults that do not drink excessively or smoke, the single most important personal choice influencing one's long-term health is what one eats. The *Surgeon General's Report on Nutrition and Health* established the fact that two-thirds of all deaths (including coronary heart disease, stroke, atherosclerosis, diabetes, and some types of cancer) are related to food that we eat. In fact, approximately 35% of all cancer deaths in America may be related to diet.

The link between diet and the leading causes of death in America is suggested by research that reveals diets high in fat, saturated fat, and cholesterol, and low in fruits and vegetables significantly increase a person's chances of developing cancer, heart disease, and other chronic diseases. There is consensus among nutrition experts and relevant health organizations—including the U.S. Surgeon General, the National Cancer Institute (NCI), and the American Cancer Society—that the typical American diet contains too much fat and not enough fruits, vegetables, and whole grains. Therefore, health authorities recommend that Americans eat a diet rich in fruits and vegetables. The U.S. Department of Agriculture with support from the U.S. Department of Health and Human Services has developed the Food Pyramid (Figure 2.7) to help consumers make better health choices. A diet with plenty of fruits and vegetables also assists in the prevention of obesity and promotes a healthy digestive tract.

Based on the strength of the scientific data and the relatively low consumption rates in the population, the NCI began the National 5 A Day for Better Health Program in the fall of 1991 to encourage Americans to eat five or more servings of fruits and vegetables every day. The program is a public-private partnership between the NCI, U.S. Department of Health and Human Services, and the Produce for Better

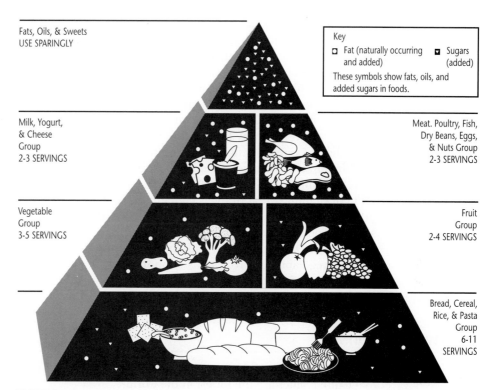

Fats, Oils, & Sweets
USE SPARINGLY

Key
☐ Fat (naturally occurring
and added)
◼ Sugars
(added)
These symbols show fats, oils, and
added sugars in foods.

Milk, Yogurt,
& Cheese
Group
2-3 SERVINGS

Meat. Poultry, Fish,
Dry Beans, Eggs,
& Nuts Group
2-3 SERVINGS

Vegetable
Group
3-5 SERVINGS

Fruit
Group
2-4 SERVINGS

Bread, Cereal,
Rice, & Pasta
Group
6-11
SERVINGS

FIGURE 2.7   Food Pyramid Guide. The Food Pyramid Guide is an outline developed by
health organizations on what to eat each day for a healthy diet.
Source: U.S. Department of Agriculture and the U.S. Department of Health and Human
Resources.

Health Foundation (PBH), a nonprofit consumer education foundation representing the fruit and vegetable industry. Examples of industry participants are supermarket chains, independent grocery stores, merchandisers, branded product companies, suppliers, commodity boards, fruit and vegetable marketers, and food service operations.

The major components of the National 5 A Day Program are retail, media, community, and research. Retailers (supermarkets and food service) advertise the program in local media and provide consumers with brochures, recipes, and interactive events such as food demonstrations. NCI and PBH work together to develop a comprehensive media campaign to obtain regular and special coverage of the 5 A Day Program. The community component of the program is vital to its success in changing behavior. State health agencies have been licensed by NCI to coordinate efforts within their states to reach consumers through all community channels. These state agencies provide a forum for the cooperative efforts of health, educational, agricultural, and voluntary agencies and the private sector in promoting the 5 A Day message.

A more recent NCI survey indicates that since the beginning of the 5 A Day program, the percentage of people able to identify the need for five daily servings of fruits and vegetables has more than tripled, from 8 to 29%. There was a corresponding decline from one in three Americans to one in 10 Americans who felt one or fewer servings each day is sufficient.

More research is needed to elucidate the role of fruits and vegetables in cancer and disease prevention. However, from a public health perspective, there is abundant evidence to suggest that substantial health benefits could be achieved by increasing the population's consumption of a variety of fruits and vegetables. As a result, the health benefits of eating vegetables are becoming more and more an important consideration in the purchase of food. Such constituents as beta-carotene and fiber entice consumers to buy certain vegetables. Many people cite beta-carotene as a reason why they choose carrots to eat. Others choose broccoli for its reputation of supplying fiber in the diet.

Vitamin C is often associated with fruits, but certain vegetables actually contain more vitamin C. For example, compared to the 120% of the recommended daily value of vitamin C that oranges contain, broccoli has 200% of the daily value, bell peppers have 150%, and cauliflower contains 100%. Also, cantaloupe is a good source of beta-carotene.

The 5 A Day for Better Health message is producing an increasingly strong impact on vegetable and fruit consumption. Consumers are working on ways to fit more fruits and vegetables into their diets.

Review Questions

1. What are some of the reasons why consumer expenditures for food in the United States are the lowest in the world?
2. What was the U.S. vegetable industry like during the following historical periods?
   a. Colonial period
   b. Post–Civil War
   c. Pre–World War II
   d. Post–World War II
   e. 1970s to the present
3. What characterized the first and second American agricultural revolutions?
4. What is the value of the entire U.S. agricultural industries and what percentage of that is the U.S. vegetable industry?
5. What are some of the characteristics of the following vegetable industries of the United States?
   a. Truck farms
   b. Market gardens
   c. Processing vegetables
   d. Vegetable forcing
   e. Niche markets
6. What is the current trend in fresh vegetable consumption?
7. Describe the 5 A Day for Better Health Program.

Selected References

Brewer, T., J. Harper, and G. Greaser. 1994. Fruit and vegetable marketing for small-scale and part-time growers. Penn State Coop. Ext. Serv. (Agricultural alternatives)

Cook, R. L. 1992. The dynamic U.S. fresh produce industry: An overview. In A. A. Kader (Ed.), *Postharvest technology of horticultural crops.* University of California, Division of Agriculture and Natural Resources Publication 3311.

The Packer. 1997. Fresh Trends: A profile of the fresh produce consumer.

Snyder, R. G. 1996. Greenhouse vegetables—introduction and U.S. industry overview. *Proc. Natl. Ag. Plastics Congress* 26:247–252.

VanSickle, J. 1998. 1998 Vegetable outlook. *American Vegetable Grower* (January):20–21.

Selected Internet Sites

www.aces.uiuc.edu/~sare/history.htm/   A Brief History of Agriculture, University of Illinois College of A.C.E.S.

www.cancer.org/   American Cancer Society.

www.cdc.gov/nccdphp/nccdhome.htm   National Center for Chronic Disease Prevention and Health Promotion, Centers for Disease Control and Prevention, U.S. Department of Health and Human Services.

www.dole5aday.com/   Dole 5 A Day Program, Dole Food Co.

www.econ.ag.gov/   Economic Research Service, USDA.

www.jan.manlib.cornell.edu/data-sets/specialty/89011/   Vegetable Yearbook, USDA, Economics and Statistical System, Cornell University.

www.nci.nih.gov/   National Cancer Institute, National Institute of Health.

www.usda/fcs/fcs.htm   Food and Nutrition Service, USDA.

www.usda.gov/cnpp   Center for Nutrition Policy and Promotion, USDA.

www.usda.gov/nass   National Agricultural Statistical Service, USDA.

www.5aday.com/   The Produce for Better Health Foundation.

chapter

3

Classifying Vegetable Crops

Importance

At least 10,000 plant species are used as vegetables worldwide. Of these plant species, only 50 or so are considered of great commercial importance. Because of a large number of different crops that are considered as vegetables, it is often useful to classify vegetables into groups or classes in order to better understand and discuss them. Placing individual vegetable crops into groups or classes with other crops that share some characteristics reduces the rather large number of crops into a smaller number of groups and results in logical associations. As with many techniques that are used to simplify, no one classification or grouping is perfect and some classifications are more useful than others depending on situations and needs.

Classification Schemes

Vegetables are often grouped according to similar characteristics, often by use, appearance, morphological features, and type of life cycle. The more common classifications include botanical, edible part, life cycle, sensitivity to temperature, and a combination of family groupings and accepted use.

Botanical Classification

The botanical classification scheme is based on similarity or dissimilarity in morphological structures, often with flower structure as the main criteria for determining relationships. The successive levels of

TABLE 3.1   *The botanical classification of some common vegetables*

Division Anthophyta

Class Monocotyledons
Alliaceae
> *Allium ampeloprasum* L. Ampeloprasum group—great-headed garlic
> *Allium ampeloprasum* L. Porrum group—leek
> *Allium cepa* L. Aggregatum group—multiplier onion
> *Allium cep* L. Cepa group—onion
> *Allium sativum* L.—garlic

Dioscoreaceae
> *Dioscorea alata* L.—yam

Liliaceae
> *Asparagus officinalis* L—asparagus

Poaceae
> *Zea mays* L. subsp. *mays*—sweet corn

Class Dicotyledon
Apiaceae
> *Apium graveolens* L. var. *dulce* (Mill.) Pers.—asparagus
> *Daucus carota* L. *sativus* (Hoffm.) Arcang.—carrot
> *Petroselium crispum* (Mill.) Nym. var. *crispum*—parsley

Asteraceae
> *Cichorium endiva* L.—endive, escarole
> *Cichorium intybus* L.—chicory
> *Cynara scolymus* L.—globe artichoke
> *Helianthus tuberosus* L.—Jerusalem artichoke
> *Lactuca sativa* L. var. *capitata* L.—head or butterhead lettuce
> *Lactuca sativa* L. var. *longifolia* Lam.—leaf or Romaine lettuce

Brassicaceae
> *Brassica napus* L. var. *napobrassica* (L.) Reichb.—rutabaga
> *Brassica oleracea* L. var. *acephala* DC—kale, collards
> *Brassica oleracea* L. var. *botrytis* L.—cauliflower
> *Brassica oleracea* L. var. *capitata* L.—cabbage
> *Brassica oleracea* L. var. *gemmifera* Zenk.—Brussels sprouts
> *Brassica oleracea* L. var. *italica* Plenck.—broccoli
> *Brassica perviridis* Bailey—spinach
> *Brassica rapa* L. var. *rapa* (DC) Metzg.—turnip
> *Raphanus sativus* L. Radicula group—radish
> *Sinapis alba* L.—white mustard

morphological relationships are a result of evolution. The successive groupings of plants in the botanical classification (from broadest grouping to most specific) are kingdom, division, subdivision, phylum, subphylum, class, subclass, order, family, genus, and species.

Most common vegetables belong to the Division Anthophyta (Table 3.1). The Division Anthophyta is generally broken down into two classes: monocotyledons and

TABLE 3.1 *The botanical classification of some common vegetables (continued)*

Division Anthophyta

Chenopodiaceae
> *Beta vulgaris* L. Cicla group—chard, Swiss chard, spinach beet
> *Beta vulgaris* L. Crassa group—beet
> *Spinacia oleracea* L.—spinach

Convolvulaceae
> *Ipomea batatus* (L.) Lam—sweet potato

Cucurbitaceae
> *Citrullus lanatus* (Thunb.) Natsum & Nakai—watermelon
> *Cucumis melo* L. Reticulatus group—muskmelon
> *Cucumis melo* L. Cantaloupensis group—cantaloupe
> *Cucumis melo* L. Inodorous group—winter, honeydew or casaba melon
> *Cucumis sativus* L.—cucumber
> *Cucurbita argyrosperma* Huber—pumpkin, cushaw pumpkin, winter Squash
> *Cucurbita mochata* Duch. ex Poir—musky pumpkin, squash or gourd, butternut squash
> *Cucurbita maxima* Duch.—pumpkin, giant pumpkin
> *Cucurbita argyrosperma* Huber—pumpkin, cushaw pumpkin, winter squash
> *Cucurbita mochata* Duch. ex Poir—musky pumpkin, squash or gourd, butternut squash
> *Cucurbita pepo* L.—summer squash, zucchini

Fabaceae
> *Phaseolus lunatus* L.—Lima bean
> *Phaseolus vulgaris* L.—common bean
> *Pisum sativum* L. ssp. *sativum*—English pea
> *Vigna unguiculata* (L.) Walp. Subsp. *unguiculata* (L.) Walp.—southern pea

Malvaceae
> *Abelmoschus esculentus* (L.) Moench—okra

Polygonaceae
> *Rhuem rhabarbarum* L.—rhubarb

Solanaceae
> *Capsicum annuum* L. Grossum group—bell pepper or pimento pepper
> *Capsicum annuum* L. Longum group—cayenne, chili or hot pepper
> *Capsicum frutescens* L.—tobasco pepper
> *Lycopersicon lycopersicum* (L.) Karsten—tomato
> *Solanum melongena* L.—eggplant
> *Solanum tuberosum* L.—Irish potato

Source: Kays, S. J., and J. C. Silva Dias. 1995. Common names of commercially cultivated vegetables. *Economic Botany* 49:115–152. Used with permission.

dicotyledons. The classes are further divided into families (with names that end in *aceae*), which are composed of individual related plant species.

The broadest grouping in which vegetables are typically discussed is family. The genus and species make up the scientific name. Scientific names are accepted worldwide and serve as positive identification, regardless of language. Plants recognized as a single vegetable, even if they have different scientific names, are said to be of one kind.

Since management systems may be governed by botanical similarities, knowledge of botanical classifications of plants is useful for producers. Also the climatic requirements of a particular family or genus are usually similar. The use of the crop for economic purposes within families is similar, and disease and insect controls are quite often similar for the related genera.

Classification Based on Edible Part

Classification by edible part provides a vegetable grower or handler with broad plant groupings that imply specific cultural or handling techniques. For example, leafy crops are very perishable and require rapid chilling after harvest to preserve quality. Root crops are similar in how they are affected by soil fertility, water management, and soil texture.

Following are some common vegetable groupings according to edible plant part, and some examples of vegetables that fit into these groupings.

1. Root
   a. Enlarged taproot (beet, carrot, radish, rutabaga, turnip, parsnip)
   b. Enlarged lateral root (sweet potato)
2. Stem
   a. Above ground, not starchy (asparagus)
   b. Below ground, starchy (Irish potato)
3. Leaf
   a. Onion group, leaf bases eaten (onion, leek, shallot, garlic)
   b. Broad-leaved plants
      1. Salad use (lettuce, cabbage, celery)
      2. Cooked [spinach, rhubarb (petiole only), kale, mustard]
4. Immature flower bud (cauliflower, broccoli, artichoke)
5. Fruit
   a. Immature (pea, snap bean, lima bean, summer squash, cucumber, okra, sweet corn, eggplant)
   b. Mature
      1. Cucurbits (pumpkin and winter squash, squash, muskmelon, watermelon)
      2. Solanum crops (tomato, pepper)

Classification Based on Life Cycle

All plants can be classified according to the time required to complete their life cycle. Annual plants complete their life cycle during a single growing season. Biennial plants require two seasons to complete their life cycle, and perennial plants grow for more than 2 years.

Most of the common vegetables are annuals. Examples of annual vegetables include spinach, lettuce, and beans. Other vegetables are biennials but are grown as annuals. These vegetables include many of the cole crops such as broccoli, cauliflower, and cabbage and root crops such as celery and parsnips. Many biennials are sensitive to temperature regulation of flowering. Other vegetables are perennials and can remain in the production for up to 15 years. Examples of these vegetables include globe artichoke, asparagus, and rhubarb.

Classification Based on Temperature

All vegetables can be separated broadly into two groups based on temperature: warm-season and cool-season vegetables. Warm-season vegetables are usually crops that are grown for and bear edible fruit. Warm-season crops are adapted to mean monthly temperatures of 65° to 85°F and are intolerant to frost. Warm-season crops include cucumber, eggplant, lima beans, muskmelons, okra, pepper, snap bean, squash and pumpkin, sweet corn, sweet potato, tomato, and watermelon (Table 3.2).

Cool-season vegetables include most root crops and crops for salads and greens. The plant growth of cool-season crops is relatively small. Cool-season crops are adapted to mean monthly temperatures of 60° to 65°F and are often susceptible to premature seeding or bolting. Harvested crops are stored at cold temperatures. Cool-season crops include artichoke, asparagus, Brussels sprouts, broccoli, cabbage, celery, garlic, kale, onion, pea, radish, and spinach.

Classification Based on Family Grouping or Accepted Use

A classification that incorporates some of the characteristics of family grouping and crop use is pragmatic, convenient, generally accepted, and widely used. Crops that are grouped together in this classification often have the same general culture and are subject to similar pests and diseases. This classification is used in Part 3 of this text.

Vegetable crops that are discussed in Part 3 are classified according to the following groupings and criteria:

1. Cole crops. Also called crucifers or *Brassicas*. These are crops that belong to the cabbage or Brassicaceae family. This family consists of 350 genera and 3,200 species of pungent herbs. The plants are indigenous to temperate and cold climates. Vegetables included in this family are cabbage, broccoli, cauliflower, and Brussels sprouts.
2. Greens. Leafy crops that are usually eaten after they have been cooked. Crops in this grouping come from the Chenopodiaceae (e.g., spinach) and Brassicaceae (e.g., collards and kale) families.
3. Salad crops. Crops used mainly for their leaves and eaten raw. Major vegetables in this group come from the Asteraceae (e.g., lettuce) and Apiaceae (e.g., celery and parsley) families.

TABLE 3.2 *Classification of selected vegetables according to their adaptation to field temperatures*

Cool-season crops		
	Hardy	Asparagus, broccoli, Brussels sprouts, cabbage, collards, garlic, kale, leek, onion, parsley, pea, radish, rhubarb, spinach, turnip
	Half-hardy	Beet, carrot, cauliflower, celery, globe artichoke, lettuce, potato
Warm-season crops		
	Tender	Southern pea, snap bean, sweet corn, tomato
	Very tender	Cucumber, eggplant, lima bean, muskmelon, okra, pepper, pumpkin, squash, sweet potato, watermelon

Adapted from A. A. Kader, J. M. Lyons, and L. L. Morris. 1974. Postharvest responses of vegetables to preharvest field temperatures. *HortScience* 9:523–529. Used with permission.

4. Perennial crops. Crops that are in the field for more than 2 years. Many members of commercial importance as vegetables have aboveground parts that are killed each year in the temperate regions but whose roots remain alive to send up shoots in the spring. Plants in this grouping include members from the Lilliaceae (e.g., asparagus), Asteraceae (e.g., globe artichoke), and Polygonaceae (e.g., rhubarb) families.

5. Root crops. Crops that have a prominent, fleshy underground structure. The underground structure may be a root or hypocotyl with a taproot forming below. Major root crops come from the Apiaceae (e.g., carrots), Chenopodiaceae (e.g., beets), Brassicaceae (e.g., turnip and rutabaga), and Convolvulaceae (e.g., sweet potato) families.

6. Bulb crops. All are species of *Allium* and members of the Alliaceae family. These plants are native to the temperate regions of South America, South Africa, and the Mediterranean region. The plants have bulbs or corms. Crops of importance include onion, garlic, and leeks.

7. Legumes or pulse crops. All are members of the Fabaceae or pea family. The pea family has about 600 genera and 12,000 species. They are primarily herbaceous plants in temperate climates but can exist as trees and shrubs in tropical climates. The fruit of members of the family is a flattened dehiscent pod called a legume. Many members of the family can fix nitrogen in their roots as a result of a symbiotic relationship with nitrogen-fixing bacteria that live in nodules in their roots. Members of this group include the common bean, English pea, Southern pea, and lima bean.

8. Sweet corn. A member of the Poaceae or grass family. It is a monocotyledon that has an inflorescence as a group of flowers called spikelets. The flowers are small and the fruit is a kernel.

9. Solanum crops. All are members of the Solanaceae or nightshade family. There are about 90 genera and about 2,200 species in this family. Many species contain alkaloids such as solanine, nicotine, and atropine. Vegetables in this family include tomatoes, peppers, eggplants, and Irish potatoes.

10. Cucurbits. All are members of the Cucurbitaceae or gourd family. This family consists of about 100 genera and 500 to 700 species. Plants in this family have tendrils, leaves that are often rough to the touch, and large fleshy fruits with many seeds. Vegetables in this family include watermelon, muskmelon and other melons, cucumbers, pumpkin, and the squashes.

Other Classifications

Other classification schemes of vegetables refer to their sensitivity to environmental factors. Included among these are those that are grouped according to sensitivity to soil pH, tolerance to nutrient levels, preference for soil moisture levels, and sensitivity to chilling damage. Other classification schemes are based on morphological features such as seed size, and depth of rooting.

Review Questions

1. Why is it important to classify vegetables into groups or classes?
2. What is the basis of the botanical classification scheme for vegetables?

3. What are the advantages to classifying vegetables according to plant part?
4. What are some of the characteristics of warm-season and cool-season crops?

Selected References

AVRDC. 1990. *Vegetable production training manual.* Asian Vegetable Research and
    Development Center, Shanhua, Tainan, Taiwan. Reprinted 1992. 442 p.

Janick, J. 1979. *Horticultural science.* 3d ed. San Francisco: W. H. Freeman.

Kays, S. J., and J. C. Silva Dias. 1995. Common names of commercially cultivated
    vegetables. *Economic Botany* 49:115–152.

Maynard, D. N., and G. J. Hochmuth. 1997. *Knott's handbook for vegetable growers.* 4th
    ed. New York: John Wiley & Sons.

Nonnecke, I. L. 1989. *Vegetable production.* New York: Van Nostrand Reinhold.
    656 p.

Peirce, L. C. 1987. *Vegetables: Characteristics, production, and marketing.* New York:
    John Wiley & Sons. 433 p.

Yamaguchi, M. 1983. *World vegetables: Principles, production, and nutritive values.* New
    York: Van Nostrand Reinhold. 415 p.

Selected Internet Sites

http://www.ashs.org/resources/plantnames/vegetablessci.html   Listing of plant
    scientific names for vegetables, the American Society for Horticultural
    Science.

chapter

4

How Environmental Factors Affect Vegetable Production

The growth and development of vegetable crops is dependent on abiotic (physical) and biotic (biological) factors. Abiotic factors include environmental conditions of weather and soil. Biotic factors include animals, insects, and diseases.

Weather and Climate

Weather is the composite of the temperature, rainfall, light intensity and duration, wind direction and velocity, and relative humidity of a specific location for a set amount of time. Climate is the weather pattern for a particular location over several years. It is the integrated effects of temperature, precipitation, humidity, sunlight, and wind. Climate can change with distance from the ground surface and with time. The soil condition is dependent on the climate.

Climatic Areas

There are four distinctive climatic areas recognized in the United States that are broadly categorized by availability of water (Table 4.1). These are arid (little to no water), semiarid (limited water), subhumid (generally sufficient water, but requires special farming practices to conserve moisture), and humid (plenty of moisture with possible drainage problems).

Topographical features such as mountains, moraines, hills, large bodies of water, and deserts all modify and create special climatic micro or regional deviations from the whole. For example, it is possible to grow cool-season crops in warm semi-tropical areas by using high elevations for crop production to escape the excessive summer heat of the lowlands.

Each crop has certain climatic requirements. To attain the highest potential yields a crop must be grown in an environment that meets these requirements. A crop can grow with minimal adjustments if it is well matched with its climate. Unfavorable climate conditions can produce a stress on plants resulting in lower yields. In such

TABLE 4.1   *Climatic areas of the United States*

Climatic categorization	Characteristics
Arid	Little or no water naturally available for crop production. These areas include the coastal valleys of California and the southwestern intermountain regions that include Colorado, Utah, western and southern California, Arizona, parts of Oklahoma, New Mexico, and the Texas Panhandle.
Semiarid	Limited water for most crops (less than 45 cm per year). These areas include the Great Plains, western Oregon, southwestern Idaho, parts of Colorado, Utah, and New Mexico.
Subhumid	Requiring special farming practices to conserve moisture. These areas include the eastern part of Texas, Oklahoma, Kansas, North and South Dakota, Iowa, western Minnesota, parts of North Carolina, Illinois, and the northeastern half of Florida.
Humid	Adequate rainfall, with a possible problem of drainage. These include areas east of the subhumid region in the United States.

Adapted from Nonnecke, I. L. 1989. *Vegetable production.* New York: Van Nostrand Reinhold. Reprinted with permission of International Thomson Publishers, Inc.

cases the environment can be artificially modified using production practices of row covers, mulches, and greenhouses to meet the crop requirements. Modifying the crop environment can be expensive, and profitability is usually determined before the environmental factor is artificially modified.

While climate determines what crop can be grown best in a particular location, the rate of growth and development largely depends on the weather. Cloudiness, amount of rainfall, and wind movement all influence how well a crop will grow at a particular time. Weather also determines when some farm operations can occur such as weeding, fertilization, harvesting, and irrigation.

Environmental Factors

Temperature

Most plants function in a relatively narrow range of temperatures. The extremes of this range may be considered killing frosts at about 32°F and death by heat and desiccation at about 105°F.

Optimum Temperature   Each kind of vegetable grows and develops most rapidly at a favorable range of temperatures. This is called the optimum temperature range. For most vegetables the optimum functional efficiency occurs mostly between 50° and 75°F.

Vegetables can be classified according to the temperature requirements of their optimum temperature range. However, they're generally grouped into whether they require low or high temperatures for growth. Temperature requirements are usually based on night temperature. Those that grow and develop below 65°F are the cool-season crops, and those that perform above 65°F are the warm-season crops. Crops that originated in temperate countries usually require low temperature, while those that originated in the Tropics require warm temperature.

Soil Temperature   Soil temperature has direct dramatic effects on microbial growth and development, organic matter decay, seed germination, root development, and water and nutrient absorption by roots. In general, the higher the temperature the faster are these processes. The size, quality, and shape of storage organs are also affected by soil temperature.

Dark-colored soils absorb more solar energy than light-colored soils. The capacity of water to move heat from one area to another (conduction) is greater than that of air. Heat is released to the surface faster in clay soils than in dry, sandy soils. The lower the air temperature, the more rapid the loss. Thus, although light-colored sandy soils absorb less solar energy, less heat is also released to the atmosphere because of the low water-holding capacity of the soil.

Chilling Injury   Most vegetables are injured at temperatures at or slightly below freezing. Tropical or subtropical plants may be killed or damaged at temperatures below 50°F but above freezing. This latter type of injury is called chilling injury.

Susceptibility to cold damage varies with different species and there may be differences among varieties of the same species. The susceptibility to cold damage varies to some degree with stage of plant development. Plants tend to be more sensitive to cold temperatures shortly before flowering through a few weeks after anthesis.

Heat Stress   When temperatures rise too high (in the range of 113° to 122°F), cell death results as the protoplasts in the plant cells are destroyed. In tomatoes, fruits exposed on vines to high temperatures and high solar radiation can reach 120° to 125°F. If green fruits are exposed to these temperatures for an hour or more, they become sunburned; and ripe fruits become scalded.

As with cold resistance, the plant cells can become gradually acclimated, to a certain extent, to heat by slowly raising the temperature and lengthening the exposure daily. Transpiration from the leaf stomata helps cool leaves. It has been calculated that transpiration can reduce heating by about 15 to 25%.

Heat stress at fruit set or during late fruit development may cause defects that render the product unmarketable. In tomatoes, cracks at the stem end (catface) may appear or the fruit may be puffy. Onions and radishes become more pungent at high temperatures. If high temperatures occur for long periods, the leaves might develop chlorosis or show scalding effect.

Symptoms of heat injury are the appearance of necrotic (dead) areas on young leaves. Heat injury occurs over a wide range of vegetables depending on the species or tissue.

Vernalization   Vernalization is the induced or accelerated flowering (bolting) that occurs in certain plants to low temperatures. The biennials and some of the cool-season vegetables (e.g., *Allium*, carrot, celery, the crucifers, and spinach) initiate flower formation after extended (several weeks or months) exposure to low temperature. The required length of low-temperature exposure varies with species.

Premature flowering is called bolting, and bolting can cause substantial yield losses in certain crops. This is particularly true for crops that require little cold exposure, like heat-tolerant Chinese cabbage.

In some species, seedlings and young plants still in the juvenile stage are insensitive to conditions that promote flowering in older plants. In some species, seeds can be vernalized. The seeds must have sufficient water to allow the vernalization process to occur. Certain tubers, corms, and bulbs require low temperatures following moderately high temperatures before growth occurs.

Light

All life on earth is supported by the radiant energy of the sun. Light is one of the most important and variable components of the plant environment. Unfiltered sunlight entering the earth's ionosphere has a brightness or intensity of 1.39 kW/m^2 between the wavelengths of 225 and 3,200 nm. The ozone layer in the stratosphere absorbs a proportion of the ultraviolet radiation while water vapor, carbon dioxide, and oxygen in the troposphere absorb the wavelengths of 1,100 to 3,200 nm. As a result, of the radiation emanating from the sun and available to the earth, only 47% reaches the earth's surface. In addition clouds and particulates in the air reflect, scatter, or absorb the sun's radiation. The amount of sunlight also varies depending on cloud cover, latitude, and altitude.

Light is used differently by plants than by humans and animals. Light to humans and animals is the wavelengths of radiant energy in the electromagnetic spectrum that activates the light receptors in the eyes. When these light receptors are activated, the brain interprets the impulses and vision is experienced.

Light for the plant is used for producing food through the process of photosynthesis. The autotrophic plant is directly influenced by the intensity of light, which drives photosynthesis. Besides affecting the photosynthetic rate, solar radiation also affects plant temperature and photomorphogenic responses.

Light Quality   Sunlight is often referred to as white light and is composed of all colors of light. A color (or spectral distribution) of light would be the relative distribution of wavelengths from a radiation or reflective source. The characteristics of direction and spectral composition of light in the plant's environment is transferred to the plant through the interception and activation of pigment systems. This information affects the morphological development (root and shoots) of the plant, hopefully imparting to the plant some type of ecological or physiological advantage for survival. Plants also use light for sensing and detecting competitors and keeping track of time.

Light Intensity   Light intensity is a major factor governing the rate of photosynthesis. The intensity of the incident (incoming) light and the length of the day affect the quantity or amount of light received by plants in a particular region. The intensity of light changes with elevation and latitude. The amount of sunlight also varies with the season of the year and time of day. Other factors include clouds, dust, smoke, or fog.

Light intensity is an important factor in determining the rate of photosynthesis of the plant. Vegetable crops can vary as to their preference for light intensity. The plant's light saturation point determines the relative light requirement of plants. The light saturation point is the point at which increased light intensity does not result in an increase in photosynthetic rate. Crops such as corn, cucurbits, legumes, potato, and sweet potato require a relative high level of light for proper plant growth while onions, asparagus, carrot, celery, the cole crops, lettuce, and spinach can grow satisfactorily with lower levels of light.

Light Duration   Due to the tilt of the earth's axis (approximately 66°) and its travel around the sun, the length of the light period (also called photoperiod or daylength) varies according to the season of the year and latitude. It varies from a nearly uniform 12-hour day at the equator (0° latitude) to continuous light or darkness throughout the 24 hours for a part of the year at the Poles.

Some plants change their growth in response to daylength and exhibit photoperiodism (Table 4.2). One important photoperiodic plant response in some

TABLE 4.2   *Photoperiodic responses of vegetables*

Response	Short-Day	Day-Neutral		Long-Day	
Flowering	Sweet potato	Corn	Pepper	Spinach	Lettuce
	Southern pea	Cucumber	Eggplant	Onion	Radish
		Sweet pea	Artichoke	Cabbage	Spinach
		Tomato		Carrot	Potato
Bulbing				Onion	
Tuber initiation	Potato				
Root enlargement	Sweet potato				

Modified from AVRDC, 1990; and Yamaguchi, 1983. Used with permission.

plants is flowering. Some vegetables flower when a specific daylength minimum has been passed. Short-day plants flower rapidly when the days get shorter and long-day plants flower fast when days get longer. Plants that are not affected by daylength are called day-neutral plants. These plants can flower under any light period.

Light Energy Capture by Plants

Photosynthesis   One of the main roles of light in the life of plants is to serve as an energy source that plants can capture through the process of photosynthesis. Using water and carbon dioxide through photosynthesis, plants produce the foodstuffs (photosynthates) necessary for growth and survival. Carbohydrates (starches and sugar) and chemical energy are produced during this biochemical process in plants.

Plants capture the energy in light using a green pigment called chlorophyll. In the research laboratory, chlorophyll can easily be extracted from plant tissue using chemical solvents. Chlorophyll can also be extracted by abrasion, as anyone who has ever pruned tomato plants by hand or had grass stains on their clothes can attest.

Photomorphogenesis   Photomorphogenesis is defined as the ability of light to regulate plant growth and development, independent of photosynthesis. Plant processes that appear to be photomorphogenic include internode elongation, chlorophyll development, flowering, abscission, lateral bud outgrowth, and root and shoot growth.

Photomorphogenesis differs from photosynthesis in several major ways. The plant pigment responsible for light-regulated growth responses is phytochrome. Phytochrome is a colorless pigment that is in plants in very small amounts. Predominantly, the red (600 to 660 nm) and far-red (700 to 740 nm) wavelengths of the electromagnetic spectrum are important in the light-regulated growth of plants (Figure 4.1). The wavelengths involved in generating photosynthesis are generally broader (400 to 700 nm) and less specific.

FIGURE 4.1   The effect of daily short duration red *(left)* and far-red *(right)* light on tomato plant growth.

Photomorphogenesis is considered a low-energy response—meaning that it requires very little light energy to get a growth-regulating response. Plants generally require a greater amount of energy for photosynthesis to occur.

Water

Water is the prime necessity for life and is cycled from the earth's atmosphere to its surface through the hydrologic cycle. The water that falls to the earth as rain, snow, hail, fog, dew, or frost comes primarily from the evaporation of water from land and water surfaces and transpiration from plants as water vapor.

Moisture in the atmosphere is often measured as relative humidity. Relative humidity is the amount of water present in air as a percentage of what could be held at saturation at the same temperature and pressure. High humidity generally increases the incidence of many diseases and insects on plants. Dew point is another common term used to quantify the atmospheric moisture environment around plants. Dew point is the temperature of the air at which the water vapor is at the saturation point. Dew point varies with the amount of water vapor in the air.

Soil water is critical to plant growth, as it is the solvent in which soil nutrients are dissolved before they can be absorbed by plant roots. Soil moisture is often expressed in inches of water per foot of soil and 1 acre-inch is equivalent to approximately 27,000 gallons of water. The field capacity of a soil is the maximum amount of moisture that is retained after the surface water is drained and after the water that passes out of the soil by gravity is removed. The field capacity takes into account the physical condition of the soil.

Soil water can also be expressed according to the availability of the water for the plants. Permanent wilting point or percentage is soil moisture that is no longer readily absorbed by the roots of the plant, and the plant shows signs of moisture stress and irreversible wilting occurs. The difference between soil moisture at field capacity and at the permanent wilting point is described as available water.

The total amount of soil moisture available to a plant is affected by the type and depth of soil, depth of rooting of the crop, the rate of water loss by evaporation and transpiration, the temperature, and the rate at which supplemental water is added. Generally as soil texture decreases, its ability to hold water increases. Therefore sands hold the least amount of water, while silt loams retain the most. Organic matter in the soil affects the soil structure and often increases the soil's capacity to retain soil moisture.

Waterlogging   Under waterlogged conditions, all pores in the soil are filled with water. As a result, plant roots cannot obtain oxygen for respiration to maintain their activities for nutrient and water uptake. Plants weakened by lack of oxygen are much more susceptible to diseases caused by soilborne pathogens. Waterlogging due to lack of oxygen in the soil causes death of root hairs. This death of root hairs reduces absorption of nutrients and water, increases formation of compounds toxic to plant growth, and finally retards growth of the plant.

The extent of flooding damage depends on the susceptibility of species or cultivar, level of water constantly present in the soil, soil texture, air temperature and presence, and type of microorganisms. Most vegetables are sensitive to flooding.

Water Balance   Water is essential to photosynthesis, plays a key role in transpiration, regulates the stomata, and is crucial to growth and leaf expansion of vegetables.

When water is in balance, the optimum performance of all components results in steady active growth. However, when the balance of water is affected either because there is insufficient available moisture in the soil or the transpiration of water through the stomata exceeds the plant's capacity to compensate for the internal loss, the plant comes under stress.

Most vegetable crops have differing critical growth periods and, if water stress occurs during critical stages of growth, yield is directly affected. When moisture requirements are not met during this critical phase, permanent damage usually is the result.

Drought   Drought is generally considered to be a meteorological term and is defined as a period without significant rainfall. Droughts may lead to water stress for field-grown vegetables and growth may be impacted. Periods of even short drought stress can reduce crop yields. Plants may adjust to short-term water stress by closing stomates and thereby reducing water loss through the leaves. When stomates are closed, photosynthesis is reduced or stopped and growth is slowed.

Wind

A slight wind is necessary to replenish CO_2 near the plant surface. CO_2 can be rapidly depleted at the leaf surface.

The use of windbreaks can minimize damage by a relatively slow wind. All vegetable crops are very susceptible to harsh wind speeds. The deeper the root system of the crop, generally the more resistant the vegetable crop is to strong winds.

Soil

Soil Components   Soils are made up of mineral matter, organic matter, water, and air. The mineral matter comes from the breakdown of parent material and organic matter comes from the breakdown of plants and animals.

Soil Types   Soils are classified according to their texture or makeup. Soils are composed of sand (2.0- to 0.02-mm diameter), silt (0.02 to 0.002 mm), and clay (< 0.002 mm). Different soil classifications will have differing mineral fractions (Figure 4.2).

Soils are generally classified into four groups: sands, loams, clays, and mucks. Sands have very low moisture-holding capacity and are low in plant nutrients, whereas clays have very high moisture-holding capacity and usually high mineral availability. Sands increase pore space, which improves aeration of the soil. Clay soils tend to drain slowly or poorly.

Soil Fertility   Soil is the main source of nutrients for plants. Plant macronutrients found in the soil include nitrogen (N), phosphorus (P), potassium (K), calcium (Ca), magnesium (Mg), and sulfur (S); and plant micronutrients include iron (Fe), copper (Cu), manganese (Mn), zinc (Zn), boron (B), cobalt (Co), molybdenum (Mo), and chlorine (Cl). Excessive amounts can cause toxicity and insufficient amounts can cause poor or abnormal growth.

Organic Matter   The organic composition of a topsoil is from 1 to 6%. Organic matter consists of plant and animal residue that is in various stages of decomposition.

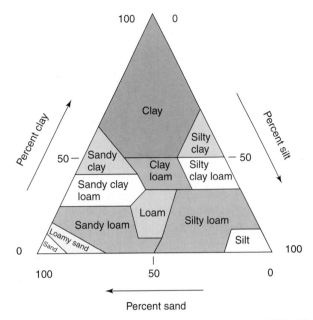

FIGURE 4.2 The USDA soil texture triangle. Source: Acquaah, G. 1999. *Horticulture: Principles and Practices.* Upper Saddle River, NJ: Prentice Hall. Used with permission from Prentice Hall.

Humus is the dark organic material that is produced through decaying action of the organic residue by the action of microorganisms and chemical reaction.

Adding organic matter to the soil will improve its physical characteristics and increase fertility and biological activity. The addition of organic matter generally improves soil structure, which enhances soil aeration and promotes penetration of water and roots into the soil.

Review Questions

1. What is the difference between weather and climate?
2. What factors influence the growth and development of plants in a particular location?
3. What affect does soil temperature have in the vegetable crop ecosystem?
4. What is chilling injury?
5. What are the differences in how humans and plants perceive light?
6. What affect does light intensity have on photosynthesis and plant growth?
7. What is produced by the plant during photosynthesis?
8. Define photomorphogenesis and how does it differ from photosynthesis?

Selected References

Acquaah, G. 1999. *Horticulture: Principles and practices.* Upper Saddle River, NJ: Prentice Hall.

AVRDC. 1990. *Vegetable production training manual.* Asian Vegetable Research and Development Center, Shanhua, Tainan, Taiwan. Reprinted 1992. 442 p.

Hale, M. G., and D. M. Orcut. 1987. *The physiology of plants under stress.* New York: John Wiley & Sons.

Janick, J. 1979. *Horticultural science.* 3d ed. San Francisco: W. H. Freeman.

Nonnecke, I. L. 1989. *Vegetable production.* New York: Van Nostrand Reinhold. 656 p.

Yamaguchi, M. 1983. *World Vegetables: Principles, production, and nutritive values.* New York: Van Nostrand Reinhold. 415 p.

Selected Internet Sites

www.nrcs.usda.gov/   National Research Conservation Center, USDA.

www.nrcs.usda.gov/wcc.htm   National Water and Climate Center, USDA.

www.nws.noaa.gov/   National Weather Service, National Oceanic and Atmospheric Administration.

PART
2

COMMON PRACTICES USED TO GROW VEGETABLES

chapter
5

Plant Selection Considerations
Choosing Which Vegetable Crops to Grow

I. Crop Considerations
II. Grower Considerations
III. Labor Considerations
IV. Marketing Considerations

One of the most important decisions made by vegetable growers is deciding which crops to grow. The environment often determines what can be grown. What should be grown is determined by grower preference as influenced by market demand or need and available resources.

Crop Considerations

Producers usually consider the hardiness of the vegetable, the length of the growing season and time required for the vegetable to reach maturity, and the adaptability of the crop to the soil, temperatures, rainfall, and daylength of the area when choosing vegetable crops to grow. Each of these factors may determine if, how much, and to what quality a crop is produced and available for marketing.

Grower Considerations

Growing and marketing characteristics of the grower affect the types of vegetables grown. Many commercial growers become specialists at producing large amounts of a few types of vegetables. These growers are often called truck farmers.

Other growers are more diversified and grow small plantings of a relatively large number of vegetables. These farmers may be supplying farm stands, roadside markets, and farmer's markets. These growers are called "market gardeners." However, even market gardeners may specialize in popular produce items such as melons and sweet corn.

FIGURE 5.1   Labor may be available for a five-person planting crew for tomatoes, but the cost of the labor may reduce the potential of any profit at harvest.

Part-time farmers may choose to grow and harvest a single crop for a specific market. However these growers must be certain that a market exists for their specialty, and that they do not overproduce their single commodity.

Labor Considerations

The availability of labor will have a direct effect on the type of crops that a vegetable grower might grow (Figure 5.1). Crops with multiple harvests, such as cucumbers and tomatoes, require more labor than crops picked only once or twice, such as beets, cabbage, and corn. For this reason some growers find it advantageous to produce higher-value, labor-intensive crops and to sell them on a pick-your-own basis.

Vegetable producers are bound to abide by the Worker Protection Standard (WPS). The WPS is a set of regulations issued by the U.S. Environmental Protection Agency designed to protect agricultural workers (including owners or managers) from exposure to pesticides used in the production of agricultural crops. The WPS requires that the employer provide specific protection to the workers. For example, information must be displayed if a pesticide has been recently used and if there is a restricted-entry interval before the field can be entered. The employer is also required to provide safety training for using pesticides, emergency assistance in the case of an accident or pesticide exposure, and decontamination areas for workers within a quarter mile of where they are working.

Marketing Considerations

The goal of any business enterprise is to be able to sell for a profit. Marketing in the vegetable industry differs from other businesses in that vegetable growers are limited in their ability to differentiate their products from those of their competitors. Growers are also less certain about the ultimate quality and quantity of what they can grow for sale. Reducing risk and managing uncertainty is how vegetable growers can improve profits.

Yet, while achieving a profit culminates the crop production practice, the determination of potential market preempts any manual practice of crop growing. Before a grower places a crop in the field, it should have already been determined that the crop is one that is anticipated as being needed and the market is sufficiently sound to handle the quantity that the grower may produce. Growers must decide on product development, product design and packaging, prices or fees that will be demanded, transportation and storage policies, product advertisement and selling procedures, and even disposal policies.

Some growers may grow some crops speculating that a market will develop for the crop when it becomes available. Such decisions constitute a risk, but any risk with potential chance for high profit of return may be deemed suitable depending on the grower. Such risk taking must be balanced by stable marketing decisions. Risk-taking ventures by successful growers constitute the exception rather than the norm.

Choosing Cultivars

    I. Factors to Consider in Making Decisions
   II. Open-Pollinated versus Hybrid Cultivars
  III. Sources of Information
       A. Seed Companies
       B. State Cooperative Extension Services
       C. Grower Evaluations

Cultivars are cultivated varieties of agriculturally important plants. A cultivar is a group of plants that share common qualities that are distinct from other groups of plants of the same species. Within the vegetable industry the terms *cultivar* and *variety* are often used interchangeably. The cultivar designation in vegetable production is important in predicting or knowing the plant's growing requirements and in the marketing of the harvested product.

Cultivars perform according to their genetic potential and the environmental conditions and cultural practices to which they are exposed. Choosing the best cultivars for an individual situation is often difficult for vegetable growers, partly because environmental conditions can vary considerably. Good cultivars as designated by growers are those cultivars that perform well under a range of environmental conditions.

Factors to Consider in Making Decisions

In selecting cultivars to grow, vegetable growers take into account the preference of their particular market, the times at which the cultivars can be expected to mature, the method of culture, disease problems that they are likely to encounter, and the adaptability of the cultivar to their soils or climate. Some other factors that influence the performance of a cultivar are climate, culture, method of harvest, and intended use.

The first consideration that vegetable growers generally use in choosing a cultivar is to select one that is adapted to the climate and soils of the area. Successful cultivars perform well under the range of conditions usually encountered on the individual farm.

The next consideration is to choose a cultivar that is high yielding and with the horticultural characteristics that the market desires. Cultivars chosen for growing should have the potential to produce crop yields either equal to or greater than cultivars that the grower may already be using. Acceptable horticultural attributes often include size, shape, color, flavor, and nutritional quality.

Another consideration taken into account by growers is uniformity of maturity. Some vegetables and cultivars have uniform maturity of the harvested product. This demands one, often labor-intensive, harvesting of the crop. Other cultivars are still blooming and producing young fruit after the initial harvest and require multiple, later harvests.

Earliness is also important in choosing a cultivar. Some cultivars are ready for harvest a week or two before another cultivar planted at the same time. This can be used by the grower to extend the harvest season by planting several cultivars at the same time.

Disease resistance of the plant germ plasm is extremely important in choosing a cultivar that will grow satisfactorily in a region. Choosing a cultivar with good disease and/or insect resistance is an important integrated pest management (IPM) grower practice that often reduces the number of pesticide sprays required and determines if a crop can grow in an area.

Open-Pollinated versus Hybrid Cultivars

Open-pollinated cultivars produce their seed by natural pollination through successive generations. A hybrid cultivar is the result of the careful crossbreeding of two parental lines (or cultivars) that are different in at least one but usually several important characteristics. The resulting plant from hybrid seed grows more vigorously and has higher yields than open-pollinated lines. Commercial growers often find that hybrids are superior to older, open-pollinated cultivars. Hybrid cultivars for certain crops tend to have more uniform plant and fruit type and uniform maturity. They also tend to have greater disease resistance, better quality, and more vigor. Hybrid seed is usually more expensive than seed for an open-pollinated cultivar.

Sources of Information

Seed Companies

Each year seed companies release new, promising cultivars. These are often advertised in colorful seed catalogs or in displays at retail stores, often with information on growth habit and disease resistance or tolerance.

State Cooperative Extension Services

Each state in the United States has an extension service located in that state, often at the land-grant university. New cultivars of selected crops are evaluated in controlled

FIGURE 5.2   A cabbage cultivar trial at a North Carolina State University experiment station.

field tests (Figure 5.2) conducted by university faculty or extension service personnel. These tests generally compare new cultivars with established cultivars in a nonbiased manner. Each state extension service generally publishes recommendations on cultivars that perform satisfactorily in that state, often with the crop's intended purpose (i.e., commercial production vs. home gardens, or fresh market vs. processing).

Grower Evaluations

Vegetable growers often evaluate new cultivars on a trial basis on their farms to observe plant performance in their particular situation. While this is the best way to determine the selection of the appropriate cultivar, it can be time consuming.

Review Questions

1. What role does the market play in determining the crops that a vegetable grower should grow?
2. What are some of the grower concerns with growing specialty crops?
3. How might labor affect which crops are grown by a commercial vegetable producer?
4. Why is knowing the cultivar important in growing vegetable crops?
5. What factors must be considered in choosing a vegetable cultivar to grow?
6. What are the differences between open-pollinated and hybrid cultivars?

Selected References

Anonymous. 1998. Varieties and kinds of vegetables. Fort Valley State Univ. Coop. Ext. Program.

Gerber, J. M., J. W. Courter, H. J. Hopen, B. J. Jacobsen, and R. Randell. 1985. *Vegetable production handbook for fresh market growers.* Univ. of Illinois Ext. Service Circular 1241.

Maynard, D. N. 1996. *Variety selection.* Univ. of Florida Coop. Ext. Serv. Cir. HS 712.

Nesheim, O. N., and T. W. Dean. 1996. *The worker protection standard.* Univ. of Florida Coop. Ext. Serv. Publ. PI 19.

VanSickle, J. J. 1996. *Marketing strategies for vegetable growers.* Univ. of Florida Coop. Ext. Serv. Publ. FRE 144.

Selected Internet Sites

www.bbg.org/gardening/kitchen/tomatoes/resources.html   Resource list of seed and plant sources, Brooklyn Botanic Garden.

www.csf.colorado.edu/sustainability/plants/vegies.html   Resource list of Internet sites for unusual vegetables, Communications for a Sustainable Future, University of Colorado.

www.usda.gov/oce/oce/labor-affairs/wpspage.htm   Environmental Protection Agency Worker Protection Standards, USDA.

www.webdesk.com/seeds/catalogs/index.html   Resource list of on-line Internet seed catalog sites, Webdesk.com.

www.aggie-horticulture.tamu.edu/plantanswers/publications/specveg1.html   Listing of seed companies that offer specialty vegetable varieties, Texas A&M University.

www.bbg.org/gardening/kitchen/tomatoes/resources.html   Resource list of seed and plant sources, Brooklyn Botanic Garden.

www.csf.colorado.edu/sustainability/plants/vegies.html   Resource list of Internet sites for unusual vegetables, Communications for a Sustainable Future, University of Colorado.

www.ianr.unl.edu/pubs/nebfacts/nf274.htm   Seed sources for commercial vegetable production, University of Nebraska Cooperative Extension Service.

www.webdesk.com/seeds/catalogs/index.html   Resource list of on-line Internet seed catalog sites, Webdesk.com.

chapter

6

Preparing the Field

Choosing the Field

Several important factors are typically considered before a field is chosen to grow vegetables. First, the environment and topography must be appropriate for the vegetable that is grown. Second, the water needs of most vegetables necessitate that the field have access to water for irrigation to grow the crop to the quality required by the market. Third, it must be determined that the field is accessible by suitable roads to facilitate transportation of machinery and harvested product.

Exposure to full intensity sunlight is generally required to satisfactorily grow most vegetables in the field. The exception to this is that some of the leafy crops can grow in partial shade. Sunlight provides the energy for photosynthesis and the signal for photoperiodic responses, such as flowering and tuber initiation, in some crops.

Variations in air temperatures can occur within a field and are often associated with topography and direction of exposure to the sun. For example, cold air flows into the low areas creating frost pockets and making this area more susceptible to frosts. Fields that have a frost pocket are usually avoided for growing vegetables. Some slope in the field may be beneficial to assist in draining cold air from the field, but these slopes cannot be so steep that erosion results. Fields with southern exposures also warm up sooner in the spring and are generally preferred when earliness of production is desired.

Many different soils can be used to grow the common vegetables. Loams are generally considered the best soil type for most vegetables, but that usually depends on the particular vegetable and potential marketing strategies (such as earliness to market). Sandy soils are easy to till and warm the soonest in the spring. As a result, they are usually chosen when earliness of production is desired. Sands hold very little water compared to the other soil types and, as a result, growers are required to be more attentive to irrigation needs when crops are grown on sands. Sandy soils have the disadvantage of leaching of some important nutrients by excessive rain or irrigation. The leachable nutrients of concern are usually nitrogen and potassium. Many of the more common vegetables can be successfully grown on sands.

Organic soils, such as mucks, have large water- and nutrient-holding compositions; however, they can be a problem for managing water and nutrients. Organic soils are often used to grow leafy vegetables such as lettuce, celery, sweet corn, and various root crops (Figure 6.1).

Clay soils hold water and nutrients but have poor aeration and drainage. Poor, high-content clay soils can be improved by incorporating organic matter. A poorly drained soil also has the disadvantage of warming slowly in the spring, as compared to soils with good drainage.

Preparing the Soil

Proper field preparation is important for optimum vegetable production. This is especially important for establishing uniform stands of vegetable crops. Geographic location and climate strongly influence field preparation practices.

Drainage

Most vegetables will not grow where surface water accumulates for long periods of time. Seeds planted in wet soils are more susceptible to seed decay and emerging

FIGURE 6.1   Lettuce grown in a field with an organic soil.

FIGURE 6.2   Raised field beds for growing vegetable crops.

seedlings in wet soils become more susceptible to soilborne diseases such as damping-off. It is also difficult to make satisfactory seedbeds in poorly drained soils.

Many vegetables are grown on raised beds (Figure 6.2) to facilitate drainage and also increase soil temperatures. Raised beds are essential with furrow irrigation or in areas with excessive rainfall. Raised beds encourage wet soils to dry quicker and can help to prevent waterlogging of plant roots.

Soil pH Testing

Soil testing determines the pH and nutrient content of the soil. Soil testing is done before vegetables are planted in the fields. Soil-testing laboratories often supply the instructions for sampling and a container for holding the soil sample for shipping to the lab. The top 6 in. of soil in several locations of a field is generally sampled. One composite sample is often formed from the several individual samples. If the soil in a field differs from location to location, more than one sample from that field is analyzed. Results from the lab indicate pH and fertility of the soil. Corrective recommendations on fertilizer and lime/gypsum applications are often included and are based on these results.

Soil pH Adjustment

Soil pH measures the concentration of hydrogen ions, H^+, in the soil. Soils are acidic if they have a soil pH below 6. Neutral soils have a soil pH of 6 to 7. Soils are alkaline or saline if they have a soil pH above 7. Soils that are either acidic or alkaline typically have their soil pH adjusted to the neutral range to ensure the proper growth of most vegetables, since most plant-essential elements reach maximal availability and most plant toxic elements become nontoxic in this pH range.

Acid Soils   Many vegetables will not grow in highly acidic soils. Vegetable growers generally apply lime when the soil pH is below 5.5 to raise the pH to a range of 6 to 7. Several types of materials that vary in their ability to neutralize soil acidity are used for liming (Table 6.1).

Alkaline Soils   While most vegetables are somewhat tolerant of alkaline soil, it is still beneficial for the vegetable grower to reduce the soil pH to avoid mineral deficiencies and toxicities. Methods to reduce pH of alkaline soil involve the field application of sulfur-containing compounds. The amount of sulfur to reduce soil pH depends on soil type and desired soil pH.

TABLE 6.1   *Commonly used sources of liming materials*

Liming materials	Relative neutralizing effectiveness
	(%, as compared to $CaCO_3$)
Calcitic lime	100
Burned lime	178
Hydrated lime	134
Magnesium carbonate	119
Dolomitic limestone	95–108
Ground shells	80–88
Basic slag	67–71
Flue dust	96
Rock phosphate	7
Wood ashes	40

Modified from: Hochmuth, G. J. 1996. (See citation on page 93.)

FIGURE 6.3 Subsoiler used to deep-till soils in which a hardpan layer may be present.

Conventional Tillage

Tillage (plowing) increases soil aeration, water penetration, nutrient availability, and the oxidation of chemical compounds in the soil. In addition, tillage improves the physical characteristics of the soil and eliminates competition from weeds. If a field has a hardpan soil layer then it is usually tilled to a deeper depth using a deep plow or subsoiler (Figure 6.3).

Vegetables are grown in a variety of soil types and climatic conditions. These factors govern to some degree the time and type of tillage. Proper field preparation helps to ensure optimum seed germination and root development.

Primary Tillage   Primary tillage involves the initial movement of soil in a field. This often eliminates surface vegetation and deepens the root zone. A moldboard plow is the most commonly used implement for primary tillage. The moldboard plow tills soil to depths of 12 in. or more (depending on soil conditions).

A disk can also be used for primary tillage, especially where a fall or winter crop was planted. The existing vegetation is usually mowed followed by a disking of the field. Disking cuts and incorporates the surface vegetation into the soil (Figure 6.4). The disking of the field is done at least 30 days prior to planting, often followed by a second disking. A shallow disking or harrowing before planting produces a smooth, clod-free surface.

A chisel plow (Figure 6.5) can be used to break up the soil. A chisel plow does not place the surface vegetation as deep in the root zone as a moldboard plow, but a chisel plow is more energy efficient than a moldboard plow and requires almost 50% less time.

Handheld, gas-powered rototillers are common tillage implements in home gardens, but are seldom used for production of vegetables on large acreage. A tractor-powered rototiller is often used in commercial fields to incorporate herbicides into the soil before planting. Repeated rototilling can compact the soil at the base of the tilled area and excessive beating of the tines can destroy soil aggregation.

FIGURE 6.4   Disks are important implements for cultivation of soil.

FIGURE 6.5   Chisel plow.

Secondary Tillage   Secondary tillage is designed to break into relatively small soil aggregates the plow ridges that are formed as a result of the primary tillage with a moldboard plow. This also smoothes the field in preparation for planting. The goal of secondary tillage is to have fine soil particles in the seed zone and coarse particles in the root zone.

FIGURE 6.6   Disking the field to smooth any plow ridges prior to planting.

FIGURE 6.7   Spring-toothed harrow.

Disks may be used in secondary tillage (Figure 6.6), but often a spring-toothed harrow (Figure 6.7) or a cultipacker (Figure 6.8) follows the disking. Relatively shallow disking reduces the chance of compaction in the root zone, especially in heavier soils. To minimize the number of passes over a field, a drag cultivator may be attached in tandem behind a disk.

Raised Beds   Raised beds are typically 3 to 8 in. higher than the normal field level. A raised bed can be produced in the field using commercially available bed shapers or disk hillers. Rotovators (Figure 6.9) also rototill and shape beds. Home gardeners can plant on flat ground and then pull soil with a hoe to form a bed.

FIGURE 6.8   Cultipacker.

FIGURE 6.9   Rotovators are used to rototill and shape plant beds prior to planting.

The benefits of raised beds in vegetable production are that they improve the drainage and increase the depth of the root zone. Using raised beds results in a warming of the soil sooner in the spring so planting may occur up to 2 weeks earlier. Raised, shaped beds are also essential for precision seeders and mechanical harvesters.

Conservation Tillage   Contour planting, strip cropping, and terracing are used for growing vegetables in cultivated areas with excessive slope. No-till planting uses the previous crop residue or a planted cover crop in place throughout the subsequent growing season. The crop debris captures surface water, and the decaying roots provide channels for water penetration. Seeds are sown in a no-till system with a slot-type seeder that places the seeds through the stubble of the previous crop. In some locations, no-till fields have cooler soil temperatures than conventionally tilled fields and these cooler temperatures have delayed seed germination and plant maturity. Crop debris in no-till fields can also increase disease and insect problems.

Cultivation   The most important effect of cultivation is weed control. By eliminating the transpiring leaf surface of weeds, soil moisture is conserved and nutrients are retained for crop growth. Repeated cultivation can increase soil compaction caused by tractor weight. Cultivation equipment includes duckfoot (flared) shovels or sweeps, chisel tines, rotary hoes, rototillers, and similar devices. Cultivation is most effective with young weeds.

Windbreaks

Severe physical damage to plants in the field can occur if winds reach 20 to 25 mph. Lesser winds can result in blowing soil particles that can adversely affect plant growth through the abrasion of leaves and stem tissue by the soil particles. Young plants are most susceptible to abrasion by soil particles moved by wind or by erosion. Abrasion slows the growth of plants or may destroy the plant completely. Damage from erosion and abrasion is most severe in dry, sandy soils.

Windbreaks are used to reduce the effects of winds on plant development. In many areas trees and brush may completely surround fields and provide adequate wind protection. In areas that are not protected from the wind, growers often plant small grains such as wheat or rye. The typical situation is to seed a couple of rows of grain for every three or four vegetable rows (Figure 6.10). This is usually done when beds are formed in early spring or during the fall. The windbreak row of small grains is often mowed later in the summer and used as a drive row during harvest. Windbreaks are usually killed before the plants become senescent, to reduce the risk of insect infestations.

FIGURE 6.10   A row of small grains serving as a windbreak for young bean plants.

Using windbreaks can also change the plant microclimate. Reduction in wind speed reduces evaporation from the leaf surface. Also, increases in air temperature, humidity, and altered CO_2 levels can occur with the use of windbreaks.

A recent report indicated that using windbreaks with peppers in the field reduced plant height, reduced time to flower bud initiation, and increased the number of flowers and fruit set. This resulted in more early and total marketable yields of the peppers. The increase in yield was also attributed to a decrease in the incidence of bacterial leaf spot on those plants that were protected by windbreaks. It was suspected that those plants that were not protected by wind suffered more wind abrasion and that this abrasion provided entry points for the disease.

Plant Nutrition and Fertilizing

At one time it was thought that the only substance needed by plants for satisfactory growth in a field was water. With time it was discovered that certain "earth substances" were beneficial for plant growth. In the 1800s, Justus Von Liebig suggested that certain elements had an essential role in plant growth and that growth was limited to the extent that an essential element was lacking.

Essential Plant Nutrients

Living plants require 16 essential elements to survive (Table 6.2). Each of the essential elements has at least one specifically defined role in plant growth so that the plants fail to grow and reproduce normally in the absence of the element. Essential elements can be broken down into macronutrients and micronutrients.

Macronutrients   Macronutrients are those elements needed by plants in relatively large quantities. While there are ten elements that are macronutrients, only nitrogen, phosphorus, and potassium are likely to be deficient in most vegetable-producing areas and soils. The following elements are considered macronutrients:

Carbon (C)   Carbon is available from CO_2 and is assimilated by plants during photosynthesis. It is a component of organic compounds such as sugars, proteins, and organic acids. These compounds are used in structural components, enzymatic reactions, and genetic material, among others.

Hydrogen (H)   Hydrogen is derived from H_2O and is also incorporated into organic compounds during photosynthesis. Hydrogen ions are also involved in electrochemical reactions and maintain electrical charge balance.

Oxygen (O)   Oxygen is derived from CO_2 and is also a part of organic compounds such as simple sugars. Oxygen is necessary for all oxygen-requiring reactions in plants including nutrient uptake by roots.

Phosphorus (P)   Phosphorus plays a major role in several energy transfer compounds in plants. Phosphorus is also very important in the structure of nucleic acids, which serve as the building blocks for the genetic code material in plant cells. Phosphorus promotes early maturity and fruit quality.

Potassium (K)   Potassium is an activator in many enzymatic reactions in the plant. Turgor in the guard cells of the stomates is also controlled by K movement in and out of these cells. Potassium is also important in cell growth primarily through its effect on cell extension. With adequate K, cell walls are thicker

TABLE 6.2   *Nutrients essential to plant growth*

Element	Chemical symbol	Form(s) taken up by plant
Macronutrients		
Carbon	C	CO_2
Hydrogen	H	H_2O
Oxygen	O	H_2O, O_2
Nitrogen	N	NH_4^+, NO_3^-
Phosphorus	P	$H_2PO_4^-$, HPO_4^{2-}
Potassium	K	K^+
Calcium	Ca	Ca^{2+}
Magnesium	Mg	Mg^{2+}
Sulfur	S	SO_4^{2-}
Iron	Fe	Fe^{2+}, Fe^{3+}
Micronutrients		
Zinc	Zn	Zn^{2+}, $Zn(OH)_2$
Manganese	Mn	Mn^{2+}
Copper	Cu	Cu^{2+}
Boron	B	$B(OH)_3$
Molybdenum	Mo	MoO_4^{2-}
Chlorine	Cl	Cl^-

From Bennett, W. F. 1993. Plant nutrient utilization and diagnostic plant symptoms. In W. F. Bennett, (Ed.), *Nutrient deficiencies and toxicities in crop plants.* St. Paul, MN: APS Press. Used with permission.

and provide more tissue stability. Potassium is also referred to as a quality element because vegetables grown with adequate K seem to have a longer postharvest shelf life.

Nitrogen (N)   Nitrogen is a component in many compounds including chlorophyll, amino acids, proteins, nucleic acids, and organic acids. A large part of the plant body is composed of N-containing compounds. Because nitrogen is contained in the chlorophyll molecule, a deficiency of N will result in a chlorotic condition of the plant.

Sulfur (S)   Sulfur is a component of sulfur-containing amino acids such as methionine. Sulfur is also contained in the sulfhydryl group of certain enzymes. In addition, sulfur is present in glycosides, which give the odor characteristics of onions, mustard, and garlic.

Calcium (Ca)   Calcium is a component of calcium pectate, a constituent of cell walls. In addition, Ca is a cofactor of certain enzymatic reactions. It is involved in cell elongation and cell division.

Iron (Fe)   Iron is used in the biochemical reactions that form chlorophyll and is a part of one of the enzymes that is responsible for the reduction of nitrate-nitrogen to ammonical-nitrogen. Other enzymes such as catalase and peroxidase also require Fe.

Magnesium (Mg)   Magnesium plays an important role in plant cells as it appears in the center of the chlorophyll molecule. Certain enzymatic reactions require Mg as a cofactor. Magnesium aids in the formation of sugars, oils, and fats.

Micronutrients   Micronutrients are essential nutrients that are needed by plants in smaller amounts than the macronutrients. The amount of micronutrients assimilated by vegetables is small in comparison to the total quantity that may be present in the soil. The availability of micronutrients is influenced by soil conditions such as pH, moisture content, aeration, and the presence and amounts of other elements. Deficiencies in micronutrients are relatively rare and are usually caused by overliming, underliming, or other poor management practices. Boron is the most widely deficient micronutrient in vegetable crop soils. The following elements are considered micronutrients:

Molybdenum (Mo)   Molybdenum is a constituent of two enzymes involved in N metabolism. The most important of these is nitrate reductase, the enzyme involved in the reduction of nitrate-nitrogen to ammonical-nitrogen. It is also a structural component of nitrogenase, which is involved in the fixation of N_2 into ammonium form in a symbiotic relationship with legumes.

Boron (B)   Boron appears to be important for meristem development in young plant parts such as root tips. Boron is involved in the transport of sugars across cell membranes and in the synthesis of cell wall material. Because of its impact on cell development and on sugar and starch formation and translocation, a deficiency of B will retard new growth and development.

Manganese (Mn)   Manganese functions in several enzymatic reactions that involve the energy compound adenosine triphosphate (ATP). Manganese also activates several enzymes and is involved in the process of the electron transport system in photosynthesis.

Zinc (Zn)   Zinc is involved in the activation of several enzymes and is required for the synthesis of indoleacetic acid, a plant growth regulator.

Chlorine (Cl)   Chlorine has a possible role in photosynthesis and may function as a counter ion for K fluxes involved in cell turgor. Chlorine is involved in the capture and storage of light energy through its involvement in photophosphorylation reaction in photosynthesis.

Crop Removal Values

Crop removal values are estimated by analyzing plants and fruits for their nutrient content and then expressing the results on a per-acre basis (Table 6.3). Crop removal values can be used to estimate fertilizer needs, and are important for information purposes and comparisons among crops. This is generally not an accurate method of determining fertilizer needs by the plants, since these values are usually determined on crops grown on well-fertilized land and crops in this situation will continue to take up nutrients in excess of their needs. As a result, crop removal values may overestimate the true nutrient content of the crops.

Fertilizers

Fertilizers are applied to a field to increase the fertility of the soil to avoid nutrient deficiency levels in the plant and attain optimum plant growth and production. Fertilizers from inorganic and organic sources are available for use by vegetable growers.

Inorganic Fertilizers   Commercial inorganic fertilizers are added to a soil to directly increase the amount of specific nutrients available to the plant (Table 6.4). They are

TABLE 6.3 *Approximate accumulation of nutrients by some vegetable crops*

Vegetable	Yield (cwt/acre)	Nutrient absorption (lbs/acre)		
		N	P	K
Broccoli	100 heads	20	2	45
	Other	145	8	165
		165	10	210
Brussels sprouts	160 sprouts	150	20	125
	Other	85	9	110
		235	29	235
Carrot	500 roots	80	20	200
	Tops	65	5	145
		145	25	345
Celery	1,000 tops	170	35	380
	Roots	25	15	55
		205	50	435
Honeydew melon	290 fruits	70	8	65
	Vines	135	15	96
		205	23	160
Lettuce	350 plants	95	12	170
Muskmelon	225 fruits	95	17	120
	Vines	60	8	35
		155	25	155
Onion	400 bulbs	110	20	110
	Tops	35	5	45
		145	25	155
Pea, shelled	40 peas	100	10	30
	Vines	70	12	50
		170	22	80
Pepper	225 fruits	45	6	50
	Plants	95	6	90
		140	12	140
Potato	400 tubers	150	19	200
	Vines	60	11	75
		210	30	275
Snap bean	100 beans	120	10	55
	Plants	50	6	45
		170	16	100
Spinach	200 plants	100	12	100
Sweet corn	130 ears	55	8	30
	Plants	100	12	75
		155	20	105
Sweet potato	300 roots	80	16	160
	Vines	60	4	40
		140	20	200
Tomato	600 fruits	100	10	180
	Vines	80	11	100
		180	21	280

From Maynard, D. N., and G. J. Hochmuth. 1997. *Knott's handbook for vegetable growers.* 4th ed. New York: John Wiley & Sons. Reprinted by permission of John Wiley & Sons, Inc.

TABLE 6.4   *Some common nutrient fertilizer sources*

Nutrient	Source	Content	
Nitrogen		% N	
	Ammonium sulfate	21	
	Ammonium nitrate	33	
	Ammonium phosphate	10–18	
	Urea	46	
	Sodium nitrate	16	
	Blood meal	13–15	
Phosphates		% P	% P_2O_5
	Superphosphates	7–22	16–50
	Ammonium phosphate	21	48
	Diammonium phosphate	20–23	46–53
	Steamed bone meal	10–13	23–30
	Rock phosphate	11–13	25–30
	Phosphoric acid	24	54
Potassium		% K	% K_2O
	Potassium chloride	40–50	48–60
	Potassium sulfate	40–42	48–50
	Sul-Po-Mag	19–25	25–30
	Potassium nitrate	37	44

not used to improve the physical condition or make soil reserves available. Commercial fertilizers furnish limited elements in the most economic manner and maintain proper ratio of nutrients for the crop being grown.

Commercial fertilizers always bear a tag that indicates the nutrient analysis for available nitrogen, phosphorus, and potassium. For example, a labeled 5-10-10 fertilizer means 5% by weight nitrogen, 10% phosphate (P_2O_5), and 10% potash (K_2O). The percent phosphorus in the fertilizer is obtained by multiplying by 0.44 and the percent potassium is obtained by multiplying by 0.83. Therefore, a 5-10-10 fertilizer is 5% nitrogen, 4.4% phosphorus, and 8.3% potassium. The fertilizer label also tells what materials are used to make the units of available fertilizers. As an example, for a 5-10-5 fertilizer it may say 2.5% nitrogen as 20% ammonium sulfate; 2.5% nitrogen as 15% sodium nitrate; 10% phosphorus as 20% triple superphosphate; and 5% potassium as 40% potassium chloride.

Timing and Application of Fertilizers

Preplant Fertilizer or Broadcast Application   A preplant fertilizer or a broadcast application (Figure 6.11) uniformly applies fertilizer with a spreader prior to planting. The fertilizer is broadcast or spread on the soil surface and then incorporated with a plow, disk, or power tiller. Fertilizer is generally not broadcast to and incorporated into a field too far in advance of planting, since considerable loss of nitrogen and potassium could occur as a result of leaching due to rains that may occur before the field is planted.

FIGURE 6.11   Broadcasting fertilizer on the surface of the soil.

Band Application   Band application is often used with direct-seeded crops such as beans and corn. In banding, fertilizer is placed 2 in. to the side and 2 in. below the level of the seed in a row at planting time. This method is especially effective when placing phosphorus in either cold or calcareous soils or soils high in hydrous oxides of iron and aluminum. Since banding is placing fertilizer where it is needed and likely to be taken up by the plant, smaller amounts of nitrogen and potassium may be lost due to leaching.

Starter Solution   Starter fertilizer solutions are commonly used with transplanted crops. Water is often applied to the recently planted transplants and small amounts of fertilizer may be included in this water to encourage early plant growth and development. Starter fertilizers provide only small amounts of the actual fertilizer requirements of the plants, but they are important in establishing crops in cool, damp soils. Transplanted crops produce little aboveground growth during the first 2 weeks after planting, often because they undergo shock at transplanting and have a limited root system. An application of starter solution provides phosphorus as well as nitrogen and potassium directly to the plant roots and could encourage early growth and increase the early yield, although it rarely affects the total yield. They can be applied at planting in a band to the side and below the seed or transplant, or dissolved in the transplant water and applied in a furrow.

Side Dressing or Split Application   Although fertilizer is needed for early growth, the greatest quantities are taken up during the second half of the season. For this reason, much of the fertilizer applied at planting time may be leached out of sandy soils before being utilized by the plant. To correct this problem, a side dressing of fertilizer (usually nitrogen) can be applied 4 to 8 weeks after planting and 6 to 10 in. from the base of the plant, and then lightly incorporated into the soil. N and K are the nutrients of the most concern because of their potential to be leached through the soil. Split applications are often used on sandy soils or with long-season crops.

Foliar Application or Foliar Spray   Foliar applications are used to rapidly enhance the green (chlorophyll) color in plants. Nitrogen foliar fertilizers usually contain low-biuret urea because the uncharged urea molecule more easily penetrates the cuticle of the leaf. Foliar sprays are widely used to furnish trace elements to vegetables. Foliar applications are usually considered a last resort for correcting nutrient deficiency problems.

Slow-Release Fertilizers   Slow-release fertilizers include sulfur-coated urea or isobutylidene-diurea that supply a portion of the N that the plant requires. Although they are more expensive, slow-release fertilizers can be useful in reducing fertilizer nutrient leaching and in supplying adequate fertilizer for long-term crops.

Nutrient/Chemical Application through Irrigation Systems   Most plant nutrients can be applied through trickle or sprinkler irrigation systems using a siphon or a metering pump. Nitrogen is soluble and moves readily in the soil. Although phosphorus and potassium move more slowly through the soil than nitrogen, their movement is greater with trickle irrigation than with other types of application techniques. Micronutrients are often insoluble in water but they can become soluble if they are applied with chelates, thereby reducing the possibility of clogging the irrigation emitters. Clogging can be a problem when insoluble or slightly soluble compounds lodge in the small openings of trickle lines.

Chemicals such as fungicides or insecticides can be applied through a trickle system in the same manner as nutrients. However, only soilborne organisms are affected by the application. Regulations regarding application of chemicals may prohibit application through an irrigation system. Herbicides are never applied through an irrigation system because of the possibility of contamination of the system.

Tissue Sampling   Vegetable growers often use tissue sampling to determine the effectiveness of a fertilizer management program. Timely tissue sampling can also diagnose nutritional problems.

Organic Fertilizers   Organic fertilizers generally used in commercial vegetable production include animal manures, green manures, and compost. Nutrient composition and decomposition rates of manures and compost vary according to source. Proper application of manure to fields takes into account the nutrient composition and decomposition rate for the manure used. The carbon to nitrogen ratio (C:N) of manure is also important since those with a low C:N decompose more quickly than those with a high C:N. Because many manures do not supply adequate nitrogen at the time of application (animal manure) or at planting (green manure), a side dressing of nitrogen is also required to provide the proper amount of nutrients for early plant growth.

Animal Manures   Animal manures are important sources of organic nutrients and can be used as fertilizers (Table 6.5). Many vegetables generally show a favorable response to animal manures. Disadvantages with the use of manures may include the presence of excess amounts of salt, introduction of weed seeds, and a poor

TABLE 6.5 *Nutrient content of some commonly used manure sources*

Source	Dry matter (%)	N	P_2O_5	K_2O	C/N (ratio)
		approximate composition (% dry weight)			
Dairy	15–25	0.6–2.1	0.7–1.1	2.4–3.6	8
Horse	15–25	1.7–3.0	0.7–1.2	1.2–2.2	2
Swine	20–30	3.0–4.0	0.4–0.6	0.5–1.0	4
Sheep	25–35	3.0–4.0	1.2–1.6	3.0–4.0	16
Poultry	20–30	2.0–4.5	4.5–6.0	1.2–2.4	—

Adapted from Maynard, D. N., and G. J. Hochmuth. 1997. *Knott's handbook for vegetable growers.* 4th ed. New York: John Wiley & Sons. Reprinted by permission of John Wiley & Sons, Inc.

cost-to-benefit ratio. Also, some manures may need to be properly fermented or aged before applying to the fields.

Green Manures   Green manures supply organic matter, prevent erosion, and aid in conserving soluble nutrients in the soil. Green manures can either be nitrogen-fixing legumes or nonlegume crops. The added organic matter in the soil retains more moisture and stores plant nutrients for crops and microorganisms. A 6- to 8-week period between plowing down of green cover crops and crop establishment is recommended to allow decay of the green cover crop plant material. Also, freshly incorporated plant material can encourage certain plant diseases such as damping-off.

Composts   Composts have relatively low amounts of nitrogen, phosphorus, and potassium as compared to inorganic fertilizers. Compost can be made from domestic sources, such as food scraps, or from municipal or commercial sources. Municipal sources of compost include yard wastes and common types of commercial composts include chicken and turkey litter. Advantages with the use of compost are that the composting process allows the safe use of materials such as sawdust that normally chemically binds soil nitrogen if applied without undergoing the composting process. Also, composting reduces pathogen populations within the materials when temperatures reach 150°F and destroys weed seeds when temperatures reach 175°F. Disadvantages are that composts generally are not as effective as raw organic matter in improving soil structure and can lose nitrogen as NH_3 when the compost is turned.

Rotations

Crop rotation is the growing of two or more crops in a sequence on the same land during a period of time. Crop rotations increase soil sanitation, fertility, and structure. Proper crop rotations can also reduce disease and insect pressures. Vegetable crops are often rotated either every year or every other year depending on the climate and previous field history.

A variety of rotations can be used as long as the ensuing planted crops are not susceptible to the same root diseases and insects as the crop that is currently being grown. Certain crops and families of crops can be grouped together according to their susceptibility to the same diseases (Table 6.6). Crops within individual groups are rotated with crops from other groups. Vegetables from one group are generally not planted in the same location more than once every 3 to 5 years.

TABLE 6.6   *Crop grouping for rotation to control soilborne diseases*

Crop grouping	Diseases aided by rotation
Cantaloupe, cucumber, honeydew, melon, pumpkin, squash, watermelon	Fusarium wilt, gummy stem blight, anthracnose, scab, belly rot, pythium, nematodes
Brussels sprouts, cabbage, cauliflower, collard, lettuce, mustard, radish, rutabaga, spinach, Swiss chard, turnip	Black rot, black leg, club root, alternaria, yellows
Eggplant, Irish potato, okra, pepper, tomato	Potato scab, bacterial canker and wilt, early blight, verticillium wilt, nematodes
Beet, carrot, garlic, onion, shallot, sweet potato	Scurf, black rot, soil rot, wilt, nematodes
Sweet corn	Smut
All beans, Southern peas, peas	Anthracnose, rhizoctonia and fusarium root rot, bacterial and halo blight, nematodes

Modified from Johnson, K. E. 1993. *Crop rotation in vegetable production.* Univ. of Tennessee, Agricultural Extension Service. Vegetable and Small Fruit Facts, no. 3.

Critical Nutrient Concentrations

Normal plant growth is achieved when all the essential elements are provided in a suitable general nutrient concentration range. This is called the adequate or sufficient nutritional concentration range. A critical concentration for a nutrient occurs at the point where plant growth is reduced by 10% because of a shortage of the element in question. The critical concentration is the borderline between elemental sufficiency and deficiency.

A deficiency of an element essential for plant growth will result in a decrease in the normal growth of the plant and will affect the yield of a crop (Table 6.7). The deficient range occurs at elemental concentrations lower than those in the transitional zone, and is accompanied by a drastic restriction in growth. Plants may begin to show deficiency symptoms. Symptoms of nutrient deficiencies have been traditionally used to diagnose growth problems.

Essential elements can be toxic when they are in excess and detrimentally affect growth and yield of the crop. In the toxic zone, tissue elemental concentrations are greater than those in the adequate zone. A gradual decrease in plant growth occurs in the toxic zone. As the tissue concentration rises further, toxicity symptoms, often necrosis, can occur. Such symptoms can be used to determine the source of the problem.

Most deficiency or toxicity symptoms can be categorized into one of five types: (1) chlorosis, which is a yellowing of plant tissue due to a reduction in chlorophyll formation; (2) necrosis, or death of plant tissue; (3) lack of new growth or terminal growth resulting in rosetting; (4) an accumulation of anthocyanin and an appearance of a reddish color; and (5) stunting and reduced growth with either normal or dark green coloring or yellowing.

Deficiency symptoms are only used as guidelines for determining nutritional needs. Symptoms plus plant and soil analyses, together with a general knowledge of crop needs and the chemistry of the soil, are all used in determining crop nutrient needs.

TABLE 6.7 *Fertilizer nutrients required by plants*

Nutrient	Deficiency symptoms	Occurrence
Nitrogen (N)	Stems thin, erect, hard; leaves small, yellow; on some crops (tomatoes) undersides are reddish; lower leaves affected first	On sandy soils especially after heavy rain or after overirrigation. Also on organic soils during cool growing seasons
Phosphorus (P)	Stems thin and shortened; leaves develop purple color; older leaves affected first; plants stunted and maturity delayed	On acid soils or very alkaline soils. Also when soils are cool and wet
Potassium (K)	Older leaves develop gray or tan areas on leaf margins. Eventually a scorch appears on the entire margin	On sandy soils following leaching rains or overirrigation
Boron (B)	Growing tips die and leaves are distorted. Specific diseases caused by boron deficiency include brown curd and hollow stem of cauliflower, cracked stem of celery, blackheart of beet, and internal browning of turnip	On soils with pH above 6.8 or on sandy, leached soils, or on crops with very high demand such as cole crops
Calcium (Ca)	Growing-point growth restricted on shoots and roots. Specific deficiencies include blossom-end rot of tomato, pepper and watermelon, brownheart of escarole, celery blackheart, and cauliflower or cabbage tipburn	On strongly acid soils, or during severe droughts
Copper (Cu)	Yellowing of leaves, stunting of plants. Onion bulbs are soft with thin, pale scales	On organic soils or occasionally new mineral soils
Iron (Fe)	Distinct yellow or white areas between veins on youngest leaves	On soils with pH above 6.8
Magnesium (Mg)	Initially older leaves show yellowing between veins, followed by yellowing of young leaves. Older leaves soon fall	On strongly acid soils, or on leached sandy soils
Manganese (Mn)	Yellow mottled areas between veins on youngest leaves, not as intense as iron deficiency	On soils with pH above 6.4
Molybdenum (Mo)	Pale, distorted, narrow leaves with some interveinal yellowing of older leaves, e.g., whiptail disease of cauliflower	On very acid soils
Zinc (Zn)	Small reddish spots on cotyledon leaves of beans; light areas (white bud) of corn leaves	On wet, cold soils in early spring, or where excessive phosphorus is present
Sulfur (S)	General yellowing of younger leaves and reduced growth	On very sandy soils, low in organic matter, especially following continued use of sulfur-free fertilizers and especially in areas that receive little atmospheric sulfur
Chlorine	Deficiencies very rare	Usually only under laboratory conditions

From Hochmuth, G. J. 1996. *Soil and fertilizer management for vegetable production in Florida*. Univ. of Florida Coop. Ext. Serv. Cir. HS-711.

Many deficiency symptoms can be caused by any one of several nutrients or by growing conditions. The growers use knowledge of soil pH and general soil conditions to accurately diagnose nutrient imbalances. Typical nutrient deficiency symptoms can also be caused by many other conditions. Certain herbicides, diseases, and insects can cause chlorosis in plants. Waterlogged or droughty soils and mechanical or wind damage can often create problems that mimic deficiencies.

Review Questions

1. What factors must be considered in choosing a field to grow vegetables?
2. What types of information are received as a result of soil testing?
3. How does conservation tillage differ from conventional tillage?
4. Why are raised beds used to grow many vegetables?
5. What are the macronutrients and micronutrients?
6. Why aren't crop removal values accurate predictors of fertilizer needs by plants?
7. What are some of the concerns with using animal manures for growing vegetables?
8. What are the benefits of using crop rotation?
9. Define the critical concentration for a nutrient.

Selected References

Bennett, W. F. 1993. Plant nutrient utilization and diagnosing plant symptoms. In W. F. Bennett (Ed.), *Nutrient deficiencies and toxicities in crop plants.* St. Paul, MN: APS Press, 1–7.

Decoteau, D. R., D. Ranwala, M. J. McMahon, and S. B. Wilson. 1995. *The lettuce growing handbook: Botany, field procedures, growing problems, and postharvest handling.* Oak Brook, IL: McDonald's International.

Gerber, J. M., J. W. Courter, H. J. Hopen, B. J. Jacobsen, and R. Randell. 1985. *Vegetable production handbook for fresh market growers.* Univ. of Illinois Ext. Service Circular 1241.

Hochmuth, G. J. 1996. *Soil and fertilizer management for vegetable production in Florida.* Univ. of Florida Coop. Ext. Serv. Cir. HS-711.

Johnson, K. E. 1993. *Crop rotation in vegetable production.* Univ. of Tennessee, Agricultural Extension Service. Vegetable and Small Fruit Facts no. 3.

Maynard, D. N., and G. J. Hochmuth. 1997. *Knott's handbook for vegetable growers.* 4th ed. New York: John Wiley & Sons.

Michigan State University. 1996. *Sites for vegetable gardens.* Michigan State University Extension, Home Horticulture Publ. 01701337.

Naegely, S. K. 1998. Windbreaks. *American Vegetable Grower.* (January): 52–53.

Parnes, R. 1990. *Fertile soil, a grower's guide to organic and inorganic fertilizers.* Davis, CA: AgAccess.

Peet, M. 1996. *Sustainable practices for vegetable production in the south.* Newburyport, MA: Focus Publication.

chapter

7

Planting the Field

The selection of an appropriate planting method for a particular vegetable crop depends on many factors, including the size or shape of the seed, soil characteristics, total acres to be planted, and personal preference of the grower.

Small-sized, irregular-shaped seeds are often difficult to separate and handle during planting and establishing a uniform stand of plants in the field. To assist the vegetable grower in establishing a stand in the field, vegetable seeds can be sized for uniform growth, pelleted for precision planting, or greenhouse germinated for later transplanting. Pelleting or coating the seeds makes them round and easy to singularize. Transplants are used with some crops to ensure stand establishment and to promote earliness of the crop. Transplanting is not appropriate or cost-effective with all vegetable crops.

Direct Seeding

Direct seeding is the least expensive method of planting and establishing vegetable crops in the field. While many vegetable crops can be successfully planted by direct seeding, environmental and biological factors can affect resulting stand establishment. Cold soil temperatures, a crusted soil surface, seeds planted too deep or too shallow, and nonviable seed can all result in poor germination and uneven stands. Poor stands will not only lower yields but also waste fertilizer, pesticides, and time.

There are many types of commercially available planters for vegetable seeds and these planters can be generally categorized as random or precision seeders. Random seeders sow a metered flow of seeds without exact placement or spacing of seeds (Figure 7.1). Precision seeders systematically dispense a seed or a group of seeds at a preselected spacing.

Random Seeding

A seed drill is the most common type of random seeder. It has a rotating feed wheel that agitates the seed over an orifice that causes the seed to fall at a controlled rate. The seeding rate is adjusted by selecting a specific orifice size. Different-sized holes in a rotary plate in the bottom of the seed hopper make selection simple. Multiple plates provide a wide range of choices for different seed sizes and planting rates.

The seed drill opens a small furrow and allows seed to flow through a hole to a specified depth and then closes the furrow. The seed drill can be fitted with a regular single-row shoe or mounted with a scatter shoe to spread seed within a narrow band. Plates with holes of various sizes are available for most vegetables.

A disadvantage of direct seeding with a random seeder is the need to thin the emerging seedlings to desired plant spacings (Figure 7.2). To reduce plant competition, thinning is done while the seedlings are small and when the initial leaves are formed. Thinning too late in the season can cause root damage and slow the growth and development of the plants that were not removed during the thinning operation.

Precision Seeding

Precision seeding places the seeds at a defined spacing (Figure 7.3) and at a specific depth in the soil. Seeding rates, however, can be decreased as seed quality and performance are improved. Precision seeders (Figure 7.4) are more expensive than seed drills but using them reduces or eliminates costs associated with the need for thinning after the seedlings become established.

FIGURE 7.1   A tractor-mounted Planet Jr. seeder.

FIGURE 7.2   Hand-thinning emerging seedlings to a desired plant spacing.

FIGURE 7.3   Young cabbage seedling at a defined plant spacing without thinning as a result of using a precision seeder.

The distance between the rows varies with respect to cultivation, irrigation, fertilization, and cultivar. Bed height is an additional factor that varies with soil type, drainage, and season. Sprinkler irrigation is often needed in some locations for successful seedling emergence and plant establishment.

FIGURE 7.4   A precision seeder used for vegetable crops.

There are several types of precision seeders. Some of those that are commercially available include the following:

Belt Type   Circular holes are punched in the belt of a belt-type planter to accommodate the seed size, and holes are spaced along the belt at specific intervals to regulate the distance between seeds planted in the row. Usually, coated seed improves the uniformity of this type of seeder.

Plate Type   Seed in a plate-type planter is dropped into a notch in a plate and is transported to a drop point. Spacing is achieved by gearing the rate of turn of the plate.

Vacuum Type   Seed in a vacuum-type planter is drawn against holes in a vertical plate by a vacuum and is agitated to remove excess seed, leaving one per hole; various spacings are achieved.

Spoon Type   Seed in a spoon-type planter is scooped up out of a reservoir by small spoons (sized for the seed) and then carried to a drop shoot where the spoon turns and drops the seed. Spacing is achieved by spoon number and gearing.

Pneumatic Type   Seed in a pneumatic-type planter is held in place against a drum until the air pressure is broken, then it drops into tubes and is blown to soil.

Gel Seeders   A gel planter is unique in that the seed is not handled directly but is suspended in a gel that is metered and dispensed into the row. This enables imbibed and/or partially germinated seed to be planted with minimal damage. One of the major advantages of gel seeding is that the seed can be imbibed and/or pregerminated under ideal conditions prior to planting. By eliminating nongerminated seeds, emergence and stand are improved.

Growing Transplants and Transplanting

For some vegetable crops and under certain environmental and economical situations, planting using transplants is the preferred method of crop establishment in the field. Vegetables are transplanted to ensure an earlier harvest, enable warm-season

TABLE 7.1   *Relative ease of selected vegetables for transplanting*

Easy	Moderate	Require special care (containerized transplants recommended)
Beet	Celery	Sweet corn
Broccoli	Eggplant	Cucumber
Brussels sprouts	Onion	Muskmelon
Cabbage	Pepper	Summer squash
Cauliflower		Watermelon
Chard		
Lettuce		
Tomato		

Adapted from Maynard, D. N., and G. J. Hochmuth. 1997. *Knott's handbook for vegetable growers.* 4th ed. New York: John Wiley & Sons. Reprinted by permission of John Wiley & Sons, Inc.

crops to be grown in regions too cold for direct seeding, allow plants to develop better root systems, and result in a uniform stand of plants. Only the early crops of many vegetables are transplanted and not all crops are suitable for transplanting at any time.

In general, crops that are easy to transplant such as broccoli, cabbage, and cauliflower are efficient in water absorption and rapidly form new roots (Table 7.1). Crops that are moderately easy to transplant such as celery, eggplant, and peppers do not absorb water as efficiently as crops that are easy to transplant, but they form new roots relatively quickly. Difficult-to-transplant crops such as sweet corn, cucumber, and summer squash have root systems that are easily injured during transplanting and the transplants are not easily stored or "held" until the proper time for field planting if they become too large during the transplant development stage.

Using transplants reduces the amount of time the plant is in the field and may increase uniformity of the crop and harvest as compared to fields that were established by direct seeding. Using transplants also eliminates the need for thinning. The costs for these operations vary from location to location, and whether a grower uses transplants depends on the climate and expected economic returns.

Growing Structures

Vegetable transplants can be produced in greenhouses, hotbeds and cold frames, or in the field in protected areas using plastic or fabric materials for coverings. Greenhouses (Figure 7.5) typically provide the most regulated growing conditions but can be expensive to construct and maintain. Heating, ventilation, and irrigation systems in a greenhouse can be either manually or automatically operated.

Hotbeds and cold frames are more traditional structures used by home gardeners and some commercial growers. They are similar in construction to greenhouses but are much smaller, with a lower level of sophistication for controlling environmental factors. Hot beds are heated with manure, hot air, hot water, steam, or electricity, while cold frames are unheated. Cold frames (Figure 7.6) are used mainly to harden plants that are grown in heated structures, but temperature control in cold frames is more difficult than in greenhouses.

Some vegetable growers practice direct seeding in the field in close plant spacings, often under a low-lying plastic or fabric covering, to produce transplants. When

FIGURE 7.5   Growing vegetable transplants in a polyethylene-covered greenhouse.

FIGURE 7.6   Cold frames used for growing vegetable transplants.

the emerging seedlings are of suitable size the covering is removed and the plants are "pulled" from the plant bed and transplanted to the field. These transplants are called bare-root transplants (Figure 7.7). Field-grown bare-root transplants are generally less costly to grow or buy than greenhouse-grown plants in trays, but they are not as uniform or as quick growing. Another disadvantage of bare-root plants is that they tend to have a high mortality rate in the field.

FIGURE 7.7 Bare-root transplants of tomatoes grown in the field at close plant spacings and prior to removal for transplanting to the field.

FIGURE 7.8 Styrofoam flats used for growing transplants in a greenhouse.

Containers and Growing Media

Vegetable seeds for transplants can be started in a variety of containers ranging from molded trays to pots. Styrofoam and/or plastic trays are considered the standard in many areas of the United States. Styrofoam flats (Figure 7.8) are often preferred to plastic trays because they encourage more uniform root zone temperatures and moisture.

Cell shapes vary within the seedling flats and available shapes include inverted pyramid, round, and hexagonal. While it is not known if cell shape affects field performance of the transplants, cell size (Figure 7.9) does appear to have an effect, and

FIGURE 7.9   Melon plant growth as affected by the size of the cell of the transplant flat.

TABLE 7.2   *Recommended container sizes for growing vegetable transplants*

Crop	Container size (in.)	Optimum size for early production (in.)
Asparagus	1.5 to 3.0	2.0
Broccoli	1.0 to 3.0	2.0 to 2.25
Cabbage	1.0 to 2.5	1.5
Cauliflower	1.0 to 3.0	2.0 to 2.25
Cucumber	1.5 to 4.0	2.0
Eggplant	1.5 to 4.0	3.0
Lettuce	1.5 to 2.5	1.5
Muskmelon	1.5 to 4.0	3.0
Okra	1.0 to 3.0	1.5
Pepper	1.0 to 4.0	3.0
Tomato	1.5 to 6.0	4.0
Watermelon	1.5 to 4.0	3.0

Adapted from Sanders, D. C. 1991. Container production. In D. C. Sanders and G. R. Hughes (Eds.), *Production of commercial vegetable transplants*. North Carolina State Univ. Coop. Ext. Serv.

recommended cell size to use for producing the transplants will vary with the crop (Table 7.2). In general, larger cell sizes, especially for longer cycle crops (tomatoes and watermelons), often result in larger yields in the field. Smaller cells for short cycle crops (leafy greens and cucurbits) may be necessary as seedling root growth may not completely fill a large cell. This could lead to excessive root damage when the transplant is removed or pulled from the cell. The standard cell size in the industry is 1 in. by 1 in. or approximately 200 plants per tray. Economics and duration of stay in the greenhouse also help determine the cell size that is used for a particular crop.

TABLE 7.3   *Advantages and disadvantages of containers used to grow transplants*

Container	Advantage	Disadvantage
Peat pot (single and strip)	Easy root penetration; available in large sizes; easy to handle in field	Difficult to separate; dries out easily; may act as wick and dry out in field
Clay pot	Long life	Slow to work with; pot dries out; pots are heavy
Prespaced peat pellet	No media preparation; can be handled as a unit of 50	Limited to small sizes
Plastic pot	Reusable; good root penetration	Requires handling as a single plant
Plastic flat with unit	Easily handled; reusable	Requires storage during off season; may be limited in sizes
Plastic cell pack	Easily handled	Roots may grow out of container; limited in sizes; requires some set up labor
Polyurethane foam flats	Easily handled; requires less media than similar sizes of other containers	Requires regular fertilization; plants grow slowly at first
Expanded polystyrene	Lightweight; easy to handle; variable cell sizes and shapes; reusable; automation compatible	Needs sterilization; moderate investment
Injection-molded trays	Variable cell sizes; reusable; long life; automation compatible	Large investment; needs sterilization between uses

Adapted from Sanders, D. C. 1991. Container production. In D. C. Sanders and G. R. Hughes (Eds.), *Production of commercial vegetable transplants.* North Carolina State Univ. Coop. Ext. Serv.

Clay and plastic pots are excellent containers for growing a small number of transplants. Peat pots are also good containers and are made from peat or paper waste fibers. They are porous and provide excellent drainage and air movement. Expandable peat pellets are popular with home gardeners. When the pellets are dry they are about the size of a silver dollar, and when placed in water they swell to form a cylindrical container filled with peat moss, ready for seeding or transplanting.

While each type of container has its advantages and disadvantages (Table 7.3), any container can be used so long as it is clean, sturdy, and fits into the space available for growing the plants. Containers used for growing transplants must have good drainage and be able to hold soil or soil mixes and water.

Soil mixes are generally used for growing transplants. Topsoil is expensive and even good topsoil often contains weed seed and diseases and requires sterilization before it can be used. Artificial mixes do not require sterilization, are easy to handle, and can be bought in large or small volumes. Most mixes will include micronutrients and a surfactant to improve wetting.

Seeds and Seeding

The seeds selected for sowing for transplants must be genetically true to type (be of the correct cultivar) and of good quality (high germination rate and do not contain seeds of other crops or weeds) (Figure 7.10). Seeds are purchased fresh for each growing season and are typically sold in packages or containers that indicate crop, cultivar, germination percentage, and chemical seed treatments, if any. When grown in a greenhouse, seeds are usually planted at a shallow depth into the potting mix, with the actual depth of planting dependent on the crop and the size of the seed (Table 7.4). Usually two to three seeds are sown per cell or container. After the seeds germinate during the transplant production process, the seedlings are thinned to one per cell.

Growing Conditions

The role of temperature on seed germination, early seedling growth, and recommended temperatures for transplant production are often dependent on the vegetable crop (Table 7.5). The amount of time needed to grow a suitable seedling for

FIGURE 7.10 Germination tests are routinely used by seed companies to ensure that the seeds that are sown are of good quality.

TABLE 7.4 *Depth of seeding guide and seed size for selected vegetables*

Vegetable	Seed per ounce (approx. #)	Good plants per ounce[1] of seed (approx. #)	Depth to plant seed in flat (in.)
Broccoli	9,000	5,000	0.5
Cabbage	8,500	5,000	0.5
Celery	70,000	15,000	0.13
Collard	8,000	5,000	0.5
Eggplant	6,000	2,500	0.5
Onion	9,500	4,000	0.5
Pepper	4,500	1,500	0.5
Tomato	11,000	4,000	0.5

[1]Based on outdoor conditions. A larger number can be expected if grown indoors.

Adapted from McLaurin, W. J., D. M. Granberry, and W. O. Chance. 1990. *Home garden transplants.* Univ. of Georgia Coop. Ext. Serv. Leaflet 128.

transplanting also varies among the vegetables. Generally, properly grown plug plants in a greenhouse have a very high survival rate in the field.

Watering and Fertilizing

The plant's water requirement during transplant production increases as the plant grows larger, the weather warms, and the amount of sunlight increases. Developing transplants are watered only when needed and the soil mix is thoroughly wet during each watering. This reduces fluctuations in moisture and washes salts through the bed or container. The soil media should be maintained in a moist condition but not excessively wet. Overwatering can cause plant tissue to become excessively succulent, resulting in plants that are more susceptible to diseases. Underwatering often results in poor germination, plant death, or stunted growth.

Watering transplants in the morning is usually sufficient while plants are small, but watering 2 to 3 times a day may be necessary during later stages of transplant growth. Water pH for irrigation should be 6.0 to 7.0.

Fertigation or the use of water-soluble fertilizers at the time of each watering is one way of feeding young transplants. The usual concentration is 100 to 220 ppm of an appropriate fertilizer mix (Table 7.6) per feeding.

TABLE 7.5   *Temperature and time requirements for germination and growth of vegetable transplants*

| | Temperature requirements (°F) | | | | | |
| | Germination | | | Growth | | Time to grow |
Crop	Min.	Opt.	Max.	Day	Night	(weeks)
Asparagus	50	75	95	65–80	55–70	8–10
Broccoli	45	85	95	65–75	55–60	5–7
Brussels sprouts	45	85	95	65–75	55–60	5–7
Cabbage	40	85	100	55–75	45–55	5–7
Cauliflower	40	80	100	55–75	45–55	5–7
Celery	40	70	85	60–75	60–65	9–12
Collard	40	80	100	55–75	55–60	4–8
Cucumber	60	95	105	70–85	65–70	2–4
Eggplant	60	85	95	70–85	65–70	6–10
Lettuce	35	75	85	50–70	45–55	5–7
Muskmelon	60	90	100	70–85	65–70	2–4
Onion	35	75	95	60–75	60–75	9–12
Pepper	60	85	95	70–80	65–70	6–8
Pumpkin	60	95	100	70–85	60–70	2–4
Summer squash	60	85	95	70–85	60–70	2–4
Sweet potato	—	—	—	75–85	65–70	4–8
Tomato	50	85	95	65–80	60–65	4–8
Watermelon	60	95	105	75–85	65–70	4–8

Adapted from Sanders, D. C. 1991. Management practices. In D. C. Sanders and G. R. Hughes (Eds.), *Production of commercial vegetable transplants.* North Carolina State Univ. Coop. Ext. Serv.

TABLE 7.6   *Fertilizer rates for producing vegetable transplants*

Fertilizer	Rate (per 50 gallons of water)
20-20-20	0.75 to 1.0
15-15-15	1.0 to 1.5
15-30-15	1.0 to 1.25

Adapted from Precheur, R. J. (Ed.). 1998. Producing transplants. In *1998 Ohio vegetable production guide.* Ohio State Univ. Coop. Ext. Bull. 672.

TABLE 7.7   *Hardening temperatures for vegetable transplants*

Vegetable or type	Hardening temperature (°F)
Cool-season crops	50 to 60 (for 5 days)
Tomato, eggplant, pepper, cucurbit	Not below 55

Adapted from Precheur, R. J. (Ed.). 1998. Producing transplants. In *1998 Ohio vegetable production guide.* Ohio State Univ. Coop. Ext. Bull. 672.

Hardening and Preparing for Field Planting

Transplants are hardened prior to placing in the field to reduce shock and stress due to the transplanting process. Plants are hardened by acclimating them to the anticipated growing conditions in the field. This is accomplished by slowly removing the optimum growing conditions of the greenhouse by reducing the water, temperature, and/or fertilizer that the plants receive (Table 7.7). For example, reducing temperatures to 50°F and reducing watering may harden crops such as cabbage, cauliflower, and onions. Squash and cucumbers are hardened only by reducing moisture. In the 2 weeks prior to transplanting, nitrogen is typically withheld from the media to harden the transplants. Fertilization just prior to transplanting is often done to give the field-set plants a rapid start.

Transplanting to the Field

Transplants must be in good condition when placed in the field (Figure 7.11). They should not be too large or hardened too severe and should not have a large number of the roots removed by pulling the transplants out of the growing containers. Transplants can be planted into a field using manual labor or with a mechanical transplanter. Single-row and multirow mechanical transplanters (Figure 7.12) drawn behind a tractor are available that will set bare-root plants or rooted plants. A mechanical transplanter with a tractor driver and two additional people can plant several acres of vegetables in a day. Transplanters are also available that will punch holes into plastic mulch, set plants, and also apply starter solution after the plants are placed into the soil.

Transplants are typically planted 1 to 2 in. deeper in the field than they were in the growing containers. The soil after the transplants are placed in the field is well firmed around the roots. A starter solution may be applied as part of the transplanting process. Starter solutions are water-soluble, high phosphorus fertilizers applied to young plants at the time of transplanting. Starter fertilizers supply phosphorus in an available form even when soil temperatures may restrict phosphorus uptake.

FIGURE 7.11   A good quality transplant ready for setting to the field.

FIGURE 7.12   A multirow transplanter for vegetables.

Review Questions

1. What are some of the environmental factors that can result in poor plant stands in the field?
2. What is the major disadvantage of using a random seeder?
3. Why are some vegetables established in the field using transplants instead of direct seeding?
4. What are some characteristics that make some crops easier to transplant than others?
5. What are the advantages of using a soil mix versus topsoil as the growing medium for transplants?
6. How is the frequency of watering transplants affected by the age of the transplants?
7. Why are transplants hardened prior to field planting and how is hardening accomplished?

Selected References

Daum, D. R., and M. D. Orzolek. 1986. Planters for seeding vegetables. Penn State Univ. Coop. Ext. Serv. Agric. Eng. Fact Sheet PM-87.

Gerber, J. M., J. W. Courter, H. J. Hopen, B. J. Jacobsen, and R. Randell. 1985. *Vegetable production handbook for fresh market growers*. Univ. of Illinois Ext. Service Circular 1241.

McLaurin, W. J., D. M. Granberry, and W. O. Chance. 1990. *Home garden transplants*. Univ. of Georgia Coop. Ext. Serv. Leaflet 128.

Orzolek, M. D. 1991. Establishment of vegetables in the field. *HortTechnology* (Oct.–Dec.):78–81.

Precheur, R. J. (Ed.). 1998. Producing transplants. In *1998 Ohio vegetable production guide*. Ohio State Univ. Coop. Ext. Bull. 672.

Sanders, D. C. 1991. Management practices. In D. C. Sanders and G. R. Hughes (Eds.), *Production of commercial vegetable transplants*. North Carolina State Univ. Coop. Ext. Serv.

Sanders, D. C., and J. T. Garrett. 1993. Precision seeding. TriState Vegetable Production Fact Sheet 29.

Vavrina, C. S. 1996. *Transplant production*. Univ. of Florida Coop. Ext. Serv. Publ. HS 714.

chapter

8

The Need for Water and Irrigation

TABLE 8.1   *Water content of common vegetables*

Vegetable and plant part	Fresh weight (%)
Cucumber fruit	96
Lettuce head	94
Cabbage head	90
Potato tuber	79
Tomato leaves	85 to 95
Corn leaves	65 to 82
Carrot roots	88
Asparagus tips	88
Tomato fruit	94
Sweet corn seeds	85

Adapted from AVRDC. 1990. *Vegetable production training manual.* Asian Vegetable Research and Development Center, Shanhua, Tainan, Taiwan. Reprinted 1992. 442 pp. Used with permission.

Since many vegetables are about 80 to 90% water (Table 8.1), their yields and quality suffer very quickly from drought or water stress. Thus, for good yields and high quality, irrigation is often essential to supply the plant's water needs for the production of quality vegetables.

The Plant's Need for Water

Vegetables require a constant and somewhat abundant supply of moisture (approximately 65% field capacity) throughout the growing season. Even brief periods of water stress can reduce crop yield and quality. Irrigation may also be needed to ensure good germination and emergence and for seedling establishment.

Watering or irrigating supplements natural rainfall when soil moisture is insufficient to meet plant needs. Proper irrigation practices are important to produce good crops, control diseases and insects, reduce soil erosion, conserve water, and protect water quality. Irrigation can also be used to cool plants during hot days and for frost control during freezing temperatures.

Shallow-rooted vegetable crops are not generally allowed to experience more than 10 continuous days without water throughout the season. Other crops must receive adequate moisture, especially during critical periods in their development (Table 8.2). Most vegetables require from 1.0 to 1.5 in. of water per week during their peak growing season.

There are many factors that effect loss of moisture from the soils. These include run-off, evaporation, percolation to the water table, and plant transpiration. The largest proportion of water loss from the field is through transpiration, with the rate of loss fluctuating with duration and intensity of solar radiation, wind speed, and relative humidity.

Incipient wilting is the wilting of plants under weather conditions favoring rapid transpiration, even when field moisture is at field capacity. Such stress increases in frequency and intensity as soils dry. At the plant's permanent wilting point, water uptake ceases, and plant tissue and cells are damaged.

TABLE 8.2   *Critical water-use periods for vegetable crops*

Crop	Growth period
Broccoli, cabbage, cauliflower	Head development
Lettuce (head)	Head expansion
Carrot, radish, turnip	Seed germination, root expansion
Beet	Continuous
Sweet corn	Silking
Cucumber, eggplant, pepper, melon, tomato	Flowering, fruit set, and fruit enlargement
Bean (dry, lima, pole, snap)	Flowering
Pea (garden and southern)	Flowering and pod development
Onion	Bulbing and bulb expansion
Potato	After flowering, tuber initiation and development

Adapted from Sanders, D. C. 1997. *Vegetable crop irrigation.* North Carolina Coop. Ext. Serv. Leaflet HIL-33E.

Methods for Monitoring Soil Moisture

In order to optimize crop production in the field, soil water must be available and irrigation, when needed, must be efficiently used. There have been a number of methods developed and used by vegetable growers for monitoring soil moisture to determine when irrigation may be needed. These monitoring methods can be broadly categorized as plant-based and soil-based.

Plant-Based Monitoring

Visible Signs   The plant's appearance can be used by growers to determine when to water. While this may not be the best way to determine water needs, it is the easiest to recognize. Visible signs of water stress include wilting and a change in leaf color. Young leaves of water-stressed plants may become dull, darken, or turn grayish.

Visible signs may not always indicate what is best for the plant at a certain stage. For instance, allowing the plant to undergo mild wilting may help "harden" the plant, making it more resistant to environmental stresses. At other times, plant injury from water deprivation may occur before signs of stress are noticeable.

Use of Pressure Bombs   Pressure bombs measure water potential in the xylem of the plant. Xylem pressure is nearly equivalent to leaf water pressure and indicates plant water status. The higher the gauge reading on the pressure bomb the greater the plant water stress. Because of the diurnal nature of water relations in a plant, critical values of water pressure are usually needed for each crop at different times of the day. Pressure bombs only indicate when a plant needs water, but not how much it needs.

Use of Infrared Thermometers   Infrared thermometers measure leaf temperature without making direct contact with the leaves. Transpirational or evaporative loss of water through the stomata results in reduced leaf surface temperatures due to evaporative cooling effects on the leaf tissue. A rise in leaf temperature may indicate water stress due to reduced evaporative cooling occurring at the leaf surface.

Soil-Based Monitoring

Appearance of Soil   The appearance of the soil can give a good indication of the soil water status (at least near the soil surface). Following are some of the characteristics of the various soil types and suggestions on when irrigation is needed:

Sandy soil—dries rapidly, should be irrigated when it appears dry
Sandy loam—dries moderately fast, should be irrigated when soil appears dry
Silt loam—dries moderately slow, should be irrigated when pale color indicates no
     obvious moisture
Clay—dries slowly, should be irrigated when soil is still pliable

Soil Moisture Feel   Feeling the soil is a simple, effective method to determine when to irrigate and how much water to apply at an irrigation. A soil has a specific feel characteristic according to the amount of available soil moisture it contains. The feel test chart (Table 8.3) uses the following tests:

Ball Test   A handful of soil is squeezed into a ball. The ball in the palm of the
     hand is pressed with the thumb. The wetter the soil the tighter the ball.
Ribbon Test   A handful of loose soil is placed in the palm of the hand. The soil is
     pushed with the thumb so that the soil "ribbons" between the thumb and
     forefinger. Since coarse soil will not ribbon, it will be necessary to look instead
     at how soil particles stick to the thumb and palm of the hand.
Open-Palm Test   A handful of soil is gently rolled between open palms. This can
     determine 80 to 100% soil moisture conditions in medium- to fine-textured soil.

Gravimetric Method   The gravimetric method requires that a portion of soil is removed from the field and weighed. The soil sample is dried in an oven and reweighed. Percent water content is determined by dividing the weight of water (initial − oven-dry soil weight) by oven-dry soil weight. This method gives an estimate of water content of the soil but not the water that is available to the plant.

Tensiometer   A tensiometer is a sealed, water-filled tube with a porous ceramic tip on the lower end and a vacuum gauge on the upper end. The tube is installed in the soil with the ceramic tip placed at the desired root zone depth and with the gauge above the ground (Figure 8.1). Water passes into and out of a chamber in the tensiometer to and from the surrounding soil through a membrane, depending on the amount of water in the soil.

Tensiometers can be placed at several locations and depths to determine soil moisture throughout the root zone. Tensiometers are most effective when used for scheduling irrigation in shallow-rooted, water stress sensitive, and frequently irrigated vegetable crops. As the soil dries, water leaves the tensiometer, and vice versa as the soil is wetted. The flow of water from the tensiometer creates a vacuum that is measured as pressure. The standard unit of measure of soil water tension or soil suction is the "bar," which is a unit of pressure. Most tensiometers are calibrated in hundredths of a bar (centibar) and can operate in a range of 0 to 80 centibars.

The pressure changes in the tensiometer are related to soil moisture or the relative wetness of the soil, and minimum values have been established for many crops to determine when irrigation should be used (Table 8.4). When specific minimum tensiometer values for beginning irrigation are not known for a crop, the following general guidelines of Harrison and Tyson (1993) can be used for interpreting tensiometer readings (in centibars) under most conditions:

TABLE 8.3 *Feel test chart for estimating soil moisture*

% Available moisture	Fine soil texture (loam, silt loam, clay loam)	Coarse soil texture (sandy and loamy sand)
Below 20	Powdery, dry, will not form a ball; if soil is crusted, easy to break into powdery condition	BT: No ball forms. Single-grained soil flows through fingers with ease
35 to 40	Dry, almost powdery BT: A ball can be formed under pressure, but some soil will fall or flake away when hand is opened. The ball is very crumbly and hardly holds it shape	BT: Forms weak, brittle ball. Fingerprint outline not discernible. No soil sticks to hand RT: Few soil particles stick to thumb
50	BT: Forms a ball readily, holds its shape. No moist feeling is left on palm nor will any soil fragments cling to palm. Ball is very brittle and breaks readily. Soil falls or crumbles into small granules when broken RT: Will not ribbon; soil too crumbly	BT: Forms very weak ball. If soil well broken up it will form more than one ball upon squeezing. Fingerprint outline barely discernible. Soil grains will stick to hand RT: No ribboning. Soil particles will just cease to lay down. Patchy soil layer on thumb
60 to 65	BT: Forms firm ball; fingerprints imprint on ball. Hand feels damp but not moist. Soil doesn't stick to hand. Ball is pliable. When broken, ball shatters or falls into medium-sized fragments RT: Just barely ribbons OPT: Soil breaks down into granules and is a little crumbly. Continues to crumble until a tiny round ball is left in palm	BT: Forms weak, brittle ball. Fingerprint outline not as distinct. Soil particles will stick to hand in a patchy layer RT: No ribboning. Soil particles will stick to thumb in a patchy layer
70 to 80	BT: Damp and heavy, slightly sticky when squeezed. Forms tight plastic ball. Shatters with a burst into large particles when broken. Hand is moist RT: Ribbons out. Moist soil particles left on thumb OPT: Sample can be molded into a round ball; somewhat plastic, will not shatter readily	BT: Forms weak ball. Distinct fingerprint outline on ball. Soil particles will stick to palm RT: No ribboning. Soil particles will stick to thumb during ribboning process in a distinct layer over surface of thumb
100	BT: Wet, sticky, doughy, and slick. A very plastic ball is formed, handles like stiff bread dough or modeling clay, not muddy. Leaves water on hand. Ball will change shape and cracks will appear before breaking RT: Ribbons readily if not too wet OPT: Forms a tight ball. Will work into a round pencil-like shape	BT: Upon squeezing, no free water appears on ball but wet outline of ball is left on hand. Ball has some stickiness and a sharp fingerprint outline is left on it RT: No ribboning. Soil particles will form smooth layer on thumb

Definition of terms:

BT = Ball Test, RT = Ribbon Test, OPT = Open-Palm Test

From Larsen D. C. and M. K. Thornton. 1990. Irrigation on crop important during growth. *The Potato Grower of Idaho.* (August): 18–19. Used with permission.

FIGURE 8.1   A tensiometer monitoring soil moisture in the field.

Readings 0 to 5   This range indicates a nearly saturated soil and often occurs for 1 or 2 days following a rain or irrigation. Plant roots may suffer from lack of oxygen if readings in this range persist.

Readings 5 to 20   This range indicates field capacity. Irrigation is discontinued in this range to prevent waste of water by percolation and also to prevent leaching of nutrients below the root zone.

Readings 20 to 60   This is the usual range for starting irrigation. Most field plants having root systems 18 in. deep or more will suffer until readings reach the 40 to 50 range. Irrigation is started in this range to ensure maintaining readily available soil water at all times. It also provides a safety factor to compensate for practical problems such as delayed irrigation, or inability to obtain uniform distribution of water to all portions of the field.

Readings 70 and higher   This is the stress range for most soil and crops. Deeper-rooted crops in medium-textured soils may not show signs of stress before readings reach 70. A reading of 70 does not necessarily indicate that all available soil water is used up, but that readily available water is below that required for maximum growth.

The need for irrigation will also be affected by environmental factors such as soil texture type, air temperature, wind speed, and relative humidity. Irrigation is often started sooner on sandy soils than heavier clay soils, since sandy soils hold less water than clay soils at the same meter reading (Table 8.5). Irrigation is also started at higher soil moisture levels during hot weather, high winds, or low humidity.

Gypsum Block   Gypsum blocks measure changes in electrical conductivity of soils. The wetter the soil the greater the conductivity. Similar to tensiometers, gypsum blocks can be buried at different depths to obtain a soil profile characterization of irrigation needs.

TABLE 8.4 *Irrigation needs and preferred irrigation methods for selected vegetables*

| Crop | Preferred minimum soil moisture | | Preferred irrigation method[2] |
	Centibars	ASM[1]	
Asparagus	70	40%	a,b
Beans, dry	45	50%	a
Beans, lima	45	50%	a,b
Beans, pole	34	60%	a
Beans, snap	45	50%	a
Beet	200	20%	a,b
Broccoli	25	70%	a,b,c
Brussels sprouts	25	70%	a,b,c
Cabbage	34	60%	a,b
Carrot	45	50%	a,b
Cantaloupe	34	60%	a,b
Cauliflower	34	60%	a,b,c
Celery	25	70%	a,b,c,d
Corn, sweet	45	50%	a,b
Cucumber, pickles	45	50%	a,b,c
Cucumber, slicer	45	50%	a,b,c
Eggplant	45	50%	a,b,c
Greens (turnip, mustard, kale)	25	70%	a,b
Leek	25	70%	a,b
Lettuce (head, Bibb, leaf, cos)	34	60%	a,b
Onion	25	70%	a,b
Peas, green	70	40%	a
Peas, southern	70	40%	a,b
Pepper	45	50%	a,b,c
Potato, Irish	35	70%	a,b
Pumpkin	70	40%	a,b
Radish	25	70%	a
Rhubarb	200	20%	a,b
Rutabaga	45	50%	a,b
Squash, summer	25	70%	a,c
Squash, winter	70	40%	a,b
Sweet potato	200	20%	a,b
Tomato, staked	45	50%	a,c
Tomato, ground	45	50%	a,b
Tomato, processing	45	50%	a,b
Turnip	45	50%	a,b
Watermelon	200	40%	a,b,c

[1]ASM (available soil moisture). % of soil water between field capacity (10 centibars) and permanent wilting point (1,500 centibars).

[2]Irrigation method: a=Sprinkler, b=Traveling gun, c=Trickle, d=Furrow.

Adapted from Sanders, D. C. 1997. *Vegetable crop irrigation.* North Carolina Coop. Ext. Serv. Leaflet HIL-33E.

TABLE 8.5   *Influence of soil type on irrigation*

Soil type	Soil moisture to begin irrigating (centibar)[1]
Loamy sands, sandy loams, very fine sandy loams	25
Silt loams	40
Clay loams, silt clay loams	60

[1]Average readings will vary on crop, rooting depth, type of irrigation system, and soil type.

Adapted from Harrison, K. and A. Tyson. 1993. *Irrigation scheduling methods.* Univ. of Georgia Coop. Ext. Serv. Bull. 974.

Gypsum blocks indicate when to irrigate but not how much to apply. Depending on soil type, gypsum blocks are generally replaced every 2 to 5 years. Gypsum blocks lose their effectiveness in determining soil water status when they're used in saline or alkali soils in which salts affect conductivity.

Thermal Dissipation Sensor   Thermal dissipation sensors monitor the dissipation of heat in a porous ceramic block in contact with the soil. Water flows into and out of the ceramic block depending on the water status of the soil. The wetter the soil, the more moisture in the block. The ceramic block is a good heat conductor when wet and a poor conductor when dry. An electric current quantifies heat conduction. The sensor can detect small changes in soil moisture and can be linked by computer to an automatic irrigation system. Once calibrated, the sensor can remain in place for long periods of time. The sensor and monitors are expensive.

Types of Irrigation Systems

There are three basic types of irrigation systems: overhead systems, surface systems, and subsurface systems. The type of system used by a vegetable grower will depend on the characteristics of the field, crop, and available resources (Table 8.6).

Overhead Systems

Portable Aluminum Pipe Sprinkler System   A portable aluminum pipe sprinkler system is an aluminum pipe system designed to cover only a portion of the acreage to be covered at one time. This system can be moved one or more times per day. It may use a variety of sprinkler sizes. It has the highest labor requirement of any sprinkler system and is only used on small acreage.

Solid-Set Irrigation System   A solid-set irrigation system (Figure 8.2) is an aluminum pipe system that is placed in the field at the start of the irrigation system and left in place throughout the season. It is adapted to most soil types, topography, field sizes, and shape. It has a low labor requirement once it is placed in the field and the system can be automated. It is normally used for crops with a high cash value.

Permanent Sprinkler System   A permanent sprinkler system is an underground pipe system with only a portion of the risers and sprinklers above ground. This system has the same characteristics as a solid-set, except labor requirements are lower and the system cannot be moved to another location.

TABLE 8.6 *Advantages and disadvantages of the various irrigation systems*

System	Advantages	Disadvantages
Furrow	• Low initial cost • Disease avoidance	• Uneven water distribution • Crusting can occur • Poor water conservation • High water use • Salt may concentrate near plants • Land must be leveled
Sprinkler	• Good soil moisture • Leaches excess salts • Reduces crusting • Relatively uniform water • Can be used on unlevel land	• High initial cost • Encourages weed growth • High energy requirement • May encourage some diseases
Trickle	• Requires little water • Can cut irrigation labor cost • Can harvest or spray • Efficient water use • Can be used on sloped land • Fertilizer and some fungicides can be applied through the system	• Can be costly

FIGURE 8.2   Solid-set irrigation system in a field.

FIGURE 8.3   Side-wheel irrigation system.

Side-Wheel Move System   A side-wheel move system (Figure 8.3) is a wheel move aluminum system where the lateral line is the axle for the wheels and is designed to be used in rectangular or square fields up to 2,000 ft in length and on crops 4 ft or less in height. The lateral line is stationary while it is being used. Once a section has been irrigated, a small engine attached to the line through drive wheels is started and the unit is moved to the next irrigation set, normally a distance of 60 ft. Wheels and sprinklers are typically spaced 40 ft apart. The system is limited to normally flat terrain and uniformity of water is excellent. The system is difficult to move from field to field.

Traveling Gun   A traveling gun (Figure 8.4) is a self-propelled, continuous move sprinkler mounted on a two-, three-, or four-wheeled trailer with water being supplied through a flexible hose. The unit follows a steel cable that has one end anchored at one end of the field and the other end attached to the machine. The cable winds onto a cable drum on the machine as the machine moves through the field. It is best suited to square or rectangular fields, straight rows, and flat to rolling topography. Uniformity of water application will vary from excellent to fair. It is easily transported from field to field and farm to farm.

Center Pivot System   A center pivot system (Figure 8.5) is an electrically or hydraulically powered self-propelled lateral line on which sprinklers or spray nozzles are mounted. Towers that are spaced 110 to 180 ft apart support the lateral line. Center pivots are available as fixed-point machines and as towable machines that can be used on two or more pivots. The pattern of irrigation is a circle; however, with an end gun it is possible to irrigate corners or odd shapes of fields. The slope of land must be less than 10%. Center pivots have a high initial cost and are usually custom-designed to fit the area.

FIGURE 8.4   A traveling gun irrigation system.

FIGURE 8.5   A center pivot system in a vegetable field.

Lateral Move System   A lateral move system is a self-propelled, electric drive lateral line on which low- or very-low-pressure nozzles are mounted. The system moves in a straight line down the field. Towers that are spaced 110 to 180 ft apart support the lateral line. The base tower that controls the speed of travel is located at the end of the lateral line or at some point in the lateral line. It follows a buried guidance or aboveground guidance cable. A lateral move system is designed for flat fields with

FIGURE 8.6   Trickle irrigation used for growing staked tomatoes.

widths up to 4,000 ft and lengths up to 9,000 ft. It has a medium to high initial cost and the operational costs are low.

Surface Systems

Drip or Trickle Irrigation   A drip or trickle irrigation system (Figure 8.6) is a low-pressure system where a lateral line is placed alongside every crop row adjacent to the plant. Water is discharged through the line through emitters or orifices in the pipe or through microsprinklers. Trickle tubing or tapes are designed to be operated at low water pressure (generally 12 psi). A drip or trickle system can be used on most soil types and works best on flat terrains. The filter system is very important. With the use of trickle systems the amount of water used is reduced because of placement near the root zone. Also, limited water sources and low pressure can be used. Field operations can be performed during irrigation. A disadvantage with the system is that the emitters can become plugged with soil, which requires a high level of management.

Furrow Irrigation   Furrow irrigation (Figure 8.7) is the application of water in furrows that have a continuous slope in the direction of water movement to crops. The system is limited to fields that have natural slopes of 2% or less. Fields must be graded to provide uniform slope and water distribution. It is not designed for small applications of water. Labor requirements are fairly high and the initial costs can be high depending on the amount of grading. However, operational costs are low and it does have a good uniformity of water.

Subsurface Systems

Subsurface irrigation systems (Figure 8.8) are used in areas where there is a naturally occurring shallow water table (e.g., selected areas in Florida) and adjustment of water level is possible. Pesticides are applied very carefully with subsurface irrigation systems to prevent contamination of the water source.

FIGURE 8.7   Furrow irrigation used for growing head lettuce.

FIGURE 8.8   A subsurface irrigation system used in Florida.

Special Considerations

Water Source

Water is generally pumped from a well or an irrigation pond on the farm. This water may contain dissolved salts, grit, and/or bacteria that can cause problems. Dissolved salts can build up in the soil around the plants and cause toxicity problems. Grit can clog nozzles or tubes. Sufficient water must be applied to leach salts and alleviate salt

TABLE 8.7   *Relative maximum rooting depths of selected vegetables*

Shallow (18 to 24 in.)	Broccoli, Brussels sprouts, cabbage, cauliflower, celery, corn, garlic, leek, lettuce, onion, parsley, potato, radish, spinach
Moderately deep (36 to 48 in.)	Bean, beet, carrot, cucumber, eggplant, muskmelon, pea, pepper, rutabaga, summer squash, turnip
Deep (more than 48 in.)	Artichoke, asparagus, lima bean, pumpkin, winter squash, sweet potato, tomato, watermelon

Adapted from Maynard, D. N. and G. J. Hochmuth. 1997. *Knott's handbook for vegetable growers*. 4th ed. New York: John Wiley & Sons. Reprinted by permission of John Wiley & Sons, Inc.

buildup. Filters or strainers placed in the system ahead of nozzles or tubes prevent clogging.

Irrigation systems are useless unless there is sufficient water for the systems. One inch of irrigation to one acre of land requires 33,000 to 37,000 gallons of water.

Soil Profile

Soil characteristics of fields below the surface may be different from at the surface. These characteristics may affect irrigation requirements.

Hardpans may exist either naturally or because of repeated plowing. Hardpans interfere with water movement through the soil, preventing the water from draining. Hardpans cause salt buildup and result in too much water in the root zone.

Deep plowing or subsoiling can break hardpans. Increasing the frequency of irrigation while decreasing the amount of water applied each time can also be used to soften the hardpan in the soil.

Pockets of sand or gravel in otherwise clay-based soils prevent uniform water distribution. Steepness of sloped land also influences subsurface distribution of irrigation water.

Crop Rooting Depth

Vegetable crops vary according to their depth of rooting (Table 8.7). The depth of rooting may affect the ability of the plant to scavenge for water and survive water stress. Crops with deeper roots are more likely than shallow-rooted crops to find available water when the surface soil water status is depleted. Some crops such as watermelon have extremely deep roots, while others such as lettuce have very shallow roots.

Common Problems

Several common problems may result as vegetable growers water their crops. Growers, hoping that it will rain, may wait too long before they begin irrigating a dry field. This results in a portion of the field that gets irrigated last being in severe water stress. Another problem involves growers trying to stretch the acreage that can be reasonably covered with their available irrigation system. Also, frequent light irrigations result in shallow root systems that make the plants susceptible to short periods of water stress. On the contrary, excessive irrigation leaves crops vulnerable to leaching from rain or irrigation.

Review Questions

1. Why is irrigation so important in the production of vegetable crops?
2. What factors affect water loss from a field?
3. How can the visual appearance of the soil indicate water needs for the plant?
4. What are the advantages of using a drip or trickle irrigation system for growing vegetables?
5. How can hardpans interfere with the plant's water needs?

Selected References

AVRDC. 1990. *Vegetable production training manual.* Asian Vegetable Research and Development Center, Shanhua, Tainan, Taiwan. Reprinted 1992. 442 pp.

Decoteau, D. R., D. Ranwala, M. J. McMahon, and S. B. Wilson. 1995. *The lettuce growing handbook: Botany, field procedures, growing problems, and postharvest handling.* Oak Brook, IL: McDonald's International.

Gerber, J. M., J. W. Courter, H. J. Hopen, B. J. Jacobsen, and R. Randell. 1985. *Vegetable production handbook for fresh market growers.* Univ. of Illinois Ext. Serv. Circular 1241.

Harrison, K., and A. Tyson. 1993. *Irrigation scheduling methods.* Univ. of Georgia Coop. Ext. Serv. Bull. 974.

Larsen, D. C., and M. K. Thornton. 1990. Irrigation on crop important during growth. *The Potato Grower of Idaho* (August): 18–19.

Maynard, D. N., and G. J. Hochmuth. 1997. *Knott's handbook for vegetable growers.* 4th ed. New York: John Wiley & Sons.

chapter

9

Extending the Growing Season

Vegetable growers often try to extend the growing season for selected crops for the purpose of obtaining a marketing advantage. For example, crops that produce early in the spring or early summer often command a greater price on the market. Also, producing a crop when large quantities of the crop are not available (considered as "off-season") can command greater prices and increased demand.

Extending the growing season can be accomplished by planting the field early in the spring (before the recommended planting date for the region) with the use of protective structures such as row covers or low tunnels. Earlier production can also be accomplished with the use of synthetic mulches, which can enhance crop maturity. Later production in the fall is possible with the use of protective structures such as row covers that are placed on top of the crop when cooler outdoor temperatures prevail. Greenhouse vegetable production potentially allows crop production all year-round. Important factors to consider on whether to extend the season are the increased costs of using season extender production systems, potential increase in sale prices of the crop if produced either earlier or later, and suitability of the crop to season extender production systems.

Mulches

Mulching is a practice of covering the soil around plants to improve crop growth and development. Mulch materials may be organic (leaves, straw, grass) or synthetic (plastic). Using a mulch can improve coverage of the seed with soil and reduce soil crusting. Mulches also modify soil surface temperatures. In high rainfall areas, woven plastic weed mats reduce fertilizer leaching by allowing light rainfall to penetrate the cover while high amounts of rain drain away over the mat. Other specific types of mulches are used to repel insects and regulate plant growth by modifying the plant light environment.

Organic Mulches

Grasses, leaves, and straw are some of the materials that can be used as organic mulches. Organic mulches do not affect soil temperatures to the extent of synthetic mulches. Organic mulches do suppress weeds, reduce crusting, and preserve moisture. The gradual decomposition of organic mulches can also add organic matter to the soil.

Following are some of the more common organic mulches and their effects on soil characteristics:

Straw   Straw mulches (Figure 9.1) can be time-consuming to apply, may introduce weeds, and create an environment for pests like slugs and rodents.

Grass   Grass mulches have a low C:N ratio that may result in later release of nitrogen to the crops.

Sawdust   Sawdust has a high C:N ratio and requires nitrogen from the soil for decomposition. It is easy to apply and generally inexpensive, but should only be used in a well-rotted condition.

FIGURE 9.1   Tomatoes grown with straw mulch.

Synthetic Mulches

Synthetic (plastic) mulches have been used commercially on vegetables since the early 1960s and have been shown to substantially increase earliness and total marketable yields of some selected (primarily warm-season) crops. The plastic mulch provides a greenhouse effect during the day, warming the soil considerably as compared to unmulched soil. The plastic also modifies fluctuations in soil moisture, partially by eliminating evaporation, minimizing crusting from hard rains, and reducing leaching of nutrients. Synthetic mulches can be expensive, and must be removed and disposed of at the end of the growing season.

Black Plastic Mulch   A dark-colored mulch, such as black (Figure 9.2), can trap heat from the sun's rays and warm the soil. In the spring, this results in warmer root zone temperatures than in bare ground and often encourages earlier plant growth. The increased value of the earlier production of some vegetables with mulches can easily offset the costs of the plastic mulch. Generally, black plastic mulch culture benefits muskmelons, watermelons, cucumbers, summer squash, peppers, tomatoes, and sweet corn. Black plastic is also effective in weed control.

Clear Mulch   A clear plastic used as a mulch material warms the soil the best, but allows for weed growth. A herbicide is often needed with the use of clear plastic. In addition, clear plastic could warm the soil to such an extreme as to adversely affect plant growth. In some areas of the country such as in the southern United States, clear plastic is also used as a means of solarizing or sterilizing soil in the field.

White Mulch   White mulch (Figure 9.3) reflects more light back to the plant than black mulch, and has little or no effect on soil temperature. White mulch may repel some insects.

FIGURE 9.2   Black plastic mulch applied to a field with the use of a mechanical mulch layer.

Degradable Mulch   Degradable plastic mulches (Figure 9.4) have many of the properties and provide the usual benefits of nondegradable polyethylene mulches except that they degrade after the film has received a critical amount of sunlight. When the film has received sufficient light, it becomes brittle and develops cracks and holes. As a result, small sections of the film may tear off and be blown away. The film eventually breaks down into small flakes and disappears into the soil. The edges of the mulch covered by soil retain their strength and break down more slowly.

FIGURE 9.3   Peppers grown with a white mulch.

FIGURE 9.4   Two formulations of photodegradable mulch in the field for growing watermelons. The formulation on the left degrades more slowly than the formulation on the right.

FIGURE 9.5   A field evaluation of different colors of mulch.

Selectively Permeable or Reflective (Colored) Mulch   Other colors of plastic mulch (Figure 9.5) are available, including a new generation of selectively permeable and wavelength selective-reflective mulches. Selectively permeable mulches allow certain wavelengths of light to pass through the mulch to warm the soil without encouraging weed growth. Wavelength selective or colored mulches reflect wavelengths of light into the plant canopy to influence plant growth and production or insect populations.

Living Mulches

Grasses, legumes, and other crops that are seeded into an established crop, usually mid- to late-season, are considered living mulches. These living mulches provide erosion control and some weed control. Crops usually chosen to be used as living mulches have relatively low growth rates and vigor as compared to the established crop, so as to not compete aggressively with the crop for water and nutrients.

Row Covers

Row covers are flexible transparent coverings that are installed over single or multiple rows of vegetables to enhance plant growth by warming the air around the plants in the field. They can also warm the soil and protect the plant from hail and wind injury. Row covers are often used to promote early production in the spring or summer or to extend the production season in the fall or early winter. Row covers originated during the 16th and 17th centuries with the use of jars, cloches, and cold frames placed over the plants during periods of cool temperatures.

Row covers can be supported above the plant using hoops of wire. A clear or white plastic is stretched over the hoops and the sides are secured by placing in soil (Figure 9.6). The plastic can be vented or slitted prior to purchase or can be mechanically slitted for venting during the growing season as the temperatures increase

FIGURE 9.6   Row covers of clear plastic stretched over wire hoops in the field.

FIGURE 9.7   A row cover made of fabric material.

within the row cover environment. Other terms used for plastic row covers include cloches, tunnels, and low tunnels.

Row covers can also be made of fabric material (Figure 9.7). These row covers are often used as floating row covers, meaning that the material is placed directly on the top of the developing plants without the need for wire hoops. This fabric material is usually spunbonded or extruded and colored white with about 80% light transmission through the material. Floating row covers can be narrow and cover a single row of crops, medium wide and cover a crop bed with three to four closely spaced rows, or wide and cover multiple rows in widths of 10 to 50 feet. The advantage of floating row covers is their relative ease of application and reduction in labor as compared

FIGURE 9.8   High tunnels for vegetable production. Courtesy of W. J. Lamont, Jr.

with using plastic row covers stretched over wire hoops. A disadvantage is the potential abrasion of the leaf surface from the cover material during windy conditions.

High Tunnels

High tunnels (Figure 9.8) are increasing in popularity with vegetable growers, particularly in the northern United States, who want earliness and improved quality and yields of a wide variety of horticultural crops. High tunnels are designed similar to greenhouses, although they are not as mechanically sophisticated. High tunnels differ from low tunnels (row covers over single rows) in that they are constructed using hopped metal bows connected to metal posts, which are driven into the ground and evenly spaced. The tunnel is about 6 to 7 feet high and 14 feet wide. The end walls are minimally framed to allow easy removal to accommodate free movement of power tillage equipment. The unit is covered with 6-mil greenhouse-grade polyethylene, which is left on year-round or can be removed during the winter. Ventilation is done by rolling up the sides of the tunnel as far up as desired. Trickle irrigation tubing beneath a black plastic mulch cover is placed on the tilled soil within the high tunnel. The high tunnel is then planted with the desired crops. High tunnels generally have either very simple heating systems or none at all.

Greenhouses

Greenhouses are used to provide the greatest control over many of the environmental conditions surrounding the plant. The greenhouse structure modifies the environmental conditions and allows vegetable crop production in regions and at times when outdoor production would not be optimum. Greenhouse production also involves the growing of specialty or niche-marketed crops (e.g., greenhouse-grown tomatoes or cucumbers).

The relative size of greenhouses in the United States can be broken down into two main categories: family-run operations and corporate operations. The family-run greenhouse operation is often 2,500 to 10,000 square feet and usually less than 1 acre. The corporate operation tends to be at least 10 acres in size.

Greenhouse vegetable production may entail growing the plants in a soilless media. Most large growers raise their plants in rock wool or glass wool. Other growers use perlite, peat-lite, pine bark, or soil for their growing media. These media allow the implementation of nearly exact fertilizer recommendations with high-quality injectors and computer controls for environmental regulation.

Special cultivars have been developed for growing in the greenhouse environment. The use of outdoor cultivars for commercial greenhouse production is generally not recommended. Cultivars developed for greenhouse growing are higher yielding, more uniform, less prone to physiological disorders, and have excellent disease resistance than cultivars that were bred for outdoor field production but grown in a greenhouse.

Insects and diseases can affect greenhouse vegetable production. Control methods for insects and diseases include chemical, barrier, sanitation, biological, and others. The most severe insect pests in greenhouses are whiteflies and leaf miners.

Review Questions

1. What are some of the reasons that a vegetable grower may want to extend the growing season for a crop?
2. What are the various types of synthetic mulches that are in use in vegetable production?
3. When did row covers originate?
4. How does a high tunnel differ from a low tunnel row cover?
5. What are the two basic categories of greenhouses for vegetable production in the United States?

Selected References

Gerber, J. M., J. W. Courter, H. J. Hopen, B. J. Jacobsen, and R. Randell. 1985. *Vegetable production handbook for fresh market growers.* Univ. of Illinois Ext. Publ. Circular 1241.

Oebker, N. F., and H. J. Hopen. 1974. Microclimate modification and the vegetable crop ecosystem. *HortScience* 9:564–568.

Orzolek, M. D. 1996. Organic matter and cover crops. Penn State University, College of Agricultural Sciences, Penn Pages Doc. No. 2940165.

Orzolek, M. D., P. A. Ferreti, A. A. MacNab, J. M. Halbrendth, S. J. Fleischer, Z. Smilowitz, W. K. Hock. 1997. *1997 commercial vegetable production recommendations.* Penn State Coop. Ext. Publ.

Snyder, R. G. 1996. Greenhouse vegetables—introduction and U.S. industry overview. *Proc. Natl. Ag. Plastics Congress* 26:247–252.

Wells, O. S. 1996. Row cover and high tunnel growing systems in the United States. *HortTechnology* 6:172–176.

Wittwer, S. H. 1993. World-wide use of plastics in horticultural production. *HortTechnology* 3:6–23.

chapter

10

Dealing with Growing Problems

Common Growing Problems

Insect, disease, and weed infestations in vegetable fields can result in heavy losses through reduced yields, lower quality of produce, increased costs associated with growing and harvesting, and additional expenditures required for materials and equipment needed to control the infestations. As a result, the amount of money spent on controlling pests can be substantial and may reduce or eliminate any profits from sale of the produce.

Insects

Plant productivity generally decreases as insect populations in the field increase. Also, the visual appearance (often the primary criteria used by consumers to determine what and how much they will buy) can be adversely affected by insects, rendering the crop as reduced quality or even unmarketable.

Insects Categorized by Feeding Habit   Insects can be categorized according to their feeding habits. These include chewing insects, sucking insects, and rasping insects.

Chewing Insects   Chewing insects consume plant roots, leaves, or stems as adults or larvae. Cabbageworms, armyworms, grasshoppers, Japanese beetles, Colorado potato beetles, and fall armyworms are examples of insects that cause injury by chewing.

Sucking Insects   Sucking insects derive nutrients by puncturing plant cells and sucking plant sap. Only the internal liquid portions of the plant are swallowed, even though the insect feeds externally on the plant. Aphids, scale insects, squash bugs, leafhoppers, and plant bugs are examples of sucking insects.

Rasping Insects   Rasping insects cut plant cells and feed on exposed plant sap. Small patches of killed tissue, often not penetrating through the leaf, is indicative of injury from rasping insects. Thrips are examples of rasping insects.

Insects Categorized by Plant Type Attacked   Insects can be grouped according to the plant type that they attack. These include insects that feed on underground plant parts, chewing insects that feed on foliage and stems, piercing and sucking insects that feed on foliage and stems, and insects that feed on seeds, pods, or fruits.

Insects that Feed on Underground Plant Parts   Insects that feed on underground plant parts are generally soil insects that eat seeds and/or roots and stems of young developing plants. Damage can range from reduced plant vigor to death of the plants. Slow plant growth, yellowing of the leaves, and wilting and death of the plant are symptoms of soil insect damage.

Following are some examples of insects that feed on underground plant parts:

Wireworms   Wireworms attack carrot, corn, potato, beet, bean, lettuce, and many other plants. There are many species, and the worms (larvae) range in size from $\frac{1}{2}$ to $1\frac{1}{2}$ in. depending on growth and/or species. The larvae are slender, quite hard, and usually have a shiny appearance (Figure 10.1). They feed on underground plant parts only. The adult is known as a click beetle because it makes a clicking sound when turning from its back to its feet. The female lays

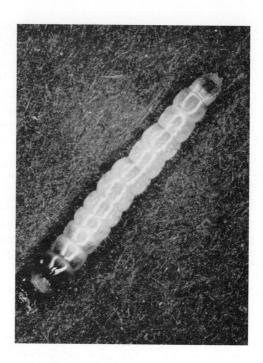

FIGURE 10.1   Wireworm larvae.

eggs in the spring. After the eggs hatch, the wireworms feed for 2 to 6 years before maturing into adult beetles.

Cutworms   Surface-feeding cutworms or caterpillars are common pests of young vegetables planted early in the spring. The worms are stout and dark colored. The cutworms eat through the stems just above ground level. Tomato, pepper, pea, bean, and members of the cole crops are particularly susceptible to cutworms. Cutworms remain in the soil during the day and feed only after sundown. The cutworm rolls into a ball when disturbed. Adult cutworms are dark, night-flying moths with bands or stripes on their forewings.

Mole crickets   Mole crickets (*Scapteriscus* sp.) tunnel through the top 1 to 2 in. of soil, loosening the soil and uprooting the plants. They also feed on roots, weakening the plants. Mole crickets feed at night and may tunnel as far as 10 to 20 feet before the sunrise. In the daytime, they return to their burrows.

Corn stalk borers   Cornstalk borers spend the winter as eggs on grasses and weeds, especially giant ragweed. After hatching in the spring, the worms feed in the leaf whorls and then bore into the sides of the stalks and burrow upward. After pupating in the soil, the grayish-brown adult moths emerge in late summer or early fall. The female moth lays eggs on grasses.

Cucumber beetle larvae   There are several species of cucumber beetles (banded, striped, and spotted) that attack vegetable crops. The adult beetles feed on the foliage, but the major source of damage is by yellowish-white larvae that tunnel and eat the roots of the plants.

Grubs   Grubs are larvae of several types of beetles that feed on roots. The grub is identified by a brown head, six prominent legs and a smooth shiny body with the rear end purple to black in color. The grubs feed on roots (Figure 10.2) 1 to 3 in. deep. In late fall, they move deep into the soil to overwinter and resume feeding in the spring.

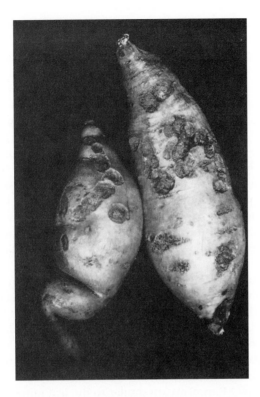

FIGURE 10.2   Grub injury on sweet potato roots.

Chewing Insects that Feed on Foliage and Stems   Foliage feeders feed on above-ground plant parts, except seeds, seed pods, and fruits. Caterpillars are the primary foliage feeders. After reaching a length of ½ in. or longer, they consume large amounts of plant tissue. The caterpillars are widely known for their foliage-feeding ability, but adult forms (moths, butterflies) do not damage the plant.

Caterpillars and Worms   Caterpillars and/or worms have chewing mouthparts that damage plants by consumption of tissue.

Armyworms   Armyworms (*Spodoptera* spp.) overwinter in cold-season areas, but the moths migrate great distances in the spring as females search for places to lay their eggs. The tan to gray adult females lay eggs on the blades of grasses and grains. A single female can lay 25 to 100 eggs. The caterpillars hatch from these eggs and feed on leaves, flowers, and fruits. Ragged holes in leaves are symptoms of armyworm feeding. They often attack plants in large numbers. After several weeks of feeding they pupate in the soil, then emerge as adult moths to resume the cycle. These pests are numerous after cold, wet spring weather that slows the development of natural parasites and diseases.

Diamondback moths   Diamondback moth (*Plutella xylostella*) worms are the larvae of gray or brown moths that fly in the evening. The adult moths are gray, about ⅓ in. long, and males have a row of three light-colored, diamond-shaped markings on the back. The adults are not injurious to the plants. The larvae feed on the underside of leaves of members of the cabbage family, chewing small holes and/or leaving transparent patches. After feeding for about 2 weeks, the larvae pupate in transparent silken

cocoons attached to the underside of leaves. Adult moths spend the
winter under plant debris.

Hornworms   Tomato hornworms (*Manduca quinquemaculata*) and tobacco
hornworms (*Manduca sexta*) feed on tomato, pepper, or eggplant fruits
and foliage. Each worm consumes large quantities of foliage and causes
extensive damage. The worms hatch from eggs laid on the underside of
the leaves, and these young feed for 3 to 4 weeks. They crawl into the soil,
pupate, and later emerge as adults to repeat the cycle. Hornworms are
characterized by having a hornlike projection on top of the rear end of
the body. The worms are quite large (3 to 4 in. long) when fully grown.
Some worms may have white sacs on their bodies. These sacs are the
cocoons of parasitic wasps that feed on and eventually kill the hornworms.

Loopers   Loopers attack members of the cabbage family as well as lettuce.
Several species of these moths or butterfly larvae feed on many vegetables.
The brownish cabbage looper (*Trichoplusia ni*) moths lay pale green eggs
on the topside of leaves in the evening. Although they are usually
nocturnal, moths can be seen flying during the twilight hours. Eggs are
usually laid with the onset of warm weather. A single female can lay up to
1,000 eggs during her lifetime. The larvae that emerge feed on leaves,
flowers, and buds for 2 to 6 weeks, depending on the weather and species.
The larvae eventually develop into green caterpillars about 0.75 to 1.0 in.
long. Mature caterpillars pupate in cocoons attached to leaves or
structures or buried in the soil.

Tomato pinworms   Pinworms are similar to the leaf miner in their attack.
Pinworms are small, yellowish gray or green, purple-spotted caterpillars
belonging to the moth order of insects. The larvae are only about ¼ in.
when grown. They tunnel through the leaf, eating away the chlorophyll
and leaving a crooked tunnel either behind or into mature tomato fruit.
Pinworms also roll and tie the leaf tips together.

Beetles

Mexican bean beetles   Mexican bean beetles (*Epilachna varivestis*) are found
throughout the United States. The adults are about ¼ to ⅓ in. long,
bronze in color with 16 black spots on their backs. The larvae are yellow
with rows of black-tipped, branched spines growing from their backs.
Mexican bean beetles feed on many of the legumes, such as lima beans,
bush beans, and cowpeas. Feeding damage by adults and larvae is often on
the underside of the leaves between the veins, resulting in a lacy
skeletonized leaf appearance. The beetles can also reduce pod
production. The adult beetles spend the winter in plant debris and
emerge in late spring and early summer. The females lay yellow eggs on
the underside of leaves. Larvae that hatch from these eggs are green at
first and then gradually turn yellow. Hot, dry summers and cold winters
reduce beetle populations.

Colorado potato beetles   Colorado potato beetles (*Leptinotarsa decemlineata*)
devastate potato, tomato, and pepper plantings. Both the adults and
larvae can damage plants by devouring leaves and stems. Small plants are
most severely affected. The female beetles lay their yellow to orange eggs
on the underside of leaves as the first plants emerge in the spring. The
larvae hatch from these eggs, feed for 2 to 3 weeks, pupate in the soil, and
emerge 2 weeks later as adults that lay more eggs.

Cucumber beetles   Cucumber beetles, both striped (*Acalymma* sp.) and
spotted (*Diabrotica* sp.) are common pests of cucumbers, melons, squash,
and pumpkins. Controlling these beetles is important because they carry
two serious diseases, mosaic virus and bacterial wilt, which damage and
may kill cucurbits. Adult beetles survive the winter in plant debris and
weeds. They emerge in early spring and feed on a variety of plants. The
beetles attack leaves and stems and may totally destroy the plants. Mature
females lay their yellow-orange eggs in the soil at the base of the plants.
The grubs that hatch from these eggs eat the roots and stems below the
soil line, causing stunting, wilting, and premature death.

Flea beetles   Flea beetles jump like fleas but are not related to them. Both
adult and immature flea beetles feed on a wide variety of vegetable plants.
The beetles are about $\frac{1}{16}$ in. long, or about the length of a big flea. They
are usually bronze, brown, or black, but may appear in many colors.
Immature beetles, legless gray grubs, injure plants by feeding on the roots
and undersides of leaves. Adults chew holes in leaves. Flea beetles are
most damaging to seedlings and young plants. Seedling leaves riddled
with holes dry out quickly and die. Adult beetles survive the winter in soil
and debris. They emerge in early spring to feed on weeds until vegetable
seedlings emerge. Grubs hatch from eggs laid in the soil, and these pests
feed for 2 to 3 weeks. After pupating in the soil, they emerge as adults to
repeat the cycle.

Leaf miners   Leaf miners belong to the family of leaf mining flies. The tiny
black or yellow adult females lay white eggs on the underside of leaves or
inside leaves. The maggots that hatch from these eggs tunnel between the
leaf surfaces, feeding on the inner tissue (Figure 10.3). The tunnels and
blotches are called mines. Damaged portions of leaves are no longer
edible. Upon reaching maturity, the maggots cut their way out of the leaf,
drop to the soil, and pupate.

FIGURE 10.3   Leaf miner injury on tomato leaf. Source: Clemson Univ. Ext. Serv.

Piercing and Sucking Insects that Feed on Foliage and Stems   Piercing and sucking insects have highly modified hollow mouthparts that function like a hypodermic needle. These needlelike mouthparts are inserted into the plant and used to remove the plant sap. Some insects in this group transmit plant viruses and diseases from plant to plant as they feed. These insects often inject certain enzymes and toxins into the plant as they feed and cause the plant to grow abnormally. Damage from piercing-sucking insects is often underestimated since the wounds they make are not readily apparent. These pests are also quite mobile and fly before they are seen. Overall plant symptoms caused by either severe or constant attack are distorted leaves, overall loss of plant vigor, spotting of leaves, and, in some cases, loss of foliage and death.

Aphids or plant lice   Aphids or plant lice are pale green, yellow-purple, or black soft-bodied insects that cluster on the underside of leaves (Figure 10.4). They come in various sizes ($\frac{1}{16}$ to $\frac{1}{8}$ in. in length) and colors (brown, yellow-pink, or black). Damage occurs when the aphids suck the juices from the vegetables. Severely infected plants may be stunted and weak, producing few fruit. Fruit yield is also reduced when aphids spread plant virus diseases. Aphids are spread from plant to plant by wind, water, and people. Aphids are unable to digest fully all the sugar in the plant sap and excrete the excess in a fluid called honeydew, which often drops onto lower leaves. Ants feed on this sticky substance and are often present where there is an aphid infestation.

Leafhoppers   Leafhoppers are spotted, pale green insects up to $\frac{1}{8}$ in. long. Most are wedge-shaped individuals, broad at the head and pointed behind. They hop, move sideways, or fly away quickly when the plant is touched. Leafhoppers feed on the underside of leaves, sucking the sap, which causes stippling. Severely infected vegetable plants may become weak and produce little edible fruit. One leafhopper, the aster leafhopper (*Macrosteles fascifrons*) transmits aster yellows, a plant disease that can be quite damaging. Leafhoppers at all stages of maturity are active during the growing season. They hatch in the spring from eggs laid on perennial weeds.

FIGURE 10.4   Aphids or plant lice. Source: Clemson Univ. Ext. Serv.

Stinkbugs   Stinkbugs remove plant juices and inject toxins, which result in limp, wilted leaves or groups of wilted leaves connected to a common fed-upon stem. The Southern green stinkbug is large, flattened, shield-shaped, bright green, and about ⅔ in. long (Figure 10.5). The nymphs are smaller, wingless, and have small red, black, and white markings on their backs. Stinkbugs produce an offensive odor when disturbed.

Spider mites   Spider mites are related to spiders. They cause damage by sucking sap from the underside of leaves. As a result of feeding, the green leaf pigment (chlorophyll) disappears, producing a stippled appearance. Although the mites do not attack the fruit directly, they do cause leaf drop, which weakens the plant and reduces fruit yield. If the plants are severely infected, flowers will not form or do not form normally. Mites are active throughout the growing season, but most are favored by dry weather with temperatures above 70°F.

Silverleaf whiteflies   Whiteflies (*Bemisia tabaci*) are tiny, winged insects about 1/12 in. long that feed on the underside of leaves. The insects are covered with white waxy powder. The four-winged adult female lays eggs on the underside of leaves. Each larvae is the size of a pinhead. White waxy filaments radiate from its body. It feeds for a month before changing to its adult form. The larval and adult forms suck sap from the leaves. The larvae are more damaging because they feed more heavily. Whiteflies cannot fully digest all the sugar in the plant sap. They excrete the excess as honeydew, which often drops onto the lower leaves. A sooty mold fungus may develop on the honeydew, causing the leaves to appear black and dirty.

Thrips   Onion thrips (*Thrips tabaci*) attack many vegetables, including onions, peas, and cabbage. Thrips are barely visible insects, less than 1/12 in. long, and are dark brown to black. The adults have narrow wings fringed with hairs. The larval stage is wingless, but resembles the adults. They reduce the quality and yield by rasping holes in leaves and sucking out the plant sap. This rasping can

FIGURE 10.5   Stinkbug.

FIGURE 10.6   Corn earworm feeding on sweet corn ear and husk.

cause white streaks on the leaves. Plants often die when thrip populations are high. Damage is most severe in leaf sheaths or areas that are protected from the elements and pesticides. Thrips are active throughout the growing season.

Insects that Feed on Seeds, Pods, or Fruits   As a group, insects that feed on seeds, pods, or fruits are probably the most damaging for vegetables. Most healthy plants can overcome some root and foliage damage; however, damage to edible parts of a plant results in direct loss of seeds or fruit. To further complicate the situation, attack at times of fruiting also means that not only is the immediate fruit lost, but the cumulative time, money, and labor expended on the crop (which is at a maximum at this time) is also lost.

Corn earworms   Corn earworms (*Heliothis zea*) are striped yellow, brown, or green worms that feed on the tip of the ear of corn inside the husk (Figure 10.6). These worms range in size from ¼ to 2 in. long. These worms are the most serious pests of corn, but they also attack other vegetables as well. Corn earworms are also known as tomato fruitworms and cotton boll worms. Corn earworms are the larvae of light gray-brown moths with dark lines on their wings. In the spring, female moths lay yellow eggs singly on corn silk and the undersides of leaves. The worms that hatch from these eggs feed on the new leaves in the whorls. More serious damage is caused when the worms feed on the silks, causing poor pollination, and on the developing kernels. Worms enter the ear at the silk end or they may bore through the husk.

Pickleworms   Pickleworms are specific pests of the cucurbits. The female moth lays eggs on tender buds, new leaves, stems, and undersides of fruit. The pickleworm larvae first feed on plant tissue and leaves, and when the worms become larger they may bore into the fruit (Figure 10.7). The whitish to green pickleworm caterpillars feed on the insides of the fruit. The fruits soon rot, sour, and mold after the interior has been exposed to the air by the burrows. Fruits of late season plants are severely attacked.

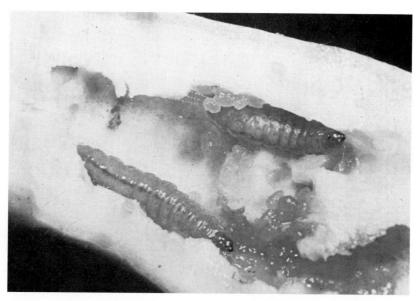

FIGURE 10.7   Pickleworm larvae feeding on fruit. Source: Clemson Univ. Ext. Serv.

Cowpea curculios   Adult weevils are about ⅛ to ⅙ in. long and are dark in color. The females drill a hole with their mouthparts through the developing southern pea and bean pods, and insert an egg into the feeding site. The tiny white grublike larvae then feed on the inside of the seed and cause its destruction.

General Control Measures   Insects must be carefully controlled during the field production of vegetables. Insects cause injury to the outer leaves and reduce vigor of the plant. Insects can become embedded within leaf clusters of some leafy and greens crops or within husks of sweet corn and not be observed during harvest, but appear during consuming or processing. Insects, once within the leaf clusters or husks, are difficult to control. Therefore, control must be exerted before insects move into these structures.

Fields are monitored regularly for insects. Control measures are usually taken when insects reach a certain threshold amount. Several cultural practices reduce insect infestations. These cultural practices include controlling weeds in nearby fields (Figure 10.8), burying plant debris that might harbor overwintering pests, and increasing populations of natural enemies of certain pests.

If insecticides are used, they are carefully chosen from an accepted list for use and are specific for the target insect. Insecticides are generally not applied in the middle of hot, dry days because coverage of wilted plants is not good. High temperatures will volatilize some insecticides before they reach the plant.

Diseases

Plant disease is the process by which living or nonliving entities interfere, over a period of time, with a plant's functions. This may cause changes in the plant's appearance and/or bring about a lower yield. There are parasitic and nonparasitic types of diseases.

FIGURE 10.8   Poor weed control along edges of a field can cause increased insect problems for the crop being grown.

Fungi, bacteria, viruses, and mycoplasmas cause parasitic diseases. All pathogens compete with plants by using metabolites produced by the host and most reduce physiological efficiency of the plant by reducing photosynthate capacity and/or water and nutrient uptake or by disrupting metabolic processes at the cellular level. The pathogens that cause parasitic diseases spread from plant to plant.

Nonparasitic diseases are the necrosis, mottling, and stunting caused by poor plant nutrition, air or water pollution, or weather-related factors. Diagnosing a nonparasitic disease is difficult because many different things can cause similar symptoms.

Fungi   Fungi are the largest group of disease organisms and abound in most environments. Fungi are plants with a threadlike structure. They enter plants through natural openings or wounds or penetrate directly through intact tissue. Fungi cannot obtain their own food and get their nutrients either from living tissues or from other organisms or from dead organic matter. Fungi produce spores, which are sexual or asexual reproductive structures capable of surviving adverse conditions and dispersing themselves. Following are some examples of fungi:

Anthracnose   Anthracnose, caused by *Marssonina panattoniana,* is also called "shothole" on some plants (Figure 10.9). The first symptoms are small, yellow, water-soaked spots on the leaves. These spots darken, enlarge slightly, and dry up. After a few days to a week the center of the spots fall out. This produces a shothole appearance that might be confused with insect damage. Affected leaves eventually wither and die. Small water-soaked spots may also appear on the midrib's underside. These spots on the midrib darken, elongate, and sink slightly giving the midrib a rough indented surface. Small, sunken, circular to elliptical lesions can also appear on mature fruit. Cool, wet periods favor anthracnose infection. The pathogen spreads when rain or sprinkler droplets splash spores from infected plants to uninfected plants. Cool temperatures are required for high spore production.

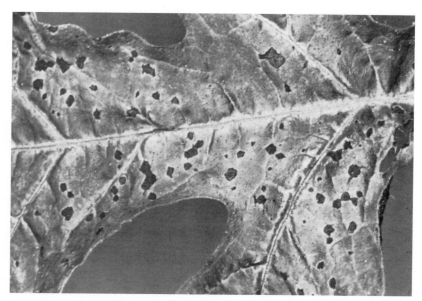

FIGURE 10.9   Anthracnose on watermelon leaf. Source: Clemson Univ. Ext. Serv.

Bean rust   Symptoms of bean rust (*Uromyces phaseoli*) are small, circular, yellow spots on leaves or pods of beans. These lesions will enlarge in diameter with time. The chlorotic halo will persist around the lesion, but the center will be "rust" red. The red color is due to a large number of red-colored urediospores produced by the pathogen. Old lesions may turn brown to black as the leaf cells die. Under conditions of heavy infection, premature defoliation can occur.

Cercospora leaf spot   Cercospora (*Cercospora capsici*) leaf spot begins as lesions that resemble small, circular, water-soaked, dark green spots. The centers of the lesions become white to gray as the fungus kills the leaf cells. These necrotic centers are surrounded by a red to dark brown border. A diffuse chlorotic halo usually surrounds the lesions during late stages of development. Lesions may enlarge and coalesce under heavy infestations. Defoliation may occur if this happens.

Downy mildew   Symptoms of downy mildew, caused by *Peronospora parasitica* on many vegetable plants, are pale, irregular, sharply angled spots on either the underside or top of leaves (Figure 10.10). The center of the lesion becomes yellow to yellow-brown and is surrounded by a diffuse area of chlorosis. These areas enlarge, often covering the entire leaf surface and extending to the upper surface. The spots are usually confined to areas of the leaf blade within the main blade, and often become covered with white fluffy spore masses. The undersurface of infected leaves appears water-soaked. The affected foliage generally loses its luster and tends to curl, turn yellow, brown, and die. Black minute spots can appear near the veins within the white mold. Downy mildew is most severe on older leaves of mature plants. If too many leaves are infected and destroyed by downy mildew, fruit formation may be impaired or prevented.

Early blight   Symptom development of early blight (*Alternaria solani*) begins on leaves as minute chlorotic spots. These lesions enlarge as the mycelium of the

FIGURE 10.10   Downy mildew on muskmelon leaf.

fungus grows between the cells of the infected leaf. The center of the lesion becomes gray to brown as the leaf cells die. Concentric rings of various shades of gray to brown tissue encircle the center of the lesion (Figure 10.11). A chlorotic halo encloses the lesion throughout the various stages of development. Several lesions may coalesce to form one to several large lesions. Large amounts of leaf surface are destroyed under this condition, and the leaf may fall off. Brown to black lesions result from infection of the fruit. Infected areas of the fruit become leathery in appearance and are covered by a black felt of sporulating mycelium.

Fusarium wilt   The initial symptom of fusarium wilt (*Fusarium oxysporum*) is a stunting of young plants and a general yellowing of the lower leaves. The leaves will die and may drop from the plant as symptoms progress. The leaves on one side of the plant may be more seriously affected than those on the opposite side. Ultimately the plant will die. The mycelium of the pathogen penetrates and grows in the vascular bundle of the root and lower stem, thus restricting the upward movement of water and nutrients. Enzymes produced by the pathogen decompose water-conducting cells of the vascular system, resulting in a chocolate brown discoloration. This is evident in a cross section of the stem as a discolored ring of tissue outside of the pith and just inside the outer layer (cortex) of the stem (Figure 10.12). Fruit will not mature on plants infected at an early stage. Some fruit may mature on late infected plants. Affected plants appear in clusters and are often in low areas of the field where there may be poor drainage.

Late blight   Late blight is caused by the fungus *Phytophthora infestans*, and infects potatoes, tomatoes, and eggplant. It was late blight that destroyed the Irish potato crop in the late 1840s, causing at least a million Irish deaths and forcing the migration of an even larger number of Irish to America. *P. infestans* reproduces asexually by producing spores, which can only survive on living plant material. The fungus is believed to live through the winter on perennial

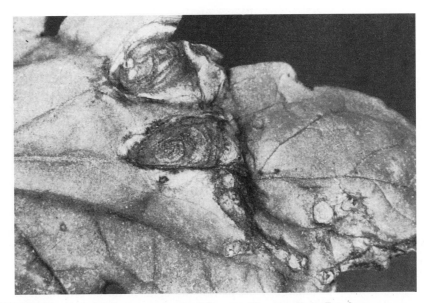

FIGURE 10.11   Early blight on tomato leaf. Source: Clemson Univ. Ext. Serv.

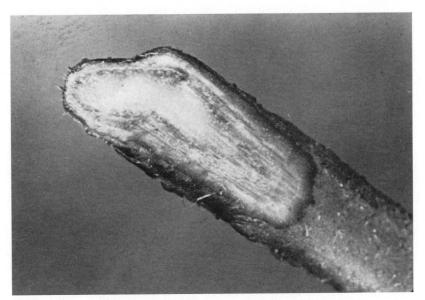

FIGURE 10.12   Fusarium wilt symptoms on an infected stem. Source: Clemson Univ. Ext. Serv.

solanaceous plants and on infected potato tubers. Even though the fungus does not move through an infected plant's tissues, its spores can wash off the leaves and down through the soil onto the tuber (soil is really rather porous). These spores cannot live in the soil. This disease needs damp conditions to propagate. With both tomatoes and potatoes, late blight often starts with infection of the leaves. Early symptoms include yellowing on the edges or sometimes in the middle. In these early stages it can resemble *Alternaria* (early blight), with the main distinguishing characteristic between the two being that *Alternaria's* lesions will not spread across the leaf's veins while late blight's lesions will. The fungus spreads outward from the initial point of infection, with the leaf turning brown at that spot, with yellow around it. The leaves may appear almost water-soaked due to the fungal hyphae breaking down the leaf's cells. Brownish-black patches on the stems, often around the leaf axils, may appear, but the stems often remain intact until the plant withers. A whitish "mold" may also be visible on the undersides of the leaves. If the fungus infects tomato fruit, the fruit may develop a characteristic known as firm rot: a hard brownish area that is smooth or sometimes corrugated. If that tomato ripens, the area immediately around the brown patch will usually remain green.

Powdery mildew   The mycelium of the fungus (*Sphaerotheca fuliginea* or *Erysiphe cichoracearum*) that causes powdery mildew develops superficially on the leaves of infected plants. This mycelium, together with the large number of chains of spores (conidia) produced on the mycelium, imparts a white, granular, or powdery appearance to infected areas on the leaf. The white, powdery areas enlarge as the pathogen begins its development on the leaf. The common name "powdery mildew" is given to this disease because of the powdery appearance of the fungus mycelium and conidia. Infected areas of leaves develop irregular, interveinal chlorotic lesions that will eventually die and become gray-brown to brown in color. Severe infection of young leaves results in leaves that are stunted, curled, and deformed. Fruit from infected plants will be small and of poor quality.

Sclerotinia drop   Symptoms of sclerotinia drop include severe wilting of mature plants, collar rot, and death. The soilborne fungi, *Sclerotinia* major and *Sclerotinia* minor, first attack leaves and stems in contact with the soil, and a cottony growth develops. The pathogen can spread downward until roots are decayed and spread upward to the leaf bases. The pathogen readily spreads in the plant. Small black spots (sclerotia) appear on leaves during the growth of the fungi. Sclerotinia is most active under moist, cool conditions.

Bacteria   Bacteria cause fewer diseases than fungi and viruses. They are single cells surrounded by a cell wall. Plant parasitic bacteria lack chlorophyll and obtain nutrients from living or nonliving sources. Bacteria enter plants through natural openings or through wounds made by insect feeding, mechanical injury, or pruning. Bacteria increases occur at an enormous rate under optimum conditions and are encouraged particularly by surface moisture from rainfall, irrigation, or high humidity and by moderate temperatures. Seeds or vegetative propagation, insects, and tillage may transmit bacteria. Some of the diseases caused by bacteria include the following:

Bacterial blight   Bacterial blight (*Xanthomonas phaseoli*) is a common disease on beans. Leaf symptoms first appear as small, water-soaked, or light green patches. The lesions enlarge as symptom development progresses, and the centers become dry and brown. The lesions are typically irregular in shape and are surrounded by an irregular yellow border. Several lesions may coalesce and result in larger lesions. Small, water-soaked, dark green lesions develop on infected pods. Individual lesions may coalesce as the bacteria continue to develop in the pod tissue to form large, irregular red to brown lesions. Pods may shrivel and die.

Bacterial leaf spot   Bacterial leaf spot (*Xanthomonas vesicatoria*) can infect tomatoes and peppers (Figure 10.13). Small, yellow lesions are the first symptoms of bacterial leaf spot. The center of the lesions turn gray to tan as the leaf cells die. A narrow black border surrounds this gray, necrotic center. A small chlorotic halo may appear around the lesion. The individual lesions will continue to enlarge, and mature lesions will grow together, resulting in large patches of brown, gray, or dead leaf tissue. Water loss from these areas will result in a buckling or curling of the leaf and leaf drop may result. Infection to other plants occurs as bacterial cells produced on the lesions of infected plants are spread by water, cultivation equipment, or insects to uninfected leaves.

Bacterial wilt   Bacterial wilt (*Pseudomonas solanacearum* or *Erwinia tracheiphila*) can infect members of the Cucurbitaceae and *Solanum* groups. The initial symptom of infection is the wilting of the tip of infected plants. The symptoms become more severe with time, especially under hot, dry conditions. Some chlorosis of the leaves may also occur. The leaves die and turn brown as the disease becomes more severe. Young plants that are infected may die before fruit production. A diagonal or lengthwise cut through the stem will reveal two tan to brown lines of the vascular cylinder (Figure 10.14). A cut across the stem frequently results in creamy-white, sticky bacterial ooze. In severe infections the central portion or pith will be brown in color. A diagnostic test for bacterial wilt can be performed in the field by cutting a wilted stem or vine close to the main stem, rejoining the cut surfaces, then slowly drawing the sections apart. A thin strand of slime that extends between the two surfaces is indicative of a positive diagnosis of bacterial wilt.

Black rot   Black rot (*Xanthomonas campestris*) infects cabbage. The disease can occur at any growth stage of the cabbage. Cells of black rot move in the translocation stream of infected plants and may move into seeds of infected plants. As a result,

FIGURE 10.13   Bacterial leaf spot on pepper leaf. Source: Clemson Univ. Ext. Serv.

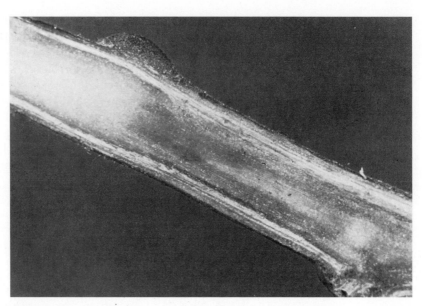

FIGURE 10.14   Bacterial wilt–infected stems of tomatoes. Source: Clemson Univ. Ext. Serv.

young plants produced from these seeds may be infected. Leaves of infected seedlings turn yellow and wilt as the bacteria invade the vascular system and prevent upward movement of water and nutrients from the roots. Vascular strands in the stem may turn black. Infection occurs through wounds or stomates. The bacteria multiply in the leaf, killing large areas. These areas are yellow, then turn brown. The lesions are V shaped when infection occurs near the veins at the edge of the leaf. The veins frequently become black. Large areas on the head develop a black, slimy symptom referred to as soft rot.

Viruses   Viruses cannot function outside a host, but most have a wide range and more than one vector. Viruses also depend on a second organism (a vector) to move about the environment and invade a plant. Viruses interfere with the biochemical reactions in plant cells and tissues. A mosaic symptom, characterized by alternating bands or patches of various shades of yellow or green colors, is the most typical and most frequently occurring symptom of virus diseases. Aphids, whiteflies, and leafhoppers are the most common vectors. Two mosaic viruses are discussed as follows:

Cucumber mosaic virus   Cucumber mosaic virus (CMV) is aphid transmitted. CMV can infect many members of the Cucurbitaceae and *Solanum* groups. Leaves of plants infected with CMV develop mottling leaf coloration, with alternating or mixed irregular patches of various shades of yellow or green. Leaves may also become curled and frequently will be dwarfed in size. Severity of disease increases with time and can be especially severe if young plants are infected. Plant stunting can occur and yields can be reduced as fruit set is diminished. Mosaic symptoms are most pronounced during periods of rapid growth, when nitrogen is ample and leaf tissue is green. Seed is the most important source of primary inoculum. The peach aphid is the most important vector.

Tomato mosaic virus   Tomato mosaic virus (TMV) can affect a wide range of crops including beet, eggplant, pepper, potato, tomato, and turnip. A wide range of symptoms can occur on susceptible crops with the most common being stunting of plants and leaves, leaf mottling and deformation, and fruit mottling and bronzing. Leaf symptoms by a common strain of TMV include mottling with raised dark areas and some distortion in younger leaves. Common spread of the virus is by mechanical means during transplanting and other cultural practices such as pruning, tying, spraying, watering, and picking. Workers who use tobacco products immediately prior to working with the plants can transmit TMV. Sanitation is important in controlling or containing the spread of the virus.

Mycoplasmas   Mycoplasmas are the smallest individual organisms and also must live close to plant cells. Mycoplasmas were once classified as viruses since they cause similar symptoms. Symptoms caused by mycoplasma include growth abnormalities, yellowing, very short internodes, and distortions of leaf and flower tissue. The most common mycoplasma is aster yellows.

Leaves affected with aster yellows become curled, yellow or white. They occasionally have small brown spots of dried latex along their margin. Center leaves fail to develop normally and are stubs in the middle of the head. Leaves often taste bitter.

A mycoplasma transmitted by a six-spotted leafhopper is the suspected causal agent of aster yellows.

General Control Measures   Methods to control diseases can be broken down into five categories: choosing resistant and tolerant cultivars, promoting vigorous plant growth, using proper water management, reducing inoculum, and applying fungicide sprays.

Choosing resistant and tolerant cultivars is often the least expensive, easiest, and most effective way to reduce losses to some diseases. For some diseases, such as soilborne vascular wilts and the viruses, the use of resistant cultivars is the only way of ensuring control. It is estimated that more than 85% of the vegetable acreage in the United States is planted with cultivars that have tolerance or resistance to one or more disease organisms.

Vigorous growth is desired for proper plant development and prevention of diseases. Stressed plants, especially those low in potash and calcium, are more vulnerable to diseases. On the other hand, plants receiving too much nitrogen may develop a succulent growth that could encourage certain pathogens. Maintaining proper crop nutrition is extremely important for long-term growth and disease prevention. Selection of cultivars adapted to the location and common pathogens is also an important aspect of disease control. Following recommended planting dates could also ensure that the plants achieve vigorous growth. Also, planting dates may be adjusted so that the crops can mature before disease strikes.

Proper water management in vegetable fields is important to prevent waterlogging conditions. In wet soils, seeds and seedlings are likely to rot. It is also important that the foliage remains dry. Wet leaves are more likely to be infected with waterborne pathogens. Inoculums spread from infected to healthy leaves by water droplets. It may be beneficial to water only early in the day so that the leaves dry before evening.

Inoculum levels of diseases are reduced through good sanitation and by controlling insect vectors. Only certified or treated seed should be used. When harvest is complete, the remains of annual crops are destroyed because stems, leaves, and roots can serve as disease and insect reservoirs for the following year.

Proper crop rotation can also reduce disease pressures. Increasing the organic matter of the soil and removing or burying diseased plant matter, which could serve as a source of inoculum, reduce the incidence of diseases.

In addition, chemical applications to the plants can be used to reduce injury due to fungal and some bacterial diseases. Chemicals that are used include seed treatment fungicides, foliar fungicides, and soil fumigants. These chemicals should be carefully chosen and properly applied.

In summary, successful control of vegetable diseases is achieved by using several steps selected on the basis of the disease organism, its life cycle, the host plants, labor availability, and cost. The degree of effectiveness that the grower achieves in controlling a particular disease will also be dependent on the growing season and the grower's level of tolerance to loss of product.

Nematodes

Nematodes are neither pathogens nor insects. They are small, unsegmented roundworms (0.7 to 1.7 mm in length) that are generally transparent and colorless. Most are slender. Nematodes reproduce by eggs that are deposited in the soil or in plant tissue. Other kinds of nematodes keep their eggs in a jellylike mass that is either attached to or inside the female's body. This mass becomes a tough protective capsule called a cyst. Under ideal conditions (80° to 86°F) as little as 4 weeks is required for many kinds of nematodes to complete their life cycle.

Plants affected by nematodes appear stunted and wilted. Leaf chlorosis and other symptoms of nutrient deficiency may also be present. Plants exhibiting stunted or declining conditions usually occur in patches of nonuniform growth rather than as an overall decline of plants within an entire field.

Root-Knot Nematodes   Root-knot nematodes (*Meloidogyne* spp.) first feed on root epidermis, penetrating with the stylet and sucking plant sap. Larvae eventually penetrate the root, taking up permanent sites and feeding on surrounding tissue. Galls (Figure 10.15) develop on affected sites, apparently in response to growth substances.

Sting Nematodes   New roots that fail to develop or appear as stubby roots are characteristics of infestation with sting nematodes (*Belonolaimus* spp.). Plants become stunted and yellow with no known resistance. In addition to the direct crop damage caused by nematodes, many of these species have also been shown to predispose plants to infection by fungal or bacterial pathogens or to transmit virus diseases, which contribute to additional yield reductions.

General Control Measures   Long rotations with grain crops can reduce nematode population levels. Because many nematode eggs can survive indefinitely, nematodes may always be a threat in fields with a history of the disease. Therefore, fungicides and nematicides may be necessary in these fields.

Weeds

A weed is a plant that is growing where it is not wanted. Weeds are competitive and aggressive, capable of surviving in competition with almost any crops. They are generally abundant, existing in dense populations and producing prolific numbers of

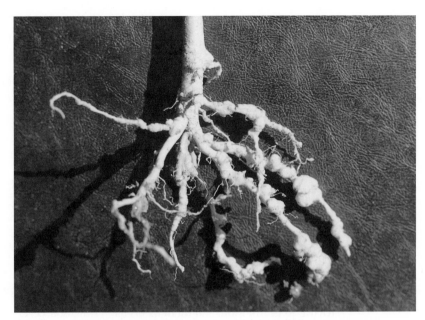

FIGURE 10.15   Galls on roots infected with root-knot nematodes.

seeds. Weeds are easily spread and the seeds have special structures that aid dispersal by wind, water, or animals.

Weeds reproduce by sexual or asexual means. Sexual reproduction occurs by seed and is the only way summer or winter annuals or biennials can reproduce. Asexual reproduction is achieved through vegetative means—runners, thistles, or rhizomes. Once established, most perennials spread by roots or rhizomes.

Weeds reduce yield and quality of vegetables through competition for light, moisture, and nutrients. Weeds also interfere with harvest operations and can harbor insects and diseases. Since early season competition is the most critical, major emphasis is placed on this time period.

Major Categories   Weeds belong to one of three major categories: grasses, broadleaves, and sedges.

Grasses   Grassy weed seeds germinate and emerge as a single leaf (monocot). Most grasses develop hollow, rounded stems and nodes that are close and hard. The leaf blades alternate on each side of the stem, are much longer than wider, and have parallel veins. Among the grass weeds are crabgrass, goosegrass, and annual bluegrass.

Broadleaves   Broadleaves have netlike veins, nodes containing one or more leaves, and showy flowers. As dicots, broadleaf weed seedlings emerge with two leaves. Broadleaf weeds include common chickweed, dandelion, and pennywort.

Sedges   Sedges are also monocots; however, they are not true grasses. Solid, triangular stems without nodes characterize sedges. Their grasslike leaves appear on the stem in clusters of three with each one extending in a different direction. Annual sedges are called water grasses. Perennial sedges such as yellow nutsedge predominate and are difficult to control. Purple nutsedge is smaller than yellow nutsedge

FIGURE 10.16   Blossom end rot of watermelon fruit. Source: Clemson Univ. Ext. Serv.

and is another perennial sedge that reproduces through a series of bulbs on radiating rhizomes called tuber chains.

General Control Measures   Prevention begins with identification and mapping of weeds in the field. Localized weed control treatments, manual removal, or chemical treatments should be used in areas with high concentrations of perennial weeds.

Successful land preparation is an effective method of weed control in fields. Early shallow cultivation reduces weed damage. Herbicides may not be necessary if the soil is properly managed during the short life cycle of some crops.

Herbicides, if used, are selected depending on weed species and soil type. For example, the effectiveness of preemergent herbicides is dependent on adequate soil moisture. Sensitive crops should not be grown in fields where persistent herbicides were applied. Other control measures also work such as establishing a good stand to be competitive with the weed species; proper crop rotation; mechanical control such as plowing, disking, cultivation, mowing, and hoeing; using polyethylene mulches; and preventing weeds from infesting or reinfesting fields.

Physiological Problems

A variety of other non-pest-associated physiological problems can occur during the growth and production of vegetables. Physiological problems in production can be associated with nutritional imbalances within the plant and/or nonoptimal environmental conditions during the growing season. Examples of physiological problems in vegetable crops are blossom end rot of fruit (a leatherlike decay of the blossom end of several vegetable fruits associated with calcium deficiency) (Figure 10.16), secondary growth of potatoes (Figure 10.17), and bolting of many biennial and perennial crops (a premature flowering that occurs when the crops are exposed to inappropriate daylengths or temperatures).

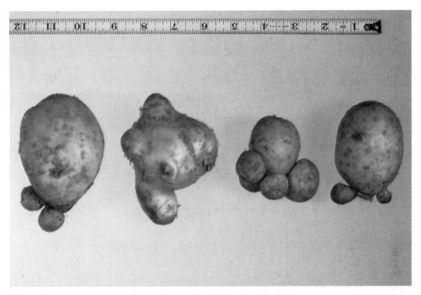

FIGURE 10.17   Secondary growth of Irish potatoes.

Symptoms associated with physiological disorders are difficult to identify and are occasionally confused with injury caused by insects and diseases. Because many physiological symptoms are crop-specific, information on their identification and potential causes are addressed according to the specific crop characteristics listed in Part 3.

Using Pesticides

Pesticides are chemical substances used to kill or control pests. Pesticides are useful when nonchemical methods fail to provide adequate control of pests and when pest populations reach a level of economic injury.

A Brief History of Pesticide Use

Pesticides in common use prior to World War II were predominantly inorganic materials such as sulfur, lead, copper, arsenic, boron, and mercury as well as plant-derived compounds such as nicotine, pyrethrum, derris, and rotenone. DDT was developed by the Swiss before the war and was declared a "miracle" insecticide. The effectiveness of DDT and related materials such as lindane, dieldrin, chlordane, and 2,4-D slowed the development of other less toxic methods of pest control (especially biological control methods). Even the gradual accumulation of evidence on the widespread environmental contamination by these chemicals failed to slow their use for many years. Today there are many chemicals on the market that are highly selective and relatively harmless to humans and other nontarget organisms. These chemicals also biodegrade rapidly.

Selecting Pesticides

Pesticides may be needed to control certain problems in the field. Nonchemical methods are usually considered before a grower applies any pesticide. Pesticides are chosen that are effective for use on the specific vegetable crop and against the specific pest problem.

The chemical label indicates among other things the brand name, and manufacturer, how the product is formulated or mixed for application, the ingredients, chemical name, sometimes a common name, and whether the chemical can be used on the crop for the intended purpose in a certain state. It also provides information on safety and other possible restrictions such as the amount of time that must pass from the time of application until it is safe to harvest the crop. Pesticides, if not properly applied, can injure the vegetable crop and farm personnel.

Pesticide Grouping According to Chemical Nature   Pesticides can be grouped according to their chemical nature. Following is a list of the major groups:

Inorganic pesticides   Inorganic pesticides are made from minerals such as copper, boron, lead sulfur, tin, and zinc. An example is the Bordeaux mixture.

Synthetic organic pesticides   Synthetic organic pesticides are man-made. They contain carbon, hydrogen, and one or more other elements such as chlorine, phosphorus, and nitrogen. Examples include Sevin (carbaryl), malathion, diazinon, and maneb.

Microbial pesticides   Microbial pesticides are microscopic organisms such as beneficial viruses, bacteria, and fungi that are cultured, packaged, and sold. Some of the pathogens multiply and spread after application while others do not. Examples of microbial pesticides are the bacterium *Bacillus thuringiensis* and the polyhedrosis virus.

Plant-derived organic pesticides   Plant-derived organic pesticides are made from plants or plant parts. Examples include rotenone and pyrethrins.

Pesticide Grouping According to Mode of Action   Pesticides can also be grouped according to their mode of action (how they work). The major groups are:

Protectants   Protectants are applied to plants, animals, structures, and products to prevent entry or damage by pests. They prevent infection from occurring by killing spores before or after they germinate. This does not prevent all infections.

Contact poisons   Contact poisons kill pests when they contact the poison.

Stomach poisons   Stomach poisons kill when swallowed.

Systemics   Systemics are taken into the blood of a host animal or sap of a host plant. Because toxic concentrations can remain in plants for more than a few days, they provide some protection against new infections.

Translocated herbicides   Translocated herbicides kill plants by being absorbed by leaves, stems, or roots and moving throughout the plant.

Fumigants   Fumigants are gases that kill when they are inhaled or otherwise absorbed by the pest.

Selective pesticides   Selective pesticides kill only certain kinds of plants or animals.

Nonselective pesticides   Nonselective pesticides kill many kinds of plants or animals.

Applying Pesticides

Pesticide application equipment is thoroughly checked and properly calibrated before it is used. Also, equipment is inspected during use to prevent misapplication of chemicals. Pesticides are not to be applied if wind speed is greater than 16 km/hour (10 mph) to avoid pesticide drift. Pesticides can be applied with a backpack sprayer (Figure 10.18) for small acreage, or a large sprayer often associated with a tractor for larger fields (Figure 10.19). Pesticide sprays can also be applied with the use of airplanes (crop dusters) (Figure 10.20).

FIGURE 10.18   A backpack sprayer for pesticides.

FIGURE 10.19   Spraying tomatoes with a large pesticide sprayer.

FIGURE 10.20   A crop duster in a tomato field.

Nonchemical Control Measures

Soil Solarization

Soil solarization is a method of soil treatment that uses energy from the sun to heat the soil to a high enough temperature to kill or reduce populations of nematodes and some other soilborne problems. Soil solarization is accomplished by covering the soil with a clear polyethylene tarp for 4 to 6 weeks during a time of the year when sunlight is near maximum. It has been effectively used to reduce damage caused by a wide range of soilborne fungi, weed seeds, and nematodes. Soil solarization appears to work by increasing soil temperatures in the normal root depths. For example, it has been reported that temperatures of tarped soil 6 in. deep in California were 14° to 23°F higher than uncovered soil.

Soil Amendments

Maintaining high organic matter in the soil usually improves the plant's ability to withstand damage from nematodes and some soilborne diseases. As organic matter decomposes in soil it forms humus. This humus increases the supply of nutrients and water available to the plants and encourages the buildup of antagonistic and competitive organisms against some nematodes and soilborne fungi. When water and nutrients are available, plants can support themselves with slightly less efficient root systems than in dry and/or infertile soils. Any organic soil amendment such as animal manure, compost, peat, and leaves will contribute to the formation of humus. Also, the use of crop rotations, manures, composts, and green manures can be beneficial for control of some pests.

Review Questions

1. How do pest problems adversely affect vegetable crop production?
2. What are some of the symptoms of piercing and sucking insects that feed on foliage and stems?
3. Why are insects that feed on seeds, pods, or fruits probably the most damaging group of insects for vegetable crops?
4. What are some general control measures to reduce insect infestations?
5. What is a plant disease?
6. If plant parasitic bacteria lack chlorophyll, how do they survive?
7. Why is choosing resistant or tolerant cultivars the cheapest, easiest, and most effective way to reduce plant losses due to diseases and insects?
8. How can proper water management in the field prevent some disease problems?
9. What are some of the symptoms of plants that are infected with nematodes?
10. Define the term weed. Can weeds in a field reduce the growth and yield of vegetables?
11. How does soil solarization reduce the populations of soilborne pests?

Selected References

Dunn, R. A. 1991. *Managing nematodes in the home garden.* Univ. of Florida Coop. Ext. Serv. Cir. SS-ENY 28.

Johnson, F. A. 1996. *Insects that affect vegetable crops.* Univ. of Florida Coop. Ext. Serv. Cir. ENY 450.

Johnson, F. A., and S. Reese. 1996. *Insect management and control in the home garden.* Univ. of Florida Coop. Ext. Serv. Cir. 563.

Kingsland, G. C., and B. M. Shepard. 1989. *Diseases and insects of fruit and vegetable crops in the Republic of Seychelles.* Chapel Hill, NC: South-East Consortium for International Development.

Kucharek, T., and B. Dunn. 1994. *Diagnosis and control of plant diseases and nematodes in a home vegetable garden.* Univ. of Florida Coop. Ext. Serv. Cir. 399-A.

Latin, R. 1993. *Diseases and pests of muskmelons and watermelons.* Purdue Univ. Coop. Ext. Serv. BP 44.

Noling, J. W. 1996. *Nematodes.* Univ. of Florida Coop. Ext. Serv. Cir. ENY 625.

Ortho Editorial Staff. 1993. *The Ortho home gardener's problem solver.* Edited by C. Smith. San Ramon, CA: Ortho Books.

Polomski, R. F. 1998. *The South Carolina master gardener training manual.* Clemson Univ. Coop. Ext. Serv. EC 678.

Relf, D. 1997. *Understanding pesticide labels.* Virginia Coop. Ext. Serv. Publ. 426–707.

Sherf, A. F., and A. A. MacNab. 1986. *Vegetable diseases and their control.* New York: John Wiley & Sons.

Stall, W. M. 1996. *Weed management.* Univ. of Florida Coop. Ext. Serv. Cir. HS 717.

chapter

11

Harvesting

Maturity

The maturity of a vegetable crop is an indication of the crop's development and its progress toward becoming a marketable product. Physiological maturity is the stage of development when maximum growth and maturation of the crop has occurred. Commercial or horticultural maturity is the stage of development required by the market. Vegetables continue to carry on physiological metabolism after they are harvested.

The maturity of the vegetable crop at harvest will affect its marketability and storage life. The nutritional content, freshness, and flavor that vegetables possess depend on the stage of maturity and the time of day at which they are harvested. Overly mature vegetables will be stringy and coarse. When possible, growers typically harvest vegetables during the cool part of the morning, and process or store them as soon as possible after harvest. If processing is delayed, the vegetables are often cooled in ice water or crushed ice and stored in a refrigerator to preserve flavor and quality.

Harvest Criteria According to Edible Plant Part

For many vegetable crops there are good indicators of when a vegetable is ready to harvest; for others there are not. Often crops that belong to the same crop grouping based on the edible plant part have similar criteria for when to harvest.

Some harvest indicators for crops based on marketable plant part include the following.

Fruit Vegetables

Fruit vegetables can be broken down into immature fruit and mature fruit. The immature fruit vegetables include the legumes, some of the cucurbits such as cucumber and soft rind squashes, some of the Solanums such as eggplant and peppers, and other crops such as okra and sweet corn. The harvest index for immature fruit vegetables is primarily based on size and color. Maturity is not a real problem unless the harvest is delayed too long and they become overmature. Immature fruit vegetables generally have tender skins that are easily damaged during harvest and handling.

The mature fruit vegetables include some of the cucurbits such as muskmelon, honeydew, watermelon, pumpkins, and hard rind squashes, and some of the Solanums such as mature-green and vine-ripened tomatoes, and ripe peppers. The harvest index for mature fruit vegetables depends on several characteristics and is dependent on the vegetable. Most mature fruit vegetables that are consumed at the ripe stage (Figure 11.1) will continue to increase in eating quality if they are allowed to fully ripen on the plant. However at this point they usually have very little shelf or storage life left, and they are too soft to withstand the rigors of harvesting, handling, and transporting to market. Some fruits may be harvested based on time and/or distance to the market rather than on maximum quality. However, fruit harvested at an immature stage will generally have very poor quality.

FIGURE 11.1   Vine-ripened tomatoes.

Leaf and Stem Vegetables

With most of the leaf crops (lettuce, cabbage, etc.) the quality and shelf life are better if harvested slightly immature than slightly overmature. Virtually all leafy vegetables are harvested by hand. The determination of horticultural maturity varies with commodity, but generally size is the principal criterion. Stem vegetables include asparagus and kohlrabi.

Floral Vegetables

All floral vegetables are hand harvested. Head size and development determine maturity of floral vegetables. Floral vegetables include artichoke, broccoli, and cauliflower.

Roots, Tubers, and Bulb Vegetables

Roots, tubers, and bulb vegetables can be both mechanical and hand harvested. Maturity indices vary with commodity. Many of these products can be harvested and marketed at various stages of development (i.e., "new" potatoes vs. storage potatoes, "baby" carrots vs. full-sized carrots).

General Guidelines on Maturity and Time to Harvest for Specific Vegetables

The following guidelines can be used for harvesting the specific vegetable crops.

Asparagus   Asparagus is typically harvested the third year after planting. Up to eight spears per plant can be harvested the second year after planting. A full harvest season, such as during the third growing season, will last 4 to 6 weeks. Spears are harvested when they are at least 6 to 8 in. tall by snapping or cutting them at ground level. Stalks of the harvested spears should be fresh and firm with compact, closed tips. Harvesting is terminated when spears show a marked decrease in diameter size (generally smaller than the size of a pencil).

Beans, lima   Lima beans are harvested when the pods are filled with the enlarged seeds. The harvested pods must be green and not show any signs of becoming yellowish.

Beans, snap   Snap beans are harvested when pods are almost full size but before seeds begin to bulge. Beans are ready to pick if they snap easily when bent in half. Harvesting usually begins 2 to 3 weeks after first bloom. The pods for marketing should be free from scars and without strings.

Beets   Early beets are harvested when about 1¼ to 2 in. in diameter. Beets that are allowed to get too large become woody, especially in warm, dry weather.

Broccoli   Broccoli is harvested when the head is dark green with a compact cluster (about 6 in. in diameter) of tight flower buds, and before any yellow flowers appear. The head is cut 6 to 7 in. below flower heads. Smaller side shoots will develop later.

Brussels sprouts   The lower sprouts of Brussels sprouts are harvested when they are about 1 to 1½ in. in diameter and firm. Buds higher up in the plant are harvested as they become firm. Lower leaves along the stem are often removed to hasten maturity, but this may also have a negative effect on ensuing growth.

Cabbage   Cabbage heads are harvested when the heads feel hard and solid but before they split. The outer leaves should be uniform green or purple color (depending on type).

Carrots   Carrots are harvested when the roots are ¾ to 1 in. in diameter. The largest roots generally have the darkest tops. Fall carrots, if mulched in certain areas, can be left in the ground all winter and harvested as needed.

Cauliflower   Sunlight is excluded (blanched) when the curds of the cauliflower are 1 to 2 in. in diameter by loosely tying together the outer leaves above the curd (head) with a string or rubber band. The curds are harvested when they are 4 to 8 in. in diameter and compact, white, and smooth. The head should be ready 10 to 15 days after tying.

Celery   Celery is harvested when the plants are 10 to 12 in. tall.

Collards   The older, lower leaves of collards are harvested when they reach a length of 8 to 12 in. New leaves will grow as long as the central growing point remains. Whole plants may be harvested, if desired.

Corn, sweet   Sweet corn silks turn brown and dry out as the ears mature. As kernels fill out toward the top, the ends become more blunt instead of pointed. Sweet corn is picked in the milk stage, when a milklike juice exudes from kernels. Harvest ranges from 18 to 21 days after the silk appears.

Cucumbers   Cucumbers are harvested when the fruits are bright deep green, before any yellow color appears. The length of the fruit should be 2 to 3 in. for sweet pickles, 5 to 6 in. for dill pickles, and 6 to 8 in. for slicing. Cucumbers are generally picked four to five times per week to encourage continuous production. Mature cucumbers left on the vine will retard the production of ensuing fruit on the cucumber plant.

Eggplant   Eggplants are harvested when the fruits are 6 to 8 in. in diameter and their color is a glossy purplish black or white (depending on cultivar). As eggplant fruits get older they become dull colored, soft, and seedy. Because the stem is woody, the fruit is cut and not pulled from the plant. A short stem is left on each fruit.

Gourds   Edible cultivars of gourds are harvested when fruits are 8 to 10 in. long, young and tender. Ornamental cultivars are harvested when the fruits are fully mature and fully colored.

Greens (general)   The outer leaves of many greens are harvested when they are 6 to 10 in. long and before they begin to yellow. Wilted or flabby leaves are not marketable.

Kale   The outer, older leaves of kale are harvested when they reach a length of 8 to 10 in. and are medium green in color. In general, heavy dark green leaves are overmature and are likely to be tough and bitter.

Lettuce   Leaf lettuce cultivars are harvested when outer, older leaves are 4 to 6 in. long. Only the outer leaves are generally harvested, as the newer leaves will continue to mature. Heading types are harvested when the heads are moderately firm and before seed stalks form.

Muskmelons (cantaloupe)   Muskmelons are harvested when the stem slips easily from the fruit with a gentle tug. Another indicator of ripeness for certain cultivars is when the netting on skin becomes rounded and the flesh between the netting turns from a green to a tan color.

Mustard   Mustard leaves and leaf stems are harvested when they are 6 to 8 in. long. New leaves will provide a continuous harvest until they become strong in flavor and tough in texture from temperature extremes.

Okra   Okra pods are harvested when they are 3 to 5 in. long and tender. They are generally harvested at least every other day during the peak growing season. Overly mature pods become woody and are too tough to eat.

Onions   Bulb onions are harvested when the tops fall over and begin to turn yellow. Ideal bulb onion diameter is 2 to 4 in. Onions are dug and allowed to dry out in the open sun for a few days to toughen the skin. The dried soil on the bulbs is removed by a gentle brushing. The stems are cut, leaving 2 to 3 in. attached to the bulb.

Parsley   Older leaves of parsley are cut when they are 3 to 5 in. long.

Peas   Edible podded cultivars of peas are harvested when pods are fully developed (about 3 in.) but before seeds are more than one-half of their full size. Regular peas are harvested when the pods are well rounded and the seeds are fully developed but still fresh and bright green. Pods are past their prime marketable maturity when they lose their brightness and turn light or yellowish green.

Peppers   Sweet peppers are harvested when the fruits are firm, crisp, and full-sized (about 4 to 5 in. long). Green peppers will turn red or yellow (depending on the cultivar) if left on the plant. This usually requires an additional 2 to 3 weeks. Hot peppers are allowed to attain their bright red color and full flavors while attached to the plant before they are cut and dried.

Potatoes (Irish)   "New" potatoes are harvested at any size greater than 2 to 3 in. in diameter. For full season potatoes, the tubers are harvested when the plants begin to yellow and die down. The tubers are stored in a cool, high-humidity location with good ventilation. Exposure of tubers to sunlight is avoided or the tubers will turn green and become nonedible.

Pumpkins   Pumpkins and winter squash are harvested when they are full size but before heavy frost. The rind of the harvested fruit should be firm and glossy and the bottom of the fruit (or ground spot) is cream to orange. The rind is tough and resists puncture from a thumbnail. A 3- to 4-in. portion of stem is left attached to the fruit.

Radishes   Radishes are harvested when the roots reach 1 in. in diameter (Chinese radishes grow much larger). The shoulders of radish roots often appear through the soil surface when they are mature. If left in the ground too long, they will become tough and woody.

Rhubarb   The harvest of rhubarb is delayed until the second year. Only a few stalks per plants should be harvested the second year. Established plantings (greater than 2 years old) are harvested for approximately 8 weeks. The quality of the stalks decreases toward the end of the harvest period. Only the largest and best stalks are harvested by grasping each stalk near the base and pulling slightly in one direction.

Rutabagas   Rutabagas are harvested when the roots are about 3 in. in diameter. The roots may be left in the ground during winter and used as needed if properly mulched.

Spinach   Spinach is harvested by cutting all the leaves off at the base of the plant when they are 4 to 6 in. long. Alternately, the entire plant may be harvested.

Squash, summer   Summer squash is harvested when the fruit is soft. Long-fruited cultivars, such as zucchini, are harvested when $1\frac{1}{2}$ in. in diameter and 4 to 8 in. long. Scallop cultivars are harvested when the fruits are 3 to 4 in. long. The rind of the harvested fruit is easily penetrated with a fingernail. The skin color often changes to a dark glossy green or yellow, depending on cultivar, when they are ready for harvest. Summer squash is harvested every 2 to 3 days during the season to encourage production.

Squash, winter   Winter squash is harvested when the fruit is mature. Mature fruit are firm and glossy and not easily punctured by a thumbnail. The portion of the fruit that contacts the soil is cream to orange when mature. To help prevent storage rot, a 2- to 3-in. portion of the vine attached to the fruit is allowed to remain on the harvested fruit. Winter squash are harvested before a heavy frost.

Sweet potatoes   Sweet potatoes are harvested late in fall but before first frost. Sweet potatoes are cured for at least 14 days in a warm, well-ventilated location.

Tomatoes   Tomato fruits are harvested at the required ripeness stage for marketing or consuming, from mature green to fully red stage.

Turnips   Turnip roots are harvested when they are 2 to 3 in. in diameter but before heavy frosts occur in the fall. The tops can be used as greens when the leaves are 3 to 5 in. long.

Watermelons   Watermelon fruits are harvested when they are ripe. Ripe watermelons produce a dull thud rather than a sharp, metallic sound when thumped. Other ripeness indicators are a deep yellow rather than white color where the melon touches the ground (ground spot), brown tendrils on the stem near the fruit, and a rough, slightly ridged feel to the skin surface.

Methods of Harvesting

The goal of harvesting is to gather a commodity from the field in a manner that preserves quality of the product at a minimum cost. For vegetable crops, that is done by hand-harvesting, mechanical (machine) harvesting, or consumer (pick-your-own) harvesting.

Hand-Harvesting

Crops grown on a small scale or for home use are usually harvested by hand. A majority of the vegetable crops grown for fresh market and crops that are susceptible to mechanical damage are also usually hand harvested (Table 11.1). The advantage

TABLE 11.1   *Percentage of selected vegetable crops that are hand harvested*

Acreage hand harvested (%)	Vegetable crop
100–76	Artichoke, asparagus, broccoli*, cabbage, cantaloupe, cauliflower, celery, cucumber*, lettuce, green onions, collards, eggplant, kale, okra, peppers, rhubarb, squash, rutabaga, turnip
75–26	Sweet potatoes, mustard greens, parsley, turnip greens, greens, dry onions, pumpkin*, tomato*
0–25	Carrots, Irish potatoes*, lima beans*, snap beans*, sweet corn*, spinach*, beets*, peas*, garlic, Brussels sprouts*

*More than 50% of crop is processed.

Adapted from Thompson, J. F. 1991. Harvesting systems. In A. A. Kader (Ed.), *Postharvest Technology of Horticultural Crops,* University of California, Division of Agriculture and Natural Resources Publ. 3311. Used with permission.

FIGURE 11.2   An economical type of harvest aid made from two old bicycles.

of hand harvesting is that the person doing the harvest can accurately select for maturity, allowing for accurate grading and multiple harvest. Handling of the vegetable in this manner results in a minimum of physical damage.

The disadvantage of hand-harvesting is the management required to maintain a harvest crew. Labor supply can be a problem for farmers who cannot offer long-term employment. Regardless of these problems, quality is so important in vegetable production that hand-harvesting is still the predominant method of harvest.

Hand-harvesting may be done on commercial farms with the assistance of harvest aids. A harvest aid (Figure 11.2) is a piece of equipment that helps expedite the harvesting process in conjunction with hand-harvesting. An example of a harvest aid is a "mule train," which is a mobile grading and packaging unit that goes down the field with the harvesters. Also, belt conveyors (Figure 11.3) are used with some crops such as lettuce and melons to move the harvested product to a central loading or in-field-handling device such as a mule train.

Mechanical Harvesting

Machine harvesting is used extensively for harvesting vegetables grown for processing. While machine harvesting can reduce costs, effective use of mechanical harvesters requires operation by dependable well-trained people. An increase in the incidence of damage to the harvested vegetable is a disadvantage for products intended for fresh market. With improvements in harvesters and cultivars adapted to machine harvesting, it is expected that in the future more vegetables will be machine harvested.

The advantages of mechanical harvesting are that the process is generally rapid, and the problems associated with hiring and managing hand labor are reduced. Disadvantages to mechanical harvesting include the great expense of the equipment,

FIGURE 11.3   A belt conveyor used for field harvesting of head lettuce. The heads are moved by the conveyor to a central location where they are boxed.

the inability to do multiple harvests, and the damage that may be done to the harvested crops.

Mechanical harvest is not presently used for most fresh market vegetables because the machines are rarely capable of selective harvest. Commodities that can be harvested at one time and are less susceptible to mechanical damage may be mechanically harvested. The rapid processing after harvest will minimize the effects of mechanical damage.

Consumer Harvesting

Consumer or pick-your-own (PYO) harvesting is becoming more popular with vegetable growers. The high cost of labor makes this a popular alternative. The grower gets the crops harvested with no grading or packaging problems, and the customer gets vegetables at a reasonable price. Success of PYOs is dependent on location near a population center. Planting and location of fields need to be carefully planned. Crops must be planted in blocks according to expected dates of harvest. The check-in and check-out station needs to be located between the parking lot and the fields that are being picked. The grower usually furnishes picking equipment and the customer takes home packages. The grower usually has packages available or will provide some other container such as bags.

Quality, Initial Grading, and Packing

Quality of fresh vegetables is the characteristic that the consumer is looking for and will judge when contemplating a decision to purchase the product. The components of quality include appearance (including size, shape, color, gloss, and defects), texture (feel), flavor (taste and smell), nutritive value, and safety. Many pre- and post-harvest factors influence the quality of fresh vegetables. These include genetic factors (cultivars), preharvest environmental factors (climatic conditions and cultural methods used), maturity at harvest, harvesting method, and postharvest handling procedures. Quality measurements can also be destructive or nondestructive and objective (based on instrument readings) or subjective (based on human judgment).

Grading is a method of determining the quality of a vegetable. The number of grades and grade names included in the U.S. standards for a given commodity varies with the number of distinct quality gradations that the industry normally recognizes with the established usage of grade names. Currently, grades include three or more of the following: U.S. Fancy, U.S. No. 1, U.S. No. 2, U.S. No. 3, U.S. Extra No. 1, U.S. Extra Fancy, U.S. Combination, U.S. Commercial, and so on.

The initial grading usually begins in the field by the harvester if the crop is hand harvested. A decision to harvest a vegetable is based on maturity and quality of the vegetable. Most nonmarketable fruit may be left in the field instead of being brought to a packing shed. Once the crop is harvested it is usually graded and packed. While a few crops are quite well suited for field packing, most commodities go through some type of packing facility.

There are four basic operations to all central packing sheds:

1. Dumping—getting the product out of the field container and onto a packing line
2. Cleaning—usually a wash but in some cases may be done with dry brushes
3. Sorting—discarding the fruit that is not marketable and separating the remainder into at least two groups based on definable quality attributes
4. Packing—placing the graded commodity into a shipping and/or marketing container

Preparing for Market

Trimming

The appearance of vegetables can be improved by removing damaged, diseased, dead, or discolored parts. In some cases, the outer leaves offer protection and may even enhance the appearance of the product. Some field trimming is desirable for most vegetables, but sufficient wrapper leaves are usually left on such crops as lettuce, cabbage, and celery for protection. Damaged or discolored leaves can be further removed as desired in the market.

Washing

Since the market demands clean produce, most vegetables are washed after harvesting (Figure 11.4). Washing removes dirt, freshens the product, and can remove spray residues. After the produce is washed, it is often cooled to prevent development of rot organisms. Most of the root crops are not washed until they are marketed. Muskmelons, cucumbers, and sweet potatoes are usually cleaned by brushing or wiping dry rather than by washing.

FIGURE 11.4   A wash tank used for mature green tomatoes.

USDA Standards and Quality Certification

The Agricultural Marketing Service branch of the United States Department of Agriculture (USDA) provides standardization and grading services that play a major role in bringing consumers the highest quality vegetables at reasonable prices. Grading is the sorting of vegetables into uniform size and/or quality (Figure 11.5). Grading makes the produce more salable. Uniformity appeals to the eye of the buyer or consumer and suggests that the vegetables have been carefully handled.

The Fresh Products branch of the Fruit and Vegetable division provides inspections of produce shipments to determine that products meet specific grade standards. Users of the service pay a fee to cover the cost of inspection. The service, generally referred to as "grading," is voluntary except for commodities that are regulated for quality by a marketing order or marketing agreement, or that are subject to import or export requirements.

The bulk of grading is conducted at the shipping point as the produce is being packed for shipment to market. A shipment also may be graded at its destination to determine its current grade, either for the receiver's use in handling, or to settle questions that may arise between shipper and receiver.

To provide grading service nationwide, the USDA maintains cooperative agreements in all 50 states and Puerto Rico. Federally licensed graders perform their work throughout the country at points of origin; often working in the fields as a crop is being harvested. In addition, the federal grading service is provided in 75 of the country's largest terminal markets.

As the basis for its fresh products grading services, the USDA has developed over 150 official grade standards for fresh fruits, vegetables, tree nuts, peanuts, and related commodities. The grade standards describe the quality requirements for each grade of commodity, giving industry a common language for buying and selling. The USDA has developed a number of specific guidelines to ensure that the standards

FIGURE 11.5   Grading tomatoes for uniformity based on weight of the fruit.

are uniformly applied throughout the country. If a request for official grading is based on U.S. grade standards, the official certificate covering the shipment will show which USDA grade the product meets. These certificates are accepted as legal evidence in all federal courts.

Review Questions

1. How does horticultural maturity of vegetables differ from physiological maturity?
2. How do the general harvest criteria for immature fruit vegetables differ from mature fruit vegetables?
3. What is the primary goal of harvesting vegetables?
4. Which vegetable crops are best suited to mechanical harvesting?
5. What is the purpose of grading vegetables?
6. Why are there official grading standards for most vegetables?

Selected References

Evans, E., and L. Bass. 1997. *Harvesting vegetables.* North Carolina Coop. Ext. Serv. HIL 8108.

Hicks, J. R. 1986. *Vegetable postharvest systems: Vegetable quality and grading.* Dept. of Fruit and Vegetable Science, Cornell University.

Kader, A. A. 1992. Quality and safety factors: Definition and evaluation for horticultural crops. In A. A. Kader (Ed.), *Postharvest technology of horticultural crops.* University of California, Division of Agriculture and Natural Resources Publ. 3311.

Steinegger, D. H., and L. Finke. 1982. *When to harvest fruits and vegetables.* Univ. of Nebraska Coop. Ext. Serv. G76-271-A.

Thompson, J. F. 1992. Harvesting systems. In A. A. Kader (Ed.), *Postharvest technology of horticultural crops.* University of California, Division of Agriculture and Natural Resources Publ. 3311.

Wagner, A., and S. Coulter. 1998. *Harvesting and handling vegetables.* Texas Agr. Ext. Serv.

Selected Internet Sites

www.ams.usda.gov/standards/vegfm.htm   Fresh Market and Fresh Vegetable Standards, Agricultural Marketing Service, USDA.

www.ams.usda.gov/standards/vegpro.htm   Processing Vegetables Standards, Agricultural Marketing Service, USDA.

www.aphis.usda.gov/   Animal and Plant Health Inspection Service, USDA.

chapter
12

Handling the Crops after Harvest

Proper handling after harvest (postharvest) is essential to maintaining the quality of harvested vegetables. After the vegetables are harvested is the time that the death process of the harvested plant part begins. Many postharvest vegetables are separated from their root system and can no longer bring water and nutrients into the plant. From the moment of harvest all the efforts of the grower, harvester, and marketer are directed toward reducing the demand of the plant for the limited reserves of water and nutrients within the plant. This prolongs the viability of the harvested crop and enhances its ability to be sold profitably.

Postharvest losses can be substantial. For example, the combined losses of lettuce at the wholesale, retail, and consumer levels have been estimated at 21.3%. General handling problems occur during harvesting, packing, transporting, grading, storage, and retail.

Maintaining Quality

Consumers generally buy only vegetables that they perceive as being top quality products. Quality is determined primarily by appearance at the time of the purchase and by texture and flavor after the purchase and during consumption. The buyers for the chain stores or other vegetable marketing industries may also use these quality factors.

Appearance

Appearance is the most important factor in products that can be seen at the time of selection for purchase by consumers. Often it is the only characteristic that the consumer or buyer may use for making decisions in purchasing. Consumers and buyers do not want vegetables that have blemishes, wilted leaves, or are dirty. They want uniform size, shape, and color of the product.

Texture

Texture may not be an important factor at the time of purchase, but it is important for the subsequent return of customers. Proper texture depends on cultivar, stage of maturity, cultural conditions, and handling.

Flavor

In comparison to appearance and texture, flavor plays a minor role in marketing of vegetables. Consumers at purchase usually have no basis by which to judge the flavor of the vegetable. Consumers also differ in their taste and usually prefer flavors that they had been exposed to previously or have grown accustomed to.

Preventing Injury

The most important key to quality maintenance of fresh vegetables is careful handling. Since no postharvest treatments exist that can overcome inferior quality resulting from poor production practices or improper handling, it is important to prevent injury of vegetables. Microorganisms enter injured areas and can cause delay and spoilage. Even slight bruises may darken the product and present an unsightly appearance. Injuries also increase respiration and loss of weight.

Temperature Control

Temperature is perhaps the single most important factor in prolonging postharvest life of vegetables, and recommended storage temperatures and cooling methods will vary depending on the crop. The higher the temperature, the faster the metabolic rate and depletion of reserves in the harvested vegetable crop. Some vegetables are kept at 33° to 36°F immediately after harvest. Auxiliary cooling is required to reach and maintain recommended storage temperatures (Table 12.1). Auxiliary cooling is often not available at the field. Proper handling procedures in the field, however, can reduce high-temperature stress of many vegetables.

TABLE 12.1   *Recommended cooling and storage conditions of some common vegetables*

Crop	Suitable cooling method(s)	Opt. temp (°F)	Freezing temp (°F)	Optimum humidity (%)	Storage life
Asparagus	H,I	36	31	95–100	2–3 weeks
Beans, lima	R,F,H	37–41	31	95	5–7 days
Beans, snap	R,F,H	40–45	31	95	7–10 days
Beets, topped	R	32	30	98–100	4–6 months
Broccoli	I	32	31	95–100	2 weeks
Cabbage	R,F	32	30	98–100	1–6 months
Cantaloupes	H,I	32–40	30	95	2 weeks
Cucumbers	F,H	45–50	31	95	2 weeks
Eggplant	R,F	46–65	31	90–95	1 week
Green onions	H,I	32	30	95–100	3–4 weeks
Leafy greens	H,I	32	30	95–100	1–2 weeks
Okra	R,F	45–50	29	90–95	7–10 days
Peas, English	F,H	32	31	95–98	1–2 weeks
Peas, Southern	F,H	40–41	30	95	6–8 days
Peppers	R,F	45–50	31	90–95	2–3 weeks
Potatoes	R,F	38–40	31	90–95	5–8 months
Squash	R,F	45–50	31	95	1–2 weeks
Sweet corn	H,I	32	31	95–98	5–8 days
Sweet potatoes	R	55	31	90	6–12 months
Tomatoes	R,F	45–50	31	90–95	1 week
Turnips	R	32	30	95	4–5 months
Watermelons	R	50–60	31	90	2–3 weeks

R = room cooling, F = forced-air cooling, H = hydrocooling, I = icing

Certain commodities may sustain chill damage at 10° to 20°F above freezing temperatures.

Source: USDA Handbook No. 66.

Harvest Procedures to Reduce Temperature Stress

Many vegetable growers harvest their crops during cool morning or evening hours or when it is cloudy. This results in less field heat in the crops that are harvested at these times than if they were harvested during the middle of the day when exposed to intense sunlight and heat.

Harvested vegetables are also kept out of direct sunlight before and after loading onto trucks. Vegetables are generally shipped in refrigerated trucks. If the trucks are not refrigerated, they are usually vented to prevent buildup of heat and detrimental gases. Although venting is important to prevent increases in temperature, excessive airflow may cause desiccation and reduce quality of some leafy vegetables such as lettuce. A loose-fitting or air-permeable, heat-reflective tarpaulin is alternatively used when transporting vegetables in open trucks during sunny conditions.

Types of Cooling Procedures after Harvest

There are several types of cooling procedures that can be used on harvested vegetables. These differ in their rate and effectiveness of cooling, compatibility with specific commodities, and cost.

Room Cooling

Many vegetables are typically packed in bins and transported to a refrigerated room. Rapid cooling is dependent on the flow of cooled air around the bins. Proper stacking of the bins is important for effective cooling. Room cooling may be used with most commodities but may be too slow for some that require quick cooling. It is effective for storing precooled produce but in some cases it cannot remove field heat rapidly enough. For bins of produce, room cooling is often inefficient because the density of the packed vegetables allows for little surface contact with the cooled air. Dehydration can also be a problem because of the constant airflow over the surface of the vegetables. Properly designed, a room cooling system can be relatively energy efficient.

Forced-Air Cooling

Forced-air cooling is similar to room cooling except the bins are vented and stacked. Cooled air is forced by a pressure gradient through the bins, allowing better contact of cool air with warm product. Pressure gradients are created by fans that draw air around and through the bins. Forced-air cooling is faster than room cooling. Although the cooling rate depends on air temperature and the rate of airflow through the packages, forced-air cooling is usually 75 to 90% faster than room cooling. Once the product is cooled, the forced air can be stopped and the product can be kept under refrigeration with little air movement and water loss. Forced-air cooling can be very energy efficient and is an effective way to increase the heat removal rate of a cooling room.

Hydrocooling

Hydrocooling is achieved by flowing chilled water over the product, rapidly removing the heat. At typical flow rates and temperature differences, water removes heat about 15 times faster than air. Hydrocooling is only about 20 to 40% energy efficient, as compared to 70 to 80% for room and forced-air cooling. Hydrocooling can be used on most commodities that are not sensitive to wetting. It is generally not recommended for leafy vegetables because excessive free moisture can encourage diseases.

Vacuum Cooling

Vacuum cooling (Figure 12.1) is the best method for cooling vegetables that have a high ratio of surface area to volume, such as the leafy vegetables. Air is pumped out of a chamber containing the product. The vacuum causes water to evaporate from the surface of the produce. As water leaves the product due to reduced air pressure, it carries heat with it. Cooling is very rapid. Vacuum coolers can be energy efficient

FIGURE 12.1   A vacuum cooler.

but are expensive to purchase and operate. Some vacuum coolers are portable and can be used in more than one location. This is especially cost-effective when the harvest moves from one location to another with the change of season.

For each 43°F of cooling, about 1% of water is lost from the product. Some vacuum coolers add a fine spray of water to reduce water loss. Water loss from vacuum cooling of lettuce is only 0.5 to 1.5% and is usually not considered a problem for the product. After cooling, the vegetable is moved to a cold room or refrigerated truck.

Evaporative Cooling

Evaporative cooling is accomplished by misting or wetting the produce in the presence of a stream of dry air. It works best when the relative humidity of the air is below 65%. At best, however, it reduces the temperature of the product only 10° to 15°F and does not provide consistent and thorough cooling. It is an inexpensive way of cooling produce.

Top or Liquid Icing

In top icing, crushed ice is added over the top of selected produce by hand or machine. For liquid icing (Figure 12.2), a slurry of water and ice is injected into the produce packages through vents or handholds without removing the package tops. Icing is very effective on dense packages that cannot be cooled with forced air and can be used on a variety of vegetables. Because ice has a residual effect, this method works best with vegetables that have a high respiration rate, such as broccoli and sweet corn. One pound of ice will cool about 3 lb of produce from 85° down to 40°F.

FIGURE 12.2   Liquid icing of broccoli.

Modified or Controlled Atmosphere

Modified atmosphere (MA) or controlled atmosphere (CA) refers to alteration of gases in the storage chamber. CA is more precise than MA. Alteration of the gases is done with normal cooling procedures. An airtight room is essential to control room gases. The primary function of modifying gases is to reduce respiration and ethylene (C_2H_4) production. This is usually accomplished by increasing carbon dioxide (CO_2) levels and reducing oxygen (O_2) levels as compared to normal atmospheric concentrations. Ordinarily, air is 78% nitrogen (N_2), 21% O_2, and 0.03% CO_2. In MA/CA for lettuce, O_2 is reduced to approximately 1 to 3% and CO_2 is increased to 15 to 18%, with N_2 making up the difference.

Carbon monoxide (CO) may be injected in a MA/CA facility at $\approx 5\%$ to inhibit some pathogens. The ability of CO to mimic C_2H_4 requires precise control of CO concentration to avoid C_2H_4 responses. Often, modified atmospheres are very hazardous to human health. Extreme precautions must be taken by workers in MA/CA facilities.

Shipping Considerations

Many vegetables are very perishable during shipping. Prompt cooling of the harvested crop before shipping is preferable (Figure 12.3). Desirable temperature for shipping will depend on the vegetable. For example, broccoli and lettuce require near freezing temperatures, while squash or tomatoes would be damaged by such low temperatures.

A refrigerated truck during transit is not a practical method of cooling vegetables from the field. The insulation provided by the containers and the usually packed load of the truck permit only slow cooling of the produce.

FIGURE 12.3   Cooled broccoli shipped with ice on top of the boxes in the truck.

Bruising and mechanical injury during shipping, which may occur from rough handling or excessively tight packing, increase susceptibility to rots. Careful handling during shipping and proper cooling temperatures provide the most effective control.

Ethylene

Ethylene (C_2H_4) is a colorless, odorless gas given off by many plants. It is a natural aging and ripening hormone. Ethylene is physiologically active in plants in trace amounts (<0.1 ppm). The response of most vegetables to ethylene is accelerated senescence. Other postharvest physiological problems can occur when plants are exposed to high concentrations of ethylene. Russet spotting is a serious physiological disorder of lettuce exposed to high concentrations of C_2H_4. Therefore, reducing exposure of harvested lettuce to C_2H_4 is essential. Air exchanges of at least once per hour in cold storage rooms will generally maintain C_2H_4 at safe concentrations.

Air exchanges are not feasible in MA/CA facilities. However, C_2H_4 can be removed by circulating air from the MA/CA unit through scrubbing material such as potassium permanganate.

Many ripening fruits give off large quantities of C_2H_4 in storage, as do some heaters and internal combustion engines. Potential sources of C_2H_4 must be identified and ethylene-sensitive vegetables must be segregated from any of these entities.

Review Questions

1. What are some of the factors that determine the quality of vegetable crops?
2. What is the most important key to maintaining the quality of harvested vegetables?

3. How do high temperatures during storage reduce postharvest life of vegetables?
4. What cooling method is the best for vegetables that have a high surface area-to-volume ratio, such as many of the leafy vegetables?
5. What type of vegetables does top or liquid icing work best on for reducing field heat and cooling the product?
6. What is ethylene and how does it affect the postharvest life of many vegetables?

Selected References

Boyette, M. D., L. G. Wilson, and E. A. Estes. 1991. *Introduction to proper postharvest cooling and handling methods.* N.C. Coop. Ext. Serv. AG-414-1. 6 pp.

Wilson, L. G., and M. D. Boyette. 1997. *Postharvest handling and cooling of fresh fruits, vegetables, and flowers for small farms. Part II, Cooling.* North Carolina State Univ. Coop. Ext. Serv. Publ. Hort. Info. Leaflet 801.

chapter

13

Marketing the Crop and Delivering It to the Buyer

The goal of successful marketing of vegetables is to sell the crops that are grown and harvested with a reasonable chance of generating acceptable profits. Marketing decisions by vegetable growers are usually considered before the crops are planted. Often, the choice of the market helps determine what crops and cultivars are grown and when they are grown.

Marketing

Marketing rather than production determines what successful vegetable producers grow. Marketing begins with determining potential customer needs. Decisions are also made about product development, product design, and packaging. Prices to charge, transporting and storage policies, when and how the vegetables are to be advertised and sold, and disposal policies are also determined at the time that marketing decisions are made.

The basic variables of marketing are determining market, product, place, promotion, and price. Deciding where to market often depends on the amount of produce to be marketed. In general, the greater the amounts of produce grown, the greater the number of alternatives for marketing. Small-volume growers are usually restricted to local or regional markets (Figure 13.1), while larger producers are able to market on the national or international level. The best market for a grower is not only the market that offers the highest price, but also the one that best matches the grower's particular situation.

Product is what is produced for the customer. It is highly dependent on the market that is chosen. *Place* is getting the right product to the target market and is the movement of the product through the distribution chain. *Promotion* is telling the market about the product through personal selling, mass selling, publicity, and sales promotion. Sales promotion may involve coupons, store signs, catalogs, novelties, and circulars.

Price takes into account the kind of competition, customer reaction, production cost per unit, and current practices as to markups, discounts, and other terms of sale. One of the more difficult marketing decisions is knowing when to accept a price and when to wait for something better. Careful attention to market trends helps the vegetable grower decide when a price is acceptable.

Characteristics of Vegetable Market Sectors

The vegetable market is made up of different sectors, often with different objectives in the variety of crops that are dealt with, the volume and the source of volume that is handled, and relationship with potential imports. The fresh market vegetable sectors for larger growers (often truck farmers) include the grower, packer/shipper, wholesaler/buyer, supermarket, and consumer. Smaller or niche vegetable growers may have similar sectors as the larger growers.

FIGURE 13.1   An indoor farmer's market in Mexico City.

Following are some of the characteristics of the various market sectors for larger growers.

Grower   The grower typically produces from 1 to 10 different crops for the market. In order to supply sufficient volume and to hold niches in the market, the grower often staggers maturity dates over the growing season. They seek reasonable quality and sell at a maximum price. Growers are constantly in competition with imports in which reduced prices can result and lower their profits.

Packer/Shipper   The packer/shipper cannot relinquish their place in the market by insufficient volume. It is generally convenient to get good produce from whatever source, including imports. They will get enough variety of crops to meet the buyer demands. Packers/shippers will buy low and try to sell high to maximize profits.

Wholesaler/Buyer   The wholesaler/buyer must have volume and flow of product. They utilize imports and domestic production to always get the volume and quality they need. The buyer needs a good variety of crops and will generally pay premium prices for top quality.

Supermarket   Supermarkets want a consistent volume of high-quality produce. They must maintain year-round sources for all vegetables, therefore relying heavily on imports. Supermarkets need a wide range of quality vegetables at competitive prices and often will pay premium prices for top quality.

Consumer   Consumers prefer to have fresh vegetables year-round regardless of source. While domestically grown crops are preferred, purchases are often made without knowledge of where the crops were grown. Consumers will not hesitate to buy imports if quality and supply are constant. Consumers generally require top quality and are willing to pay a premium price. They also like to choose from a variety of crops.

Prepackaging

There has been an increase in the quantity of fresh vegetables marketed in consumer-sized packages (Figure 13.2). Packages protect the product against dust and dirt and reduce losses due to dehydration or bruising by customers. Prepackaging is a specialized business, usually requiring elaborate equipment and cold storage. Prepackaging increases cost of retailing merchandise but reduces product loss.

Value-added Vegetables

Value-added, or minimally processed, vegetables are prepared and handled to maintain their fresh nature while providing convenience to the user. Minimal processing of vegetables generally means cleaning, washing, trimming, coring, slicing, and shredding the intact vegetable (Figure 13.3). Value-added vegetables include peeled and sliced potatoes; washed and trimmed spinach; vegetable snacks such as carrot and celery sticks and cauliflower and broccoli florets; packaged mixed salads (Figure 13.4); cleaned and diced onions; and microwavable fresh vegetable trays.

FIGURE 13.2   Consumer-sized packages of shredded lettuce.

FIGURE 13.3   Minimal processing of head lettuce.

Delivering the Crop

Containers or packages are convenient units for marketing and distributing vegetables. An important purpose of containers is to protect the contents against damage during distribution. Several modes of transportation are available to get the crops from shipping points to destination markets. These include trucks and truck trailers, refrigerated railcars, air transport, and marine transport.

FIGURE 13.4   Prepackaged cole slaw mix and shredded cabbage. Courtesy of J. Rushing.

Containers

A large number of containers for vegetables are used throughout the United States (Table 13.1). Weights of containers are only approximate, since fresh vegetables by nature vary in weight when packed in containers of the same volume. The range in weight may be due to different cultivars being grown, the amount of moisture during the growing period, different growing areas, different seasons, storage conditions, and length of time in the same container. Most packers pack containers with a weight overage to comply with minimum package weight regulations. Containers must be durable and strong and designed to facilitate rapid cooling of contents.

Shipping

Trucks, railroads, airplanes, ships and combinations, such as trailer on flatcar and container on flatcar, are used to transport vegetables in the United States. Trucks (Figure 13.5) are the major method of moving vegetables in the United States while all methods are used to ship exports (with only marine and airline carriers used for transoceanic shipments).

Trucks and Truck Trailers   Trucks are used to haul one commodity or mixed loads. Load space is intermediate in size (2,000 to 3,500 cu ft) and truck weights are limited in gross weight by state regulations (generally 80,000 lb maximum). Refrigerated units are powered by diesel engines. Microprocessor temperature controllers are used in some modern units that incorporate thermostat control, digital thermometer, fault indication, and data recording. Common problems include damaged insulation, walls, doors, air delivery shoots, and floor grooves with debris.

TABLE 13.1 *Commonly used containers for selected vegetables*

Commodity	Containers	Net weight (approx. lb)
Artichokes	Ctns, wax-treated by count and loose pack	20–25
Asparagus	Pyramid crates, loose pack	32
	½ Pyramid crates, loose pack	15–17
	½ Pyramid crates, holding 6 bchs	13½–14
	Pyramid crates, holding 12 bchs	30
Beans, snap	Bu wbd crates & bu hampers	26–31
	Ctns	25–30
Beets, bunched	½ crates, 2-dz bchs	36–40
	⅘ crates, 12s, bchd	20
Broccoli	½ ctns, wax treated, 14 bchs	23
Brussels sprouts	Ctns, wax-treated, loose pack	25
	Ctns holding 12 10-oz or 12-oz film wrpd cups or trays	8¼–10
Cabbage, green	Flat crates (1¾bu)	50–60
	Ctns, place pack	53
	Sacked, mesh	50
Carrots	Ctns, wax-treated, holding 2-dz bchs	23–27
Carrots, minus tops	Film plastic bags, mesh sacks and ctns, holding 48 1-lb film bags	55
	Burlap sacks	74–80
Cauliflower	2-lyr ctns, holding 12–16 trimmed heads, film wrpd	23
	Crates	45–50
Celery	14½-in. wbd crates, flat pack	55–60
	Ctns, flat pack	60–65
Corn, sweet	Wbd crates	50
	Ctns, packed 5-dz ears	50
	Wbd crates, 4½–5 dz	42
Cucumbers	1⅑-bu ctns & wbd crates	55
	Bu ctns	47–55
	Ctns	50
Eggplant	1⅑-bu ctns & wbd crates	33
	Ctns, wbd crates	22
Garlic	Loose pack	30
	Ctns, holding 48 mesh bags (2 cloves per bag)	10
Greens	Bu baskets, crates & ctns, 24 pack	20–25
	1⅗-bu & 1⅗-bu wbd crates	30–35
	Crates & ctns, 12–24 bchs	25
Leeks	Ctns, holding 10 1-lb film bags	10–11
	Crates, 2-dz bchs	24–30
	⅘-bu crates & ctns, packed 12 bchs	20
Lettuce, Iceberg	Ctns, packed 24	43–52
Melons, cantaloupe	½ ctns, packed 9, 12, 18, 23	38–41
	⅔ctns & crates, packed 12, 14, 18, 24, 30	53–55
	Jumbo crates, packed 18–45	80–85
	Std crates	70
	½ wbd crates, packed 9–23	38–41
Okra	Bu hampers & crates	30
	⅝-bu crates	18
	Bu crates and ctns	18–20
	12-qt baskets	15–18

(Continued)

TABLE 13.1 *Commonly used containers for selected vegetables (continued)*

Commodity	Containers	Net weight (approx. lb)
Onions, dry	Sacks	50
Onions, green	Ctns, wax-treated, 4-dz bchs	10–16
	4/5-bu crates, holding 36 bchs	11
Parsley	Jumbo crate, 5 dz	20–25
	Ctns, wax-treated, 5-dz bchs	21
	Bu baskets & 1 1/9-bu crates, 5-dz bchs	21
Peas, black-eyed	Ctns, holding 12 11-oz cello pkgs	9 1/4
	Bu crates	24
Peas, English	Bu wbd crates	30
	Bu baskets	28–30
Peppers	1 1/9-bu ctns	28
	Bu ctns	25
	Ctns	30
Potatoes	100-lb sacks	100
	50-lb ctns & sacks	50
	20-lb film bags	20
	10-lb film bags	10
Pumpkins	Various types of crates	—
Radishes, minus tops	Ctns, packed 30 6-oz film bags or 14 1-lb film bags	15
	25-lb film bags	25–26
Rhubarb	Ctns, place pack	20
	Crates, place pack	30
	10 1-lb film bags in master ctns	10
Rutabagas	Sacks and ctns	50
	25-lb film bags	25
Spinach	Ctns & wbd crates, packed 2 dz	20–22
	Film bags, holding 12 10-oz bags	7 1/2–8
	Bu baskets, ctns & crates	25
Squash, summer	Ctns	24–30
	3/4 lugs and ctns	18–22
	1/2-bu crates & ctns	21
	Bu hampers & crates	41
Squash, winter	1/9-bu crates	42
	Ctns	20–25
	Wbd crates	45–50
	Bulk bins	800–900
Sweet potatoes	Ctns & crates	50
	Ctns	40–41
Tomatoes		
mature green	Ctns and wbd crates, volume fill pack	30–31
pinks and ripes	Ctns, loose pack	30
	2-lyr flats & ctns, place pack	18–25
	3-lyr lugs & ctns, place pack	24–33
Turnips, minus tops	Film bags	25
	Bu baskets	50
	1/2-bu baskets and ctns	25
Watermelons	Various containers	—
	Bulk	45,000
	Bulk	34,000
	Bulk bin ctns	1,000
	Bulk bin ctns	1,200

Abbreviations: bch(d)(s), bunch(ed)(es); bu, bushel; ctns, cartons; dz, dozen; lb, pound; lyr, layer; pkgs, packages; pt, pint; std, standard; wbd, wirebound; wrpd, wrapped.

From Magoon, C. E. 1979. *Container net weight.* United Fresh Fruit and Vegetable Association Leaflet. Used with permission.

FIGURE 13.5   Shipping harvested vegetables in a truck.

Refrigerated Railcars   Refrigerated railcars are used primarily for long-haul shipments. Transportation time ranges from 6 to 10 days for transcontinental shipments in the United States. Common vegetables that are rail shipped include potatoes, onions, carrots, and other less-perishable commodities. Rail shipments are generally straight (or single) and mixed commodity loads. Refrigerated railcars have large load compartments (4,000 cu ft) and can carry loads up to 100,000 lb.

Air Transport   Air transport is used mainly to ship highly perishable and valuable commodities to distant domestic and overseas markets. Produce is shipped in closed (mostly nonrefrigerated) container units or in net-covered pallet loads. Air travel time can be from 6 to 18 hours or longer. Waiting time at distant markets can be as long as 1 to 2 days at ambient (often warm) temperatures, and rapid deterioration of product is common.

Marine Transport   Containerized marine transport units are common and popular. Container vans are loaded in the holds or on decks of container vessels. The containers may be refrigerated by their own refrigerated units connected to the ship's electrical power supply or by cold air supplied from a large onboard refrigeration unit. Refrigeration of a common air supply around many different products could be a problem.

Common Shipping Terms

Upon arrival at a distant market, the shipped vegetables become integrated into a regional or local distribution/marketing phase of moving the crops to their final destination. Some of the more common terms used during this phase include the following:

Arrivals   The monthly rail, piggyback (hauling truck trailers cross-country on railroad flatcars), truck, and air shipments shipped from state of origin and unloaded in a delivered city. Arrivals are used to compute season average charges.

FOB shipping point prices   The average price (by shipping container) of a commodity that shippers receive. FOB is a term of purchase in which the buyer pays the transportation charges and assumes all risks of damage and delay in transit not caused by the shipper.

Retail price   The average price paid (per unit) for a commodity in retail stores.

Retail value   The dollar value (by shipping container) of a commodity sold at retail after allowing for waste and spoilage incurred during the marketing process. Retail value is derived from retail price.

Transportation charges   The rail or truck rate received (by shipping container) from shipping point to delivered city.

Wholesale price   The average price of a shipping container for a commodity sold in a city wholesale market.

Review Questions

1. When are marketing decisions by growers usually considered?
2. What often determines where to market harvested vegetable crops?
3. What are some examples of sales promotions?
4. What are some examples of value added vegetables?
5. What are some of the roles of containers in the marketing of vegetable crops?
6. What are some ways that vegetables are shipped?

Selected References

Cantwell, M. 1991. Postharvest handling systems: Minimally processed fruits and vegetables. In A. A. Kader, (Ed.), *Postharvest technology of horticultural crops.* Univ. of California, Division of Agriculture and Natural Resources Publ. 3311.

Kasmire, R. F., and M. J. Ahrens. 1991. Transportation of fresh market horticultural crops. In A. A. Kader (Ed.), *Postharvest technology of horticultural crops.* Univ. of California, Division of Agriculture and Natural Resources Publ. 3311.

Magoon, C. E. 1979. *Container net weight.* United Fresh Fruit and Vegetable Association Leaflet.

VanSickle, J. J. 1996. *Marketing strategies for vegetable growers.* Univ. of Florida Coop. Ext. Serv. Cir. FRE 144.

Selected Internet Sites

www.ams.usda.gov/   Agricultural Marketing Service, USDA.
www.ams.usda.gov/directmarketing/   Farmers Direct Market, USDA.
www.ams.usda.gov/farmersmarkets/   Farmers Markets Home Page, USDA.

chapter

14

Uses of Vegetables

As complex and diverse as the number and types of crops that are considered as vegetables, the actual uses of these crops are even more diverse. A typical definition for a vegetable is an edible, usually succulent part or portion of a plant, eaten either as a main course or as a supplementary food in cooked or raw form. Vegetables also have other food uses such as desserts, spices or flavorings, and garnishes; as sources of vitamins, nutrients, and other phytoceuticals; and as ornamentals.

Food Uses

Vegetables are more often used as food than for any other use. Food uses could be as a main dish, as a side dish or appetizer, as a dessert, or as a garnish.

Main Dish

Vegetables effectively serve as main dishes in many situations. Salads often comprise a total meal, and consumer interest in salads has resulted in the development of some very sophisticated salad bars at restaurants and grocery stores. Soups can also be used as a main course with some typical ingredients including potatoes, tomatoes, broccoli, onions, and peppers. Casseroles, soufflés, omelets, quiches, and lasagna are often main course items with typical vegetables including cabbage, squash, broccoli, turnips, eggplant, and spinach. Baked beans are a traditional favorite meal, and more and more Mexican dishes containing beans, tomatoes, and onions are becoming mainstays in our diets.

Side Dish and Appetizer

The more traditional use of vegetables is as a side dish (often cooked) to a main dish, traditionally some type of meat. Many vegetables such as squash, Brussels sprouts, broccoli, asparagus, and cauliflower are traditional side dishes that are steamed, boiled, baked, or fried. A potato dish is a common side dish in many meals, from the home-cooked meal to the meal prepared at a restaurant. Many soups are made with vegetables and can be used as side dishes or as a main dish.

Side dishes of vegetables that are not cooked are also common. Cole slaw made from cabbage and carrots is a traditional mainstay for many meals. Cabbage is also fermented and made into sauerkraut and used as a side dish or as a condiment to another dish.

Salads can serve as a side dish or as an appetizer. Salads contain many more vegetables than just iceberg lettuce, tomatoes, peppers, or onions. Vegetable ingredients in salads are becoming more diverse and may contain broccoli, cauliflower, green or red Spanish onions, cold peas, raw spinach, celery, special lettuce types, asparagus, radishes, shredded and uncooked turnips and rutabagas, and red peppers. Cold potato salads and potato soups are also popular side dishes or appetizers.

Many vegetables are eaten raw as appetizers. Carrots, broccoli, cauliflower, cucumbers, radishes, and celery are often sliced and sometimes dipped into a sauce and eaten as an appetizer. Cherry tomatoes are often served with a dip as an appetizer.

Other vegetable appetizers are cooked. These include artichokes (served both hot and cold), which can be eaten as an appetizer or as a side dish. Some vegetables such as cucumbers, sweet corn, and cabbage are also included as a component in various pickles.

Dessert

Some fruit vegetables are often used as desserts, either sliced cold or cooked into pies, tarts, and so forth. Uncooked dessert vegetables include many of the melons (Figure 14.1). The melons can also be sliced and used as a breakfast item. Watermelon rinds can also be pickled as a preserve. Rhubarb is a component of pies and preserves and is used in tarts, sauces, puddings, punch, jams, and jellies. Sweet potatoes, squash, and pumpkins are used to make biscuits, bread, muffins, pies, custard, cookies, and cakes.

Spice or Flavoring

Some vegetables are used in cooking not only to be consumed but also to add flavoring to another component of the cooked dish. Garlic has a distinct flavor that

FIGURE 14.1   Ripe watermelon decoratively cut for display and use.

adds to the taste of many meat and vegetable dishes. Peppers are potent flavorings and the paprika pepper is the source of the spice paprika. Leeks and onions are also used to add flavor to soups and stews. Celery is used extensively in Chinese cooking to add crispness to vegetable sautés and other dishes. Celery leaves can also be dried and used as an herb and celery seed is used in soups and pickles. Fresh and dried parsley also is used as a flavoring.

Garnish

Garnishes are used in cooking not only as edible parts of the meal but also to add color or for decoration on plates and salad bars. Kale is a typical garnish used in salad bars or as a decorative feature on dinner plates with meats. Kale is used because it is less likely to wilt than other greens. Parsley is also used as a garnish to a cooked meal, with uncooked sprigs included with the main cooked items. Other vegetables used as garnish include lettuce and radishes.

Other Food Uses

Many vegetables are components of sandwiches. Lettuce, tomatoes, pickles, onions, and peppers are often added to both cold and hot sandwiches or hamburgers. Bulb onions may be sliced and used on sandwiches or dipped in batter and fried as onion rings or blooming onions. Pumpkin seeds, potato chips, and shoestring potatoes can be used as desserts, snack items, or as side dishes for a sandwich. Garlic butter has many uses in home and restaurant cooking and many restaurants offer garlic bread. Some flowers of vegetables are also eaten as a side dish. For example, certain squash flowers can be dipped into a light mixture of flour and water, fried, and eaten.

Medicinal Uses

A rapidly mounting body of scientific research has indicated that certain foods, when consumed regularly, can promote health and even prevent some chronic diseases such as osteoporosis, cancer, diabetes, and heart disease. The notion that diet can play a major role in maintaining health and staving off disease has given rise to a new class of products called "functional foods"—sometimes referred to as "nutraceuticals" or "designer foods." Functional foods generally refer to foods that provide health benefits beyond their traditional nutritional value.

Consumption of particular vegetables and fruits has long been believed to be useful in the prevention and care of disease. Until recently, the practice of Western medicine largely involved the prescription of specific plants and foods, a practice that has its origins in ancient Egypt, Greece, and Rome. The modern medical practices of China and India continue to use different plants in their prescriptions.

As early as 1933, a study in Great Britain suggested an association between higher intake of certain vegetables and lower risk of cancer at all sites. Since then over 200 studies have been conducted in many different parts of the world to investigate the role of vegetables and fruit in altering the risk of cancer in different organs of the body.

Vitamins

Vitamin C   Vitamin C is an antiscorbic vitamin that is important in immune response, wound healing, and allergic reactions. Vitamin C availability has increased since the mid-1960s because of better quality, increased variety, and year-round availability of many fresh fruits and vegetables. Most vegetables contain some vitamin C, beta-carotene, and vitamin E. These are powerful substances that seem to offer protection against some cancers. Broccoli, peppers, cauliflower, Brussels sprouts, tomatoes, and potatoes are good sources of vitamin C.

Vitamin A, Carotenes   Vitamin A is essential for vision, growth, bone development, the integrity of the immune system, and reproduction. In the early 1970s, the vegetable group of dark green and deep yellow vegetables became the leading source of vitamin A in human nutrition. The increase in carotenes is thought to be due to the development of carotene-rich cultivars of deep yellow vegetables in the mid-1960s, and the increased availability of broccoli and green peppers. Carrots, pumpkin, tomatoes, and spinach are good sources of beta-carotene.

Vitamin E   Vitamin E functions as an antioxidant at the cellular level to prevent the peroxidation of polyunsaturated fatty acids. Increased levels of vitamin E in the diet are due to the increased use of vegetable oils for salads and cooking, and to a lesser extent, the use of margarine and shortening. Increased levels of vitamin E have been reported to reduce the incidence of heart attack. Vitamin E may also work in conjunction with vitamins C and A to delay the onset of cataracts. Vitamin E has also been shown to slow the progression of Alzheimer's disease in patients with early symptoms. Watercress, parsnip, spinach, and broccoli are good sources of vitamin E.

Vitamin B_6   Vitamin B_6 is a coenzyme that aids in the synthesis and breakdown of amino acids, fatty acid synthesis, and the conversion of tryptophan to niacin. Vegetables are an important source of vitamin B_6, providing 23% of the total daily recommended allowance of vitamin B_6 intake in 1994.

Fiber

Vegetables also contain fiber, which helps in the functioning of the digestive system. Consuming plenty of vegetables and fruit can assist in protection against bowel cancer, which is one of the most common cancers. Vegetable skins have a different type of fiber, which is particularly helpful in this protective role. Other vegetables high in fiber are broad beans, peas, spinach, watercress, green beans, sweet corn, cabbage, butter beans, broccoli, and Brussels sprouts.

Phytoceuticals or Phytonutrients

Phytoceuticals describe the protective, disease-preventing compounds that are found in some plants. Phytonutrients refer to the role of some plant compounds as "quasi nutrients." In the past, the phytonutrients found in vegetables were classified as vitamins: flavonoids were known as vitamin P, cabbage factors (glucosinolates and indoles) were called vitamin U, and ubiquinone was vitamin Q. Vitamin designation was dropped for these phytonutrients because specific deficiency symptoms could not be established.

Phytonutrients can be grouped into classes on the basis of similar protective functions as well as individual physical and chemical characteristics of the molecules. Following are some of the phytonutrient classes:

Terpenes   Terpenes are found in the green vegetables and comprise one of the largest phytonutrients. Included in the terpenes is beta-carotene. The terpenes function as antioxidants, protecting lipids, blood, and other fluids from attack from free radical oxygen species. Terpenoids are dispersed widely throughout the plant kingdom.

Carotenoids   Carotenoids are responsible for many of the bright yellow, orange, and red pigments found in vegetables. There are more than 600 naturally occurring carotenoids. While some carotenoids are precursors to vitamin A, fewer than 10% have actual vitamin A activity. Of these, beta-carotene is the most active. Some carotenes seem to offer some protection against lung, colorectal, breast, uterine, and prostate cancers. Carotenes also enhance immune response and protect skin cells against UV radiation.

Phytosterols   Sterols occur in most plant species, especially in the green and yellow vegetables (and especially in their seeds). Phytosterols compete with dietary cholesterol for uptake into the intestines. They block the uptake of cholesterol and facilitate its excretion from the body. Other investigations have revealed that phytosterols block development of tumors in colon, breast, and prostate glands.

Phenols   Phenols protect plants from oxidative damage and are suspected of performing the same function in humans. Phenols also appear to have the ability to block specific enzymes that cause inflammation.

Isoflavones   This phenol subclass comes from beans and other legumes. Isoflavones block enzymes that promote tumor growth.

Thiols   Phytonutrients of this sulfur-containing class are present in garlic and cruciferous vegetables. Researchers at Johns Hopkins University have been studying a phytochemical found in broccoli and cauliflower called sulforaphane. In a 1994 study, sulforaphane was shown to be a powerful cancer fighter, since laboratory animals fed sulforaphane had a 60 to 80% reduction in cancer development. In their most recent study, researchers have shown that sulforaphane is 50 times more concentrated in young broccoli sprouts as compared to mature plants.

Glucosinolates   Glucosinolates, found in the Brassicas, are powerful activators of
    liver detoxification enzymes. They also regulate white blood cells and
    cytokines (which coordinate the activities of all immune cells).
Allylic sulfides   Garlic and onions are the most potent members of this thiol
    subclass. Leeks, shallots, and chives also contain allylic sulfides. The allylic
    sulfides in these plants are released when the plants are cut or smashed. Once
    oxygen reaches the plants' cells, various biotransformation products are
    formed. Allylic sulfides appear to possess antimutagenic and anticarcinogenic
    properties as well as immune and cardiovascular protection. They also appear
    to offer antigrowth activity for tumors, fungi, parasites, cholesterol, and
    platelet/leukocyte adhesion factors.
Indoles   Most vegetables that contain indoles also contain high levels of vitamin C.
    Indoles bind chemical carcinogens and activate detoxification enzymes, mostly
    in the gastrointestinal tract.

Folk Medicinal Uses

Some vegetable crops have a history of being used for their medicinal uses. These
uses are often handed down from generation to generation, especially in the rural
areas or by individuals wishing to use alternatives to contemporary medications. Following are some suggested folk medicinal uses of selected vegetables according to
family. Many of these uses or cures have not been verified by modern medical research and should not be considered as recommendations for use.

Liliaceae   Asparagus was given for jaundice and is recommended for disorders of
    the breast.
Alliaceae   The antibiotic effects of the alliums (onions and garlic) are due to
    sulfur compounds. Garlic was used as a cure for smallpox. This was
    accomplished by applying cut pieces of the clove to the feet about 8 days
    following onset of the disease.
Polygonaceae   Rhubarb contains large amounts of oxalates. Oxalates tie up
    calcium, preventing absorption of this element. Rhubarb has been used for
    medicinal purposes since before Christ. Dioscorides, who was a physician to
    Anthony and Cleopatra, recommended rhubarb for diseases of the liver and
    weaknesses of the stomach. He used it to cure ringworm by applications of a
    concoction of roots in vinegar.
Chenopodiaceae   Members of the Chenopodiaceae (especially spinach) have been
    used through the years for various medicinal purposes. For example, the
    English crushed spinach and used it to clean wounds and to cure warts.
Brassicaceae   Plants of the Brassicas contain *S*-methylcysteine sulfoxide, a
    compound known as the "kale anemia factor." This compound has been
    shown to lower blood cholesterol levels. The Romans considered cabbage a
    very important drug and it was often the only medicine given for many
    diseases. The Greeks recommended the juice of cabbage as an antidote for
    eating poisonous mushrooms.
Fabaceae   Roman physicians used peas boiled in seawater to cure erysipelas (a
    bacterial infection of the subcutaneous tissue). An ancient superstition
    maintained that if one ate peas too freely, one could contract leprosy.
    Pythagoras, the Greek philosopher, forbade his disciples to eat beans because
    he supposed that beans came from the same putrid matter as that from which
    man was formed.

Apiaceae   Plants of this family contain alkaloids. Carrots contain carotoxin, although not an alkaloid, which is toxic to mice. Extracts of carrot seeds have also been implicated in preventing pregnancy by inhibiting the implantation of the fertilized egg in the uterus.

Convolvulaceae   The sweet potato contains a bitter substance called ipomeamarone. In New Zealand the natives use an infusion of the sweet potato to lower fever, and apply the crushed leaf tissues for diseases of the skin.

Solanaceae   Most of the plants in the nightshade family contain some alkaloids. The sprouts of potato tubers are extremely poisonous, as are the tops. The Maori of New Zealand apply the extract from boiled potato tubers to cure pimples and soothe burns. Tomato leaves, stems, and immature fruits contain an alkaloid, tomatine. Tomatine is a natural insect repellent, and is considered poisonous.

Cucurbitaceae   Many members of the cucurbits, such as zucchini squash, contain glycosides called curcurbitacins. These compounds are very toxic to animals and humans. Pumpkins and squashes contain cholinesterase inhibitors.

Asteraceae   The bitter taste of many members of the Asteraceae is probably due to an aromatic compound called lactucopicrin. Boiled dandelion leaves were frequently prescribed for fevers and as a diuretic.

Ornamental Uses

Pumpkins are famous for their use as jack-o'-lanterns at Halloween. Various types of squashes and gourds are also used for decorative purposes. Garlic wreaths or braids are often used as ornamental features in kitchens. Ornamental kale is an outdoor plant and is not considered a vegetable. Ornamental corn is also not generally considered a vegetable crop.

Review Questions

1. What are some of the food uses of vegetables?
2. What are some of the uses of vitamin C in human nutrition and health?
3. What is the role of fiber in the diet?
4. Describe phytoceuticals and phytonutrients.

Selected References

AVRDC. 1990. *Vegetable production training manual.* Asian Vegetable Research and Development Center, Shanhua, Taiwan. Reprinted 1992. 442 pp.

Potter, J. D. 1996. Cancer prevention: Food and phytochemicals. Second International Symposium on the Role of Soy in Preventing and Treating Chronic Disease, September 15–18, Brussels, Belgium. (Abstract)

USDA. 1998. *Nutrient content of the U.S. food supply, 1909–94.* USDA Center for Nutrition Policy and Promotion. Washington, DC. Released January 22, 1998.

Yamaguchi, M. 1983. *World vegetables: Principles, production, and nutritive values.* New York: Van Nostrand Reinhold. 415 pp.

chapter

15

Nutritional Value of Vegetables

Change in Diet

Many Americans have changed their diets and are eating healthier now than they did 25 years ago (Table 15.1). For example, red meat consumption dropped 13% between 1970 and 1995. During the same period, consumption of nonred meats such as poultry rose by 86% and fish and shellfish rose by 27%. The consumption of vegetables and fruits also increased by 22%. Egg use declined by nearly 25% while cheese consumption more than doubled to 27 pounds per person.

These changes in food consumption may be due in part to an expanding scientific base relating to diet and health and a greater number of people receiving the healthy food message. Recommended dietary guidelines have been designed that promote health and prevent disease, and improved nutrition labeling helps the consumer select foods that meet these recommended guidelines.

More and more Americans are also embracing the USDA's new recommendations of consuming five fruits and vegetables a day. This increase in vegetable and fruit consumption is also occurring as the number of choices of vegetables and fruits for the consumer continues to increase. Fresh-cut fruits and vegetables, prepackaged salads, locally grown items, and exotic produce have been introduced or expanded in the last decade.

TABLE 15.1   *U.S. per capita consumption of selected major foods*

Food	1970	1980	1995
	(pounds)		
Beef	79.6	72.1	64.0
Pork	48.0	52.1	49.1
Chicken	27.4	32.7	48.8
Turkey	6.4	8.1	14.1
Fish and shellfish	11.7	12.4	14.9
Eggs (number)	308.9	271.1	234.6
Cheese	11.4	17.5	27.3
All dairy products	563.8	543.2	585.8
Fruits and vegetables	564.4	594.4	685.9
Fruits	229.0	257.9	280.9
Vegetables	335.4	336.5	405.0
Flower and cereal products	135.6	144.7	192.4

Adapted from USDA. 1997. *USDA Factbook 1997.* US Dept of Agriculture, Office of Communications, Washington DC. Also located at www.USDA.gov.factbook/contents.htm

U.S. Per Capita Consumption of Vegetables

Per capita consumption of fruits and vegetables increased during the early 1980s presumably in response to higher consumer incomes, increased ethnic diversity, and increased interest in healthful diets. A typical supermarket's produce department today now carries over 400 produce items, up from 250 in the late 1980s and 150 in the mid-1970s. Also the number of ethnic, gourmet, and natural food stores continues to increase and the number of new or designer foods offered at these institutions is also increasing.

The declining household size and aging of the U.S. population may also be contributing to the increased consumption of vegetables and fruits. Per capita expenditures on fresh vegetables are 87% greater for one-person households than for households with more than six people. Also, people in the age group 45 to 64 consume 34% more fresh fruit than the national average. This age group is projected to be the single largest segment of the population by the year 2000.

The supermarket continues to play an important role in the marketing and education of consumers about fresh vegetables. For example, 98% of American consumers report that the quality of produce has a major influence on where they shop for food. Many supermarkets have taken advantage of this and now have the produce section as the area that the consumer first goes through upon entering the store. The produce section is also the segment of the supermarket that generates the highest profitability.

Consumers continue to have more access to fresh, local produce as well. The number of farmer's markets has grown substantially throughout the United States over the last several decades. According to the USDA, farmer's markets have increased from 1,755 in 1993 to 2,116 by the end of 1995.

Recommended Dietary Allowances and Guidelines

Healthful diets contain the amount of essential nutrients and calories needed to prevent nutritional deficiencies and excesses. Healthful diets also provide the right balance of carbohydrate, fat, and protein to reduce the risk of chronic diseases, and are part of a full and productive lifestyle.

Recommended Dietary Allowances (RDAs) are the amount of nutrients that will prevent deficiencies and excesses in most healthy people. Although some people with average nutrient requirements may eat adequately at levels below the RDA, diets that meet RDAs are almost certain to ensure intake of enough essential nutrients. RDA values assist in educating consumers about healthy diets.

Such diets are obtained from a variety of foods that are available, affordable, and enjoyable. The Dietary Guidelines are USDA recommendations that describe food choices that assist in meeting nutritional needs. Like the RDAs, the Dietary Guidelines apply to diets consumed over several days and not to single meals or foods.

Foods such as vegetables contain combinations of nutrients and other healthful substances (Table 15.2). No single food can supply all the nutrients in the amounts that are needed to ensure good health and prevent disease. Choosing the recommended number of daily servings from each of the five major food groups displayed in the Food Guide Pyramid is a good way of getting a good supply of nutrients and minerals.

Nutrition Facts Labels

The Nutrition Facts Labels were designed to indicate the nutritional content of certain foods in order to help the consumer select foods that meet the recommended Dietary Guidelines. Most processed foods now carry nutrition information. However, foods like coffee and tea (which contain no significant amounts of nutrients), ready-to-eat foods like deli and bakery items, and restaurant food are not required to carry nutrition labels. Labels are also voluntary for many raw foods, but many grocers supply this information for the most commonly consumed raw fruits, vegetables, fish, meat, and poultry.

Vitamin and Mineral Intake

As Americans become more health conscious they are also improving their intake of vitamins and minerals. All vitamins (A, C, E, B_6, thiamin, riboflavin, niacin, and folate) except vitamin B_{12}, and all minerals (calcium, phosphate, magnesium, iron, zinc, copper, and potassium) have increased in per capita consumption from 1970 to 1994 (Table 15.3). A 16% increase in vitamin C consumption reflects higher vegetable and fruit consumption as a result of improvements in cultivars and year-round availability.

Fiber

Fiber is only found in plant foods such as whole-grain breads and cereals, beans and peas, and other vegetables and fruits. Because there are different types of fiber in foods, it is often recommended that a variety of foods be eaten to supply fiber. Eating a variety of fiber-containing plant foods is important for proper bowel function and for reducing symptoms of chronic constipation, diverticulosis, and hemorrhoids, as well as lowering the risk of heart disease and some cancers. Some of the health benefits associated with a high-fiber diet come from other components present in these foods, not just from fiber itself. For this reason, it is recommended that fiber be obtained from foods rather than from supplements.

Development of Supernutritious Vegetables

A vegetable's nutrient content is becoming an increasingly important measure of its value, and plant breeders are focusing on making vegetables even more healthful. For example, new cultivars of vegetables that contain uniformly high levels of vitamin C, beta-carotene, and quercetin, a compound that has properties useful in preventing certain kinds of cancers, are being bred. It is suggested that these "designer vegetables" may ward off cancer and heart disease. Other breeding programs are trying to enhance the nutritional content of foods. Some examples of these types of vegetables and their enhanced nutritional benefits follow.

Carrots and Carotene

Researchers have gradually increased carotene levels over the years. As recently as 1980, a typical hybrid carrot had 80 to 100 parts per million (ppm) of carotene. New high-carotene cultivars have twice as much, and breeders are testing cultivars that have up to 500 ppm.

Although dark, leafy greens, sweet potatoes, and winter squash are all good sources of carotene, carrots are the greatest suppliers. If these carotene-rich carrots were widely available, the positive impact would be immense since Americans rely on carrots for most of their dietary vitamin A. At Texas A&M University, researchers have developed a purple-skinned, orange-cored carrot named 'BetaSweet', with carotene levels of 180 to 220 ppm.

Other high-carotene carrots include:

'Ingot'—A 6- to 8-in.-long, blunt-ended, Nantes-type hybrid carrot that has a sweet flavor. It has carotene levels between 120 and 170 ppm.

'Beta Champ'—An Imperator-type hybrid that has 10-in.-long tapered roots. It is very suitable for juicing. It has carotene levels between 150 to 270 ppm.

'Healthmaster'—A large, 3-in.-diameter, 10-in.-long, Danvers-type hybrid carrot that requires a relatively long season to mature and is sweet tasting. It has carotene levels between 60 and 95 ppm, which are about 35% higher than levels in older, open-pollinated Danvers varieties.

Tomatoes and Vitamins

Tomato cultivars are being developed with increased levels of vitamin A and vitamin C. Significant impetus for this work on tomatoes comes from the processing industry, which sees marketing potential in using nutrient-enriched tomatoes to increase the health value of soups and vegetable juices. It is anticipated that several processing cultivars with enhanced vitamin content will be available soon.

TABLE 15.2 *Nutritional content of vegetables. Edible portion per 100 grams**

**Raw Vegetables	Water %	Food Energy Cal.	Protein gr.	Fat gr.	Carbohy- drate Total gr.	Fiber gr.	Ash gr.
Bean, green	90.1	32	1.9	.2	7.1	1.0	.7
Bean, mature seed, dry	10.9	340	22.3	1.6	61.3	4.3	3.9
Bean, wax	91.4	27	1.7	.2	6.0	1.0	.7
Beet, greens	90.9	24	2.2	.3	4.6	1.3	2.0
Beet, roots	87.3	43	1.6	.1	9.9	.8	1.1
Broccoli	89.1	32	3.6	.3	5.9	1.5	1.1
Brussels sprouts	85.2	45	4.9	.4	8.3	1.6	1.2
Cabbage, common	92.4	24	1.3	.2	5.4	.8	.7
Cabbage, red	90.2	31	2.0	.2	6.9	1.0	.7
Cabbage, Savoy	92.0	24	2.4	.2	4.6	.8	.8
Cantaloupe, casaba	91.5	27	1.2	Trace	6.5	.5	.8
Cantaloupe, honeydew	90.6	33	.8	.3	7.7	.6	.6
Cantaloupe, other netted	91.2	30	.7	.1	7.5	.3	.5
Carrot	88.2	42	1.1	.2	9.7	1.0	.8
Cauliflower	91.0	27	2.7	.2	5.2	1.0	.9
Celery	94.1	17	.9	.1	3.9	.6	1.0
Chicory	95.1	15	1.0	.1	3.2	—	.6
Chinese cabbage	95.0	14	1.2	.1	3.0	.6	.7
Collard	85.3	45	4.8	.8	7.5	1.2	1.6
Cucumber	95.1	15	.9	.1	3.4	.6	.5
Eggplant	92.4	25	1.2	.2	5.6	.9	.6
Endive	93.1	20	1.7	.1	4.1	.9	1.0
Fennel	90.0	28	2.8	.4	5.1	.5	1.7
Kale	82.7	53	6.0	.8	9.0	—	1.5
Kohlrabi	90.3	29	2.0	.1	6.6	1.0	1.0
Leek	85.4	52	2.2	.3	11.2	1.3	.9
Lettuce, butter head	95.1	14	1.2	.2	2.5	.5	1.0
Lettuce, cos. Romaine	94.0	18	1.3	.3	3.5	.7	.9
Lettuce, crisp head	95.5	13	.9	.1	2.9	.5	.6
Lettuce, loose leaf, bunching	94.0	18	1.3	.3	3.5	.7	.9
Lima bean, immature	67.5	123	8.4	.5	22.1	1.8	1.5
Lima bean, mature, dry	10.3	345	20.4	1.6	64.0	4.3	3.7
Okra	88.9	36	2.4	.3	7.6	1.0	.8
Onion, mature, dry	89.1	38	1.5	.1	8.7	.6	.6
Onion, young green, bunching	89.4	36	1.5	.2	8.2	1.2	.7
Parsley	85.1	44	3.6	.6	8.5	1.5	2.2
Parsnip	79.1	76	1.7	.5	17.5	2.0	1.2
Pea, edible podded	83.3	53	3.4	.2	12.0	1.2	1.1
Pea, green, immature	78.0	84	6.3	.4	14.4	2.0	.9
Pea, mature, dry, whole	11.7	340	24.1	1.3	60.3	4.9	2.6
Pepper, hot, chili, immature, green	88.8	37	1.3	.2	9.1	1.8	.6
Pepper, hot, chili, mature, red	80.3	65	2.3	.4	15.8	2.3	1.2
Pepper, sweet, immature, green	93.4	22	1.2	.2	4.8	1.4	.4
Pepper, sweet, mature, red	90.7	31	1.4	.3	7.1	1.7	.5

Calcium mg.	Phos-phorus mg.	Iron mg.	Sodium mg.	Potassium mg.	Vitamin A ***I.U.	Thiamine mg.	Ribo-flavin mg.	Niacin mg.	Ascorbic Acid mg.
56	44	.8	7	132	600	.08	.11	.5	19
144	425	7.8	19	1,196	—	.65	.22	2.4	—
56	43	.8	7	243	250	.08	.11	.5	20
119	40	3.3	130	570	6,100	.10	.22	.4	30
16	33	.7	60	335	20	.03	.05	.4	10
103	78	1.1	15	382	2,500	.10	.23	.9	113
36	80	1.5	14	390	550	.10	.16	.9	102
49	29	.4	20	233	130	.05	.05	.3	47
42	35	.8	26	268	40	.09	.06	.4	61
67	54	.9	22	269	200	.05	.08	.3	55
14	16	.4	12	251	30	.04	.03	.6	13
14	16	.4	12	251	40	.04	.03	.6	23
14	16	.4	12	251	3,400	.04	.03	.6	33
37	36	.7	47	341	11,000	.06	.05	.6	.8
25	56	1.1	13	295	60	.11	.10	.7	78
39	28	.3	126	341	240	.03	.03	.3	9
18	21	.5	7	182	Trace	—	—	—	—
43	40	.6	23	253	150	.05	.04	.6	25
250	82	1.5	—	450	9,300	.16	.31	1.7	152
25	27	1.1	6	160	250	.03	.04	.2	11
12	26	.7	2	214	10	.05	.05	.6	5
81	54	1.7	14	294	3,300	.07	.14	.5	10
100	51	2.7	—	397	3,500	—	—	—	31
249	93	2.7	75	378	10,000	.16	.26	2.1	186
41	51	.5	8	372	20	.06	.04	3	66
52	50	1.1	5	347	40	.11	.06	.5	17
35	26	2.0	9	264	970	.06	.06	.3	8
68	25	1.4	9	264	1,900	.05	.08	.4	18
20	22	.5	9	175	330	.06	.06	.3	6
68	25	1.4	9	264	1,900	.05	.08	.4	18
52	142	2.8	2	650	290	.24	.12	1.4	29
72	385	7.8	4	1,529	Trace	.48	.17	1.9	—
92	51	.6	3	249	520	.17	.21	1.0	31
27	36	.5	10	157	40	.03	.04	.2	10
51	39	1.0	5	231	2,000	.05	.05	.4	32
203	63	6.2	45	727	8,500	.12	.26	1.2	172
50	77	.7	12	541	30	.08	.09	.2	16
62	90	.7	—	170	680	.28	.12	—	21
26	116	1.9	2	316	640	.35	.14	2.9	27
64	340	5.1	35	1,005	120	.74	.29	3.0	—
10	25	.7	—	—	770	.09	.06	1.7	235
16	49	1.4	25	564	21,600	.10	.20	2.9	369
9	22	.7	13	213	420	.08	.08	.5	128
13	30	.6	—	—	4,450	.08	.08	.5	204

(continued)

TABLE 15.2   *Nutritional content of vegetables. Edible portion per 100 grams* (continued)*

**Raw Vegetables	Water %	Food Energy Cal.	Protein gr.	Fat gr.	Carbohy- drate Total gr.	Fiber gr.	Ash gr.
Pumpkin	91.6	26	1.0	.1	65	1.1	.8
Radish	94.5	17	1.0	.1	3.6	.7	.8
Rutabaga	87.0	46	1.1	.1	11.0	1.1	.8
Salsify	77.6	13	2.9	.6	18.0	.9	1.8
Spinach	90.7	26	3.2	.3	4.3	.6	1.5
Squash, acorn	86.3	44	1.5	.1	11.2	1.4	.9
Squash, butternut	83.7	54	1.4	.1	14.0	1.4	.8
Squash, cocozelle, zucchini, green	94.6	17	1.2	.1	3.6	.6	.5
Squash, crookneck, straightneck, yellow	93.7	20	1.2	.2	4.3	.6	.6
Squash, Hubbard	88.1	39	1.4	.3	9.4	1.4	.8
Squash, scallop, pale green, white	93.3	21	.9	.1	5.1	.6	.6
Squash, summer	94.0	19	1.1	.1	4.2	.6	.6
Squash, winter	85.1	50	1.4	.3	12.4	1.4	.9
Sweet corn, white, yellow	72.7	96	3.5	1.0	22.1	.7	.7
Swiss chard	91.1	25	2.4	.3	4.6	.8	1.6
Tomato, green	93.0	24	1.2	.2	5.1	.5	.5
Tomato, ripe	93.5	22	1.1	.2	4.7	.5	.5
Turnip, greens	90.3	28	3.0	.3	5.0	8	1.4
Turnip, roots	91.5	30	1.0	.2	6.6	.9	.7
Watermelon	92.6	26	.5	.2	6.4	.3	.3

*"Edible Portion" includes only the part or parts that humans would eat.

**The data shown in this table are values currently considered most representative of each species or horticultural type, not of any particular variety. Generally, the figures are weighted with such pertinent factors as stages of maturity, growing season, geographical location and variation in nutrient levels within samples and varieties having been taken into consideration.

***One International Unit (I.U.) of Vitamin A is equivalent to 0.3 micrograms of Vitamin A alcohol.

Source: USDA Handbook No. 8.

Some other nonprocessing types of tomatoes are available. They include:

'Caro-Rich'—A determinate tomato cultivar that produces 10- to 12-oz, globe-shaped, orange fruits that contain 10 times the vitamin A of other tomatoes. Other orange tomatoes are available, but the skin color in many of these cultivars doesn't necessarily indicate higher carotene since their color may be gained from a gene other than the high-carotene beta gene.

'Double Rich'—A midseason tomato cultivar that has 3- to 4-in. bright red fruits and contains twice the vitamin C of other tomatoes. The indeterminate plants grow 2 to 4 ft tall and require trellising.

Squash and Carotene

Winter squash, especially the Cucurbita maxima types (such as the buttercups, some pumpkins, and hubbards), are often good sources of vitamin A.

Calcium mg.	Phos-phorus mg.	Iron mg.	Sodium mg.	Potassium mg.	Vitamin A ***I.U.	Thiamine mg.	Ribo-flavin mg.	Niacin mg.	Ascorbic Acid mg.
21	44	.8	1	340	1,600	.05	.11	.6	9
30	31	1.0	18	322	10	.03	.03	.3	26
66	39	.4	5	239	580	.07	.07	1.1	43
47	66	1.5	—	380	10	.04	.04	.3	11
93	51	3.1	71	470	8,100	.10	.20	.6	51
31	23	.9	1	384	1,200	.05	.11	.6	14
32	58	.8	1	487	5,700	.05	.11	.6	9
28	29	.4	1	202	320	.05	.09	1.0	19
28	29	.4	1	202	460	.05	.09	1.0	25
19	31	.6	1	217	4,300	.05	.11	.6	11
28	29	.4	1	202	190	.05	.09	1.0	18
28	29	.4	1	202	410	.05	.09	1.0	22
22	38	.6	1	369	3,700	.05	.11	.6	13
3	111	.7	Trace	280	400	.15	.12	1.7	12
88	39	3.2	147	550	6,500	.06	.17	.5	32
13	27	.5	3	244	270	.06	.04	.5	20
13	27	.5	3	244	900	.06	.04	.7	23
246	58	1.8	—	—	7,600	.21	.39	.8	139
39	30	.5	49	268	Trace	.04	.07	.6	36
7	10	.5	1	100	590	.03	.03	.2	7

'Jade-A' - 'Jade-A', a 5- to 10-lb, dark green, heart-shaped squash, has a high total carotene content and is also very high in lutein, an antioxidant and antitumor carotene. This hybrid buttercup-type squash produces four or five 8-lb squash per plant.

Carotene-rich Cauliflower, Cabbage, and Cucumbers

There are carotene-rich cauliflower, cabbage, and cucumber cultivars. They include the following:

'Orange Bouquet' cauliflower—This hybrid produces pastel orange heads with a yellow interior. Color is best when heads are not covered.

'Orange Queen' Chinese cabbage—This upright-growing hybrid produces Napa-type heads with higher-than-normal carotene levels. When the head is cut and exposed to light, the internal color changes from lemon yellow to pale orange.

TABLE 15.3   *Nutrients per capita per day for selected years*

Nutrient	1970	1994
Vitamin A (μg)	1,500	1,520
Carotene (μg)	510	660
Vitamin E (mg alpha-TE)	13.7	16.9
Vitamin C (mg)	107	124
Thiamin (mg)	2.0	2.7
Riboflavin (mg)	2.3	2.6
Niacin (mg)	22	29
Vitamin B_6 (mg)	2.0	2.3
Calcium (mg)	890	960
Phosphorus (mg)	1,460	1,680
Magnesium (mg)	320	380
Iron (mg)	15.4	21.2
Zinc (mg)	12.2	13.2
Copper (mg)	1.6	1.9
Potassium (mg)	3,510	3,780

Adapted from USDA. 1998. *Nutrient content of the U.S. food supply, 1909–94.* USDA Center for Nutrition Policy and Promotion. Released January 22, 1998.

Peppers, Potatoes, and Vitamin C

Peppers, cabbage, tomatoes, and dark leafy greens are all excellent sources of vitamin C. Ripe peppers have three times the vitamin C of most citrus. Among these peppers, yellow wax or banana peppers (hot or sweet) have the most vitamin C. Some examples include:

'Sweet Banana'—A widely adapted cultivar that is the most popular of the banana peppers. It produces 6-in. tapered yellow peppers that turn red at maturity.
'Hungarian Hot Wax'—A cultivar that produces medium-hot, 8-in. peppers that are used at the yellow fruit color stage, or when the fruit is red and fully mature.

Though potatoes are only moderate sources of vitamin C, they can be a major source for those whose diet consists of burgers, fries, and sodas.

'Ranger Russet'—A new processing potato cultivar that was developed by the USDA. Intended for commercial growers, it has about twice the vitamin C of an average potato.
'Butte'—One of the parents of 'Ranger Russet' that has almost as much vitamin C. This late maturing, high-yielding russet baking potato has medium to large tubers and dry, white flesh. Compared with ordinary russets, it is 50% higher in vitamin C and 20% higher in proteins.

Vitamin content of most vegetables is influenced by variations in growing conditions, time to maturity, postharvest handling, and food preparation. Many of the vitamin levels often reported for vegetables are usually attained under optimum growing conditions and may be higher than actual levels in commercially available vegetables.

Review Questions

1. How has the average American diet changed in the last 25 years?
2. What are some of the characteristics of a healthy diet?
3. What are Nutrition Facts Labels and how should they be used?
4. What are "designer vegetables" and why are they important?
5. Which vegetables are good sources of carotene?

Selected References

Cook, R. L. 1992. The dynamic U.S. fresh produce industry: An overview. In A. A. Kader (Ed.), *Postharvest technology of horticultural crops.* Univ. of California, Division of Agriculture and Natural Resources Publ. 3311.

Herbs for Health Staff. 1997. Colorful new carrot may be packed with health benefits. *Herbs for Health* (November/December). Can be found on the web at http://vic.tamu.edu/Press%20Release/herbs-health.htm

USDA. 1995. *Nutrition and your health: Dietary guidelines for americans.* 4th ed. U.S. Department of Agriculture, U.S. Department of Health and Human Services. Home and Garden Bull. No. 232.

USDA. 1996. *USDA's Food Guide Pyramid booklet, 1992,* rev. ed. Home and Garden Bull. 252.

USDA. 1997. *USDA Factbook 1997.* U.S. Dept. of Agriculture, Office of Communications, Washington DC. Also located at www.USDA.gov/factbook/contents.htm

USDA. 1998. *Nutrient content of the U.S. food supply, 1909–94.* USDA Center for Nutrition Policy and Promotion. Released January 22, 1998. Washington, D.C.

Wechsler, D. 1996. *Super-nutritious vegetables.* National Gardening Association. February. Also located on the web at http://208.156.226.50/articledetails.taf?id=135&kwd=nutritious%20 vegetables&Articlesstart=21

Selected Internet Sites

www.dole5aday.com/   Dole 5 a Day Program, Dole Food Co.

www.fda.gov/   Food and Drug Administration.

www.nal.usda.gov/fnic/cgi-bin/nut_search.pl   Search the USDA Nutrient Database for Standard Reference, USDA.

www.usda/fcs/fcs.htm   Food and Nutrition Service, USDA.

www.usda.gov/cnpp   Center for Nutrition Policy and Promotion, USDA.

www.vic.tamu.edu/   Vegetable Improvement Center, Texas A&M University.

www.vm.cfsan.fda.gov/label.html   Food Labels, Food and Drug Administration.

SPECIFIC VEGETABLE CROP
CHARACTERISTICS AND GROWING
PRACTICES

chapter
16

Cole Crops

I. Broccoli
II. Cauliflower
III. Cabbage
IV. Brussels Sprouts

Cole crops, also called crucifers or Brassicas, belong to the Brassicaceae or cabbage and mustard family. Cole crops are indigenous to temperate and cold climates. Most cole crops are biennial herbaceous plants that are typically grown as annuals. A characteristic of cole crops is the odor and taste that is attributed to glucosinates within the plant tissue. Vegetables in this family include cabbage, broccoli, cauliflower, and Brussels sprouts. Other cole crops, such as collards, kale, radish, turnip, and rutabaga, are discussed in either Greens (Chapter 17) or Root Crops (Chapter 20).

Broccoli

Scientific Name

Brassica oleracea L. var. *italica* Plenck.

Common Names in Different Languages

broccoli (English); qing hua cai (Chinese); chou brocoli, brocoli (French); Brokkoli (German); cavolo broccoli (Italian); buotkorii, Italia-kanran (Japanese); brokoli (Russian); brocoli (Spanish)

Classification, History, and Importance

Broccoli (*Brassica oleracea* L. var. *italica* Plenck.) is a member of the Brassicaceae or cole crop family. Other members of the family include cauliflower, cabbage, and kale. All members of the Brassicaceae are considered cool-season crops. Most types of broccoli are biennials. Italian green, also called calabrese, is the predominant broccoli type grown in the United States for commercial production and in home gardens.

189

Members of the Brassicaceae are thought to be of European or Siberian ancestry. Wild forms are found along the Mediterranean Sea. A sprouting form of broccoli was grown by the Romans in the 1st century. Broccoli was relatively unknown in England until about 1720 and is first mentioned in the United States in 1806. In 1923, D'Arrigo Brothers Company began growing broccoli in California. This broccoli was shipped to the East Coast of the United States where large populations of Italian immigrants had settled. Advertising for this vegetable was done on Italian-speaking radio stations and in newspapers. Many consider this the beginning of the establishment of the broccoli industry in the United States.

Broccoli for fresh market was a $471 million industry in 1997 (Table 16.1). This was up from $384 million in 1996. The broccoli for processing market is considerably smaller than for fresh market with a production value of $24.3 million in 1997. Per capita consumption of broccoli in the United States continues to increase from 2.2 pounds per person in 1976 and 4.7 pounds per person in 1986 to 6.7 pounds per person in 1996 (Table 16.2). Of the 134,000 acres of broccoli grown in the United States, 123,000 are in California (Table 16.3). Arizona and Texas grow approximately 10,000 and 1,000 acres of broccoli, respectively.

TABLE 16.1   *The production value of cole crops for fresh market and processing in the United States*

	Value of production		
Crop	1995	1996	1997
		(1,000 dollars)	
Fresh market			
Broccoli	405,286	384,666	471,144
Brussels sprouts[1]	14,390	20,120	26,800
Carrots	394,356	355,829	441,193
Cauliflower	197,790	200,809	184,304
Processing[2]			
Broccoli	38,018	24,501	24,371
Cabbage (for kraut)	7,549	6,029	8,299
Cauliflower	18,758	13,174	13,652

Source: USDA, NASS.

[1]Includes processed and fresh market.

[2]Value at processing plant door.

TABLE 16.2   *U.S. per capita consumption of commercially produced broccoli for fresh market and processing*

Use	1976 (lb, farm weight)	1986	1996
Fresh	1.1	3.0	4.1
Frozen	1.1	1.7	2.6
All	2.2	4.7	6.7

Source: USDA, Economic Research Service.

TABLE 16.3   *The top broccoli producing states in the United States. Broccoli grown for either fresh market or processing*

Rank	State	Planted acres (000)	Harvested acres (000)	Yield (cwt/acre)	Production (000)
1	California	123.00	123.00	130.00	15,990
2	Arizona	10.00	9.90	125.00	1,238
3	Texas	1.30	1.30	67.00	87
	United States	134.30	134.20	129.00	17,315

Source: USDA, NASS, 1997 data.

Plant Characteristics

Height: 18 in.
Spread: 15 to 24 in.
Root Depth: 18 to 36 in.

Climatic Requirements

Broccoli is a cool-season crop that can be grown in many areas of the United States. It is generally considered hardy in cold temperatures. Preferred temperature ranges for successful broccoli production are between 60° and 68°F. If planted in the spring, young broccoli plants can withstand light frosts down to 25°F. During later plant development the plant often becomes sensitive to chilling (40°F) and may initiate flowering, rendering the plant unmarketable.

The marketable plant part is a head composed of immature flowers or florets (Figure 16.1). Rapid, uninterrupted growth is required for high-quality and good yields of broccoli. Temperatures above 80°F generally are not conducive to satisfactory broccoli growth and production and can cause broccoli heads to become loose and puffy, making them practically unmarketable. Broccoli buds will turn yellow and flower prematurely (bolting) in hot weather.

Broccoli can be grown on a wide range of well-drained soils. Since broccoli has a shallow root system, a constant supply of moisture and nutrients is required for good succulent growth. A well-drained, medium-heavy soil with a high organic content is optimal for successful broccoli growing.

Field Preparation

Broccoli fields can be established by direct seeding or by transplanting. If directly seeded to the field, broccoli requires a good seedbed to obtain a uniform depth of seeding, and loose soil to cover the seed. A plow or disk is often used to begin preparing the field during the winter for planting in the early spring. A bedshaper is used prior to planting to smooth the bed for either direct seeding or transplanting of broccoli.

Optimum soil pH for broccoli is 5.5 to 6.5 and fields are limed if the pH is too low. Adjusting the pH to the preferred range will minimize tipburn as a result of calcium deficiency, whiptail as a result of molybdenum deficiency, and hollow stems as a result of boron deficiency.

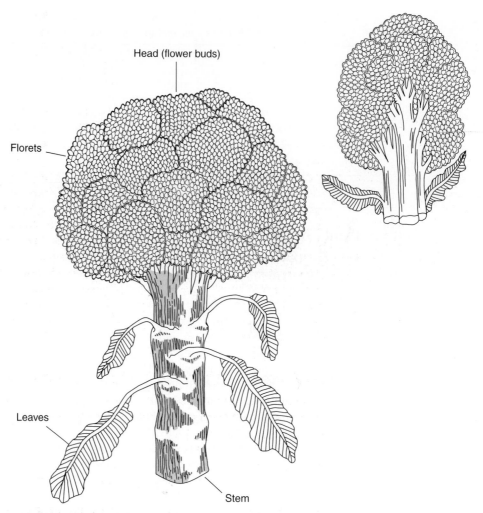

FIGURE 16.1   Diagram of a broccoli head.
Source: Iowa State University.

Broccoli can require 150 to 200 lb per acre of total N (nitrogen). Up to a third of this is often applied to the field prior to planting. The remaining N is applied as a side-dressing at about 2 to 3 weeks after transplanting and when heading begins. Adding organic matter to improve the soil is beneficial for good broccoli growth.

Selected Cultivars

Calabrese is the chief commercial broccoli cultivar in the United States. In the cooler mountain climates, such as in western North Carolina, Green Comet is often grown with good success. Other successful cultivars include Marathon, Arcadia, Green Duke, Emperor, Cruiser, Shogun, and Premium Crop.

Planting

Direct Seeding   Broccoli seeds are generally planted ¾ in. deep, 2 to 6 in. apart in rows 30 in. apart (Figure 16.2). Alternately, the planting of multiple rows (2 to 3) per raised bed is also used (Figure 16.3). Irrigation is often necessary for good seed germination and seeds germinate best at 75°F. Emerging seedlings are especially susceptible to insect injury, such as flea beetles, and must be protected.

FIGURE 16.2   An established planting of broccoli planted in single rows.

FIGURE 16.3   An established planting of broccoli with 3 rows per bed. Rye is planted in strips along the edges of the planting to serve as a windbreak.

The first cultivation of a field of broccoli is done with a disk on either side of the row as soon as the seedlings in the rows are visible from a distance. Broccoli is thinned to about a plant every 6 in. when the plants begin to crowd. A final stand may have an in-row spacing of 12 to 24 in. apart depending on the cultivar.

The final spacing of broccoli is important and can affect the length of the stalk and size of the broccoli head. A closer plant spacing generally induces the stalk to become longer and the diameter of the head smaller than a wider plant spacing.

Transplanting   Broccoli transplants can be grown in hotbeds or greenhouses. Transplants grown in hotbeds are seeded ¼ to ½ in. deep in rows 4 to 6 in. apart with two to four seeds per inch. Emerging seedlings in the hotbeds are thinned at the two-leaf stage to allow 1 in. between plants. Plants are usually watered twice a day and fertilized. It may take 6 to 8 weeks to produce broccoli transplants in hotbeds.

Broccoli transplants in the greenhouse are produced in containers (i.e., cell packs, peat pots, styrofoam trays). The seeds are sown directly into the containers containing a soil mix and thinned to one plant per cell or pot. Growing transplants of broccoli in the greenhouse may take 4 to 6 weeks before they are ready for field planting. Final field plant spacings using transplants are typically the same as with direct seeding.

Irrigation

Since broccoli receives most of its moisture from the upper foot of soil, irrigation is often necessary to supplement natural rainfall for optimum growth and production. Irrigation can be provided by furrow, overhead, or drip irrigation. As a general rule, at least 1 in. of water per week is required for optimal growth. Lighter, sandier soils may require up to 2 in. of irrigation especially during hot, dry periods.

Irrigation is often used after planting to help establish the broccoli plants in the field. Moisture stress during plant development can reduce growth and quality of the broccoli head and stalk. Tough, fibrous stalks and tipburn of broccoli can result if plants are exposed to inadequate moisture. Adequate moisture is especially necessary during head initiation to produce large quality heads required for successful marketing.

Culture and Care

Producing a uniform developing field of broccoli is a challenge to growers. It is unusual for complete uniformity of growth and maturity of broccoli in a field, mainly because no field is perfectly balanced with uniformly placed fertilizer and/or perfect water balance to meet the needs of the plant. Using transplants and uniformly sized seed can contribute to uniform plant development.

Common Growing Problems and Solutions

Insects   Several species of caterpillars attack broccoli. These include the imported cabbageworm (Figure 16.4), cabbage looper, and the diamondback moth. The cabbage aphid is also a very destructive pest on broccoli and the flea beetle can severally damage young seedlings. An insect scouting and control program is usually started early in the growth of the plants in the field and continued on a regular basis. If an insecticide is used, thorough coverage under the leaves and inside the heads is important for good control.

FIGURE 16.4   Imported cabbageworm.
Source: Clemson Univ. Coop. Ext. Serv.

Diseases   Following are some of the diseases of broccoli, their symptoms, and non-chemical methods of control.

Black Rot Yellow angular spots that progress inward from the leaf margin are
   indicators of black rot (*Xanthomonas campestris* pv. *campestris*). Leaf veins
   become dark brown to black and heads may be deformed. Nonchemical
   control measures for black rot include crop rotation, clean seeds, clean
   transplants, and resistant varieties.
Alternaria Leaf Spot Yellow, concentric spots on foliage are symptoms of alternaria
   leaf spot (*Alternaria brassicae*). Infected broccoli seedlings may be stunted or
   killed. Nonchemical controls include crop rotation and clean seed.
Downy Mildew Yellow spots on the upper surface with bluish white fungal growth
   on the lower surface of leaves may be caused by downy mildew (*Peronospora
   parasitica*). Nonchemical controls include crop rotation, old crop sanitation,
   weed management, and recommended planting dates.

Physiological Disorders   If the broccoli transplants are exposed to periods of stress,
ensuing growth in the field after transplanting may result in the formation of small,
button-shaped heads and premature flowering (bolting). It is suspected that extremes of temperature, moisture, and fertility are the causative factors of these physiological disorders.

Harvesting

Broccoli may take 45 to 60 days after transplanting before harvest can begin. The upper stem (stalk) and clusters of unopened flower buds (heads) are the marketable parts of the broccoli plant. Broccoli is harvested by hand while the clusters are still compact and before the individual flowers begin to mature and turn yellow (Figure 16.5). Overmature heads often have a woody outer stem.

FIGURE 16.5   Proper stage of maturity to harvest broccoli.

FIGURE 16.6   Small lateral side shoots developing on the main stem of a broccoli plant after the central head has been cut off.

Commercially marketable mature heads are 3 to 7 in. across and may weigh 0.3 to 1.0 lb each. The stem is generally cut to a length of 9 to 10 in. from the base of the stem to the top of the head. An average commercial field of broccoli may be harvested 4 to 6 times, at 2- to 3-day intervals (depending on the temperature). The use of hybrid cultivars that have a high uniformity in maturity is becoming more popular. These hybrid cultivars have smaller heads with a high percentage of plants harvested during the first harvest with only one or two secondary harvests.

A few days after the central head is cut, small lateral side shoots often emerge from the main stalk (Figure 16.6). These side shoots grow into small broccoli heads

FIGURE 16.7   Harvested broccoli heads placed in a shipping box prior to the addition of slurry ice in the packing shed.

FIGURE 16.8   Ice placed around the harvested broccoli in the boxes.

measuring 1 to 3 in. in diameter. These small heads are often not of commercial marketable quality but are suitable for home use or for freezing.

Postharvest Handling

Harvested broccoli is placed in shipping boxes or cartons often in the field (Figure 16.7). The broccoli after harvest must be cooled to 32°F as soon as possible and ice is often placed in the shipping boxes or cartons (Figure 16.8). Often slurry ice (40% ice and 60% water) is forced into each box by a special machine

FIGURE 16.9   Slurry ice is forced into each box of the harvested broccoli by a special machine.

(Figure 16.9). The approximate storage life of broccoli under ideal conditions is approximately 10 to 14 days.

Market Preparation and Marketing

Grading, trimming, and bunching are required for marketing commercial broccoli. Stems are trimmed to 6 to 8 in. in length and the leaves are removed. Commercial sales of broccoli require bunching two to five heads into bundles of 1 to 1.5 lb and binding with rubber bands, plastic collars, or paper-covered wires ("twist-ties"). This is often done in the field with the use of a mechanical cutter and bander (Figures 16.10 and 16.11). Small-acreage broccoli may be marketed by pick-your-own methods. Some growers harvest the largest central heads and allow customers to pick the side shoots.

An increasing percentage of the broccoli crop is being utilized for freezing. Broccoli packs are also available to the consumer in florets, bunch, broccoli cole slaw, spears, and stalk cuts.

Uses

The edible portion of broccoli is the upper stem and unopened flowers. Broccoli has many uses including soups, salad bars, and as a side entree. Other uses include raw broccoli for dipping or adding to a salad, soup, macaroni and cheese, or as major components of broccoli au gratin and cream of broccoli soup.

The main stem should be trimmed and washed in order to prepare broccoli for cooking. The crisp texture and nutrients are conserved by cooking as briefly as possible with a small amount of water. Broccoli is best steamed, and if boiled only the butt end should be placed in the water.

FIGURE 16.10 A mechanical cutter and bander of broccoli that is used in the field.

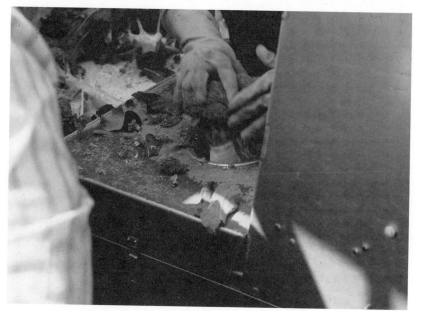

FIGURE 16.11 Two to five heads of broccoli (depending on size) placed by hand in the bander and cutter.

Nutritional Value

Broccoli is a low sodium food and is fat free, cholesterol free, low in calories, high in vitamin C, and a good source of folate. One medium stalk serving of broccoli contains only 0.5 g of fat while containing 5 g of dietary fiber (Figure 16.12). This serving of broccoli also provides 220% of the RDA of vitamin C, 15% of vitamin A, and 6% of the RDA for both calcium and iron.

FIGURE 16.12   Nutritional value of broccoli. Percent daily values are based on a 2,000 calorie diet. Used with permission from Dole Food Co., Inc.

Cauliflower

Scientific Name

Brassica oleracea L. var. *botrytis* L.

Common Names in Different Languages

Cauliflower (English); hua ye cai (Chinese); chou fleur (French); Blumenkohl (German); cavolfiore (Italian); kalifurawaa, hana-yasai (Japanese); kapusta cvetnaja (Russian); coliflor (Spanish)

Classification, History, and Importance

Cauliflower, *Brassica oleracea* var. *botrytis,* is a member of the Brassicaceae family. It is closely related to broccoli, cabbage, kale, turnips, and mustard. Cauliflower is a biennial plant producing an edible head of creamy white malformed and condensed flowers or florets (often called curds) whose stalks are short, fleshy, and closely crowded (Figure 16.13). The word cauliflower comes from the Latin terms *caulis* meaning stem, stalk, or cabbage; and *floris* meaning flower.

Cauliflower is a cultivated descendant of common cabbage. The oldest record of cauliflower dates back to the 6th century B.C. European writers mentioned cauliflower in Turkey and Egypt in the 16th century. Cauliflower was marketed in England as early as 1690 and was grown in France around 1600. It was first mentioned

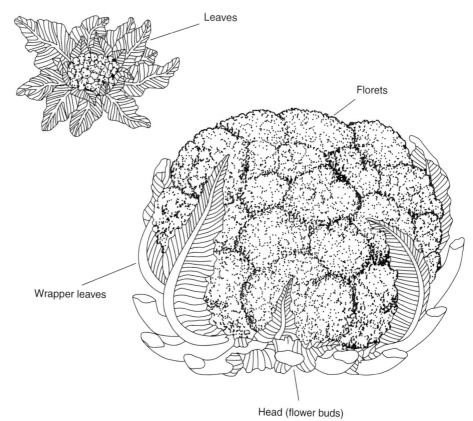

FIGURE 16.13   Diagram of a cauliflower head. Source: Iowa State University.

TABLE 16.4   *U.S. per capita consumption of commercially produced fresh and processing cauliflower*

Use	1976 (lb, farm weight)	1986	1996
Fresh	1.0	2.2	1.4
Frozen	0.6	0.9	0.5
All	1.6	3.1	1.9

Source: USDA, Economic Research Service.

in the United States in 1806. Cauliflower has been an important crop in this country since 1920.

Cauliflower for fresh market was a $184.3 million industry in 1997 (see Table 16.1). This was up from $200.8 million in 1996. The cauliflower for processing market is considerably smaller than for fresh market with a production value of $13.6 million in 1997. Per capita consumption of cauliflower has decreased from 3.1 pounds per person of both fresh market and processing cauliflower in 1986 to 1.9 pounds per person in 1996 (Table 16.4). Of the 47,600 acres of cauliflower grown in the United States, 41,200 are in California (Table 16.5).

TABLE 16.5   *The top cauliflower producing states in the United States.*

Rank	State	Planted acres (000)	Harvested acres (000)	Yield (cwt/acre)	Production (000)
1	California	41.20	41.20	130.00	5,356
2	Arizona	3.90	3.90	200.00	780
3	New York	1.10	1.00	200.00	200
4	Michigan	0.70	0.60	140.00	84
5	Texas	0.70	0.70	90.00	63
	United States	47.60	47.40	137.00	6,483

Source: USDA, NASS, 1997 data. Cauliflower grown for either fresh market or processing.

Plant Characteristics

Height: 18 to 24 in.
Spread: 24 to 36 in.
Root Depth: 18 to 36 in.

Climatic Requirements

Cauliflower tends to be more exacting in its climatic requirements than the other members of the Brassicaceae and grows best in comparatively cool temperatures with plenty of moisture, considerable humidity, and heavy applications of fertilizer. Areas bordering large bodies of water (along the Pacific Coast and near Long Island, New York) and in higher elevations such as Colorado, where natural cool conditions prevail, are conducive to successful commercial production. Cauliflower is not adapted to warm growing conditions.

Field Preparation

While cauliflower can be grown in many different types of soils, it does best in a rich, well-drained soil with a high moisture-holding capacity. A high humus content in the soil often provides better aeration and water penetration. Cauliflower grows best in a neutral or slightly acidic (pH 6.0 to 6.5) soil.

Cauliflower requires plentiful amounts of nutrients for optimum plant growth. A general fertilizer recommendation would be 80 lb of nitrogen, 80 lb of P_2O_2, and 80 to 100 lb K_2O_5 plus 15 to 20 lb of borax per acre. The fertilizer is broadcasted or mixed into the row. Side-dressing with 30 lb of N for 4 weeks after transplanting is often practiced. Cauliflower requires high magnesium levels and deficiency symptoms of magnesium readily develop when soils are too acidic or the amount of available magnesium is not adequate.

Selected Cultivars

There are several strains or types of cauliflower. One of the most important differences in strains of cauliflower is the period required for the crops to reach maturity. Some early strains of the snowball type become marketable in 50 to 55 days after transplanting to the field. Midseason types require 70 to 80 days to mature and late types typically require more than 150 days.

Cultivars of cauliflower are normally grouped by head size and density. The super snowball cultivar type is an early dwarf type with medium-sized leaves. The snowball cultivar type is later than the super snowball type and somewhat larger. Winter cauliflower types are produced during the cooler winter months in areas with mild winters. These plants tend to be larger than the other types.

Self-blanching cultivars are available that do not require their leaves to be tied above the developing head (as needed with nonblanching types) to maintain the whiteness of developing curd. These self-blanching cultivars generally do not have the purity of curds required by the market for commercial production but are popular in home gardens.

Planting

Cauliflower can be either direct seeded or transplanted. It is often transplanted in the eastern and northern United States and direct seeded in the western United States. Field spacing is 30 to 42 in. between rows, with in-row spacings from 15 to 36 in., depending on the cultivar. Cauliflower can also be planted into two rows per raised bed.

Irrigation

One to 1.5 in. of water every 5 to 7 days is generally required from either rainfall or irrigation to produce large yields of high-quality cauliflower heads. Growers in the eastern United States generally use overhead irrigation, while growers in the western United States typically use furrow irrigation.

Culture and Care

For optimum yields, cauliflower plants must be kept growing vigorously from the seedling stage through harvest. Any interruption in growth such as extreme heat, cold, drought, or plant damage can abort the development of the cauliflower head and curds.

Cauliflower is ready to blanch when it has 2 to 3 in. of white curd developed in the head. Blanching is accomplished by tying the outer leaves together over the center of the plant to protect the head from sunburn and to keep it from turning green and developing off-flavors. The leaves are tied with rubber bands, tape, or twine.

Self-blanching cultivars are named for their natural tendency to curl their leaves over their head. These self-blanching cultivars blanch better naturally under cool conditions than under warm conditions.

Common Growing Problems and Solutions

Insects Insects that may affect cauliflower include cabbage root fly maggots, cutworms, cabbageworms, cabbage looper worms, flea beetles, and aphids. Flea beetles can damage young seedlings, especially those along the edges of fields. Cabbage looper, imported cabbageworm, diamondback moth, and other caterpillars feed on foliage and infest heads. Aphids can infest young plants and cause distorted growth.

Diseases Black rot, powdery mildew, downy mildew, and alternaria leaf spot are common diseases of cauliflower.

Physiological Disorders   Cauliflower is extremely sensitive to unfavorable growing conditions such as hot weather, drought, or too low temperature. These unfavorable conditions can result in formation of premature heads or curds. These baby cauliflower heads never develop fully and are often referred to as buttons. Blindness is a physiological disorder in which no curds are formed. Blindness may be due to poor fertility, disease, insects, or cold temperatures.

Molybdenum deficiency results in "whiptailing" of leaves and is prevalent on very acid soils. Deficient levels of boron in the soil results in the plants developing hollow stems with brown discoloration. Riciness or fuzziness of the curds of the heads is often the result of high temperatures. Development of bracts or small green leaves between the segments of the curds is a result of too high temperatures or drought.

Harvesting

The cauliflower head develops rapidly under proper growing conditions. It is ready for harvest when the head has grown to 6 to 8 in. in diameter (Figure 16.14), typically within 7 to 12 days after blanching. To be marketable, the mature head should be compact and the curds firm and white. The head is cut by hand below the main stem before it becomes overmature. Overmature cauliflower heads have a coarse "ricey" appearance and are not marketable.

The harvested heads are trimmed of most of their leaves by cutting squarely across the leaves, leaving ½- to 1-in. stubs projecting above the head. These stubs protect the head from injury caused by rubbing against other heads or the shipping containers.

Postharvest Handling

During commercial production, cauliflower heads are sorted to uniform size and packed into crates. Wirebound crates usually hold 6, 12, or 24 cauliflower heads. In

FIGURE 16.14   Proper stage of maturity to harvest cauliflower.

California, the heads are typically trimmed, wrapped in perforated film (Figure 16.15), and packed in cartons (Figure 16.16). The film is perforated to prevent off-colors and off-flavors from developing after the cauliflower is cooled. Cauliflower is marketed soon after harvest or placed in 32° to 35°F storage. Cauliflower is kept in refrigeration during shipping and marketing and ice is generally not used on cauliflower as a postharvest practice.

FIGURE 16.15   Cauliflower head trimmed and wrapped in perforated film.

FIGURE 16.16   Wrapped cauliflower heads placed in shipping containers.

FIGURE 16.17   Nutritional value of cauliflower. Percent daily values are based on a 2,000 calorie diet. Used with permission from Dole Food Co., Inc.

Market Preparation and Marketing

Almost all cauliflower is marketed without its leaves (except for protective stubs). Much of western U.S.-grown cauliflower is wrapped in perforated film, then placed in cardboard cartons. Eastern growers often use wirebound crates without wrapping individual heads.

Uses

Cauliflower is cooked and served as a vegetable. Raw cauliflower is often added to green salads or cut into pieces and dipped into sauces or dips. Cauliflower is often packaged with broccoli, bok choy, bean sprouts, and snow peas in pre-made vegetable mixes.

Nutritional Value

Cauliflower is low in calories and is a good source of vitamin C. One medium head has only 25 calories, with no grams of fat and 2 grams of dietary fiber (Figure 16.17). This serving of cauliflower also provides 100% of the RDA of vitamin C, and 2% of the RDA for both calcium and iron.

Cabbage

Scientific Name

Brassica oleracea L. var. *capitata* L.

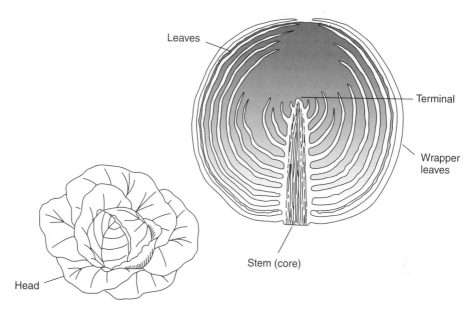

FIGURE 16.18   Diagram of a cabbage head. Source: Iowa State University.

Common Names in Different Languages

common cabbage, white cabbage (English); gan lan (Chinese); chou cabus, c. blanc, c. pommé (French); Weisskohl, Weiskraut, Kopfkohl (German); cavolo cappucio (Italian); kyabetsu, chirimen-kanran (Japanese); kapusta belokočannaja (Russian); col roja, repollo morado (Spanish)

Classification, History, and Importance

Cabbage, *Brassica oleracea* var. *capitata,* is grown for its large leafy head. It is a biennial member of the Brassicaceae family and grows best in cool, moist climates. Cabbage is typically green, but types exist with red or purple leaves. The cabbage head is comprised of numerous thick overlapping leaves with the resulting head shape round or oval (Figure 16.18).

Cabbage has been cultivated since at least 2500 B.C. and is suspected of evolving from a wild, loose, or nonheading type grown for thousands of years in Asia Minor. The eastern Mediterranean as well as Asia Minor are suggested as the places of origin for cabbage. The colewort, a weed perennial to the seacoast of Great Britain and southwestern Europe, is thought to be a wild ancestor of cabbage.

Ancient Greeks apparently held cabbage in high esteem and heading cultivars of cabbage became extremely popular in the 13th century. Cabbage was first introduced into North America by Jacques Cartier in 1541 and was planted in the United States in 1669. The popularity of cabbage grew as colonists brought cabbage with them from Europe. During the 18th century it was being grown by native American Indians as well as by the colonists. In New England, the colonists used cabbage in a boiled dinner.

Cabbage production and consumption have been relatively stable, and the cabbage industry has been an important part of the vegetable industry for many years. Cabbage was a $279 million fresh market industry and an $8.2 million processing industry (primarily for kraut) in 1997 (see Table 16.1). This was up from

$245 million and $6 million, respectively, for fresh market and processing in 1996. Per capita consumption of cabbage has slightly decreased from 10.7 pounds per person of both fresh market and processing cabbage in 1976 to 10.3 pounds per person in 1996 (Table 16.6). There are about 82,500 acres of cabbage grown for fresh market use and 5,900 acres grown for processing. New York, California, and Georgia are the leading cabbage producing states for fresh market, each producing over 10,000 acres (Table 16.7). Wisconsin and New York are the leading states growing cabbage for processing with about 2,300 acres each (Table 16.8).

TABLE 16.6   *U.S. per capita consumption of commercially produced fresh and processing cabbage*

Use	1976 (lb, farm weight)	1986	1996
Fresh	8.5	8.8	7.7
Canning (kraut)	2.2	1.6	2.6
All	10.7	10.4	10.3

Source: USDA, Economic Research Service.

TABLE 16.7   *The top cabbage producing states for fresh market in the United States*

Rank	State	Planted acres (000)	Harvested acres (000)	Yield (cwt/acre)	Production (000)
1	New York	14.00	13.60	480.00	6,528
2	California	11.50	11.50	355.00	4,083
3	Georgia	11.00	10.90	350.00	3,815
4	Florida	8.30	8.20	370.00	3,034
5	Texas	9.30	8.50	340.00	2,890
6	North Carolina	9.00	8.50	200.00	1,700
7	Wisconsin	5.00	4.80	290.00	1,392
8	Colorado	2.30	2.10	390.00	819
9	New Jersey	2.40	2.30	280.00	644
10	Ohio	1.90	1.90	320.00	608
	United States	82.55	79.83	343.00	27,395

Source: USDA, NASS, 1997 data.

TABLE 16.8   *The top cabbage producing states for processing in the United States*

Rank	State	Planted acres (000)	Harvested acres (000)	Yield (cwt/acre)	Production (000)
1	Wisconsin	2.40	2.10	38.16	80
2	New York	2.30	2.30	30.10	69
.	United States	5.87	5.49	33.46	184

Source: USDA, NASS, 1997 data.

Plant Characteristics

Height: 12 to 15 in.
Spread: 24 to 42 in.
Root Depth: 18 to 36 in. (majority of roots are confined in the upper 12 in.)

Climatic Requirements

Cabbage is grown under a wide variety of conditions and is adaptable to most areas of the United States either as a spring or fall crop. Cabbage is produced in the South mainly during the winter and early spring. It is also widely grown as a home garden crop.

Cabbage is a relatively shallow-rooted plant that requires plenty of moisture for optimum growth. Hardened plants are tolerant of frost and may be among the earliest planted vegetables in the spring. Fall cabbage can be started during the heat of midsummer, but the heads must develop during the cool weather of fall for good production.

Field Preparation

Cabbage requires a moderately well-drained fertile soil that is high in organic matter with a high water-holding capacity. Soils that dry rapidly are generally avoided for growing cabbage. Acceptable soil pH is between 6.0 and 6.8.

A good seedbed is desirable for transplanting cabbage and is essential for direct field seeding if seeds are to be planted at a uniform depth and properly covered with loose soil. Many growers in the southern United States will plow their fields in the fall in preparation of spring planting. Just before planting, plant beds are formed with a bedshaper and/or tilled to eliminate weeds and give a firm seedbed.

Fertilizer typically is applied as suggested from a soil test. A typical fertilizer application may be 50 lb of nitrogen, 100 lb of phosphoric acid per acre with another 50 lb of nitrogen applied just before the plant starts to form the head.

Cabbage fields are often broadcast-fertilized after plowing and before disking or tilling. Fertilizer may also be shanked into the bed at or just before seeding or transplanting. The band of fertilizer is placed 2 in. to the side and 2 in. below the seeds or transplants. Injury may result if the fertilizer is placed in contact with the seeds or the transplants.

Selected Cultivars

Cabbage cultivars are grown for three types of markets: fresh market, late or stored market, and the sauerkraut market. Quick heading early cabbages are usually the smallest headed (1 to 3 lb) and mature in 50 to 60 days from transplanting. Midseason types form medium-sized heads and mature in 90 to 95 days, and the late season types form the largest heads (4 to 8 lb) and require 130 days or more. Savoy cabbage cultivars are crinkled or crumpled leaf types.

Market Prize, Rio Verde, Bravo, and Gourmet are some of the more common cabbage cultivars. Gourmet is an early cultivar that produces medium-sized round heads. Market Prize is a midseason cultivar with round, firm heads. Bravo and Rio Verde are late midseason cultivars with somewhat flattened globe-shaped heads. Chieftain Savoy is a common savoy cultivar and Red Head is a red cultivar. Generally, market demand for either savoy or red types is not as great as for the smooth leafy green types.

Planting

Direct Seeding   Weeds, insects, diseases, and soil crusting must be avoided to obtain a successful crop by direct seeding. A field to be planted to cabbage should not have had cabbage or cabbage-related crops grown in it for at least 4 years in order to minimize exposure of the plantings to susceptible soilborne diseases and insects. Seeds that are direct-seeded to a field are placed 3/4 in. deep, 2 to 6 in. apart in single rows 30 in. apart or twin rows 14 in. apart on 40-in. beds.

Irrigation is often used to ensure good germination. Emerging seedlings are thinned as soon as the plants begin to crowd. Plants are generally spaced 6 in. apart in a row. Final stands may be 12 to 24 in. between plants (Figure 16.19), depending on the cultivar.

Transplanting   Six weeks are required to produce a cabbage transplant. Plant spacings for transplanted cabbage are typically similar to those used for the final stand of direct-seeded cabbage and are dependent on cultivar.

Irrigation

Irrigation is often necessary to ensure a good stand of cabbage. Irrigation is also beneficial after transplants are placed in the field and may be required when cabbage is planted at high population densities.

The critical period to provide adequate moisture for cabbage is from the time the leaves begin to cup and form a head to the time the crop is ready for harvest. Inadequate water during this period will result in reduced yields and delayed maturity.

Culture and Care

Polyethylene mulch has been used to increase the yields of cabbage in Florida. The mulch is applied to beds of moist soil that have been shaped and pressed. In this situation, black mulch is used for spring crops and white mulch is used for fall plantings.

FIGURE 16.19   An established planting of cabbage.

Common Growing Problems and Solutions

Insects Major insects affecting cabbage include cabbageworms of various types, flea beetles, diamondback moths, and cabbage aphids. Flea beetles are injurious to young seedlings, especially along the edges of fields. Cabbage looper, imported cabbageworm, diamondback moth, and other caterpillars feed on foliage and infest heads. Aphids can infest young plants causing distorted growth. They may also infest marketed heads.

Diseases There are several major diseases of cabbage. Fusarium wilt, often called yellows, is a disease found in the upper southern and the northern United States. The lower leaves of the cabbage plant become yellow and turn brown. Black rot and blackleg are diseases spread by diseased seeds or transplants or by insects. Black rot causes yellowing of the leaves and blackening of the veins. Blackleg causes dark sunken areas on the stem of young plants. The control is to use disease-free transplants and to purchase hot water-treated seeds.

Physiological Disorders The heads of early cultivars of cabbage frequently split in warm weather. This is caused by too much moisture moving into the mature head. To reduce this problem, growers try to prevent the cabbage plants from receiving excessive moisture in the field or they may prune the roots of the plants to reduce the ability of the plant to uptake water.

Harvesting

A cabbage head is typically harvested when it is firm and well sized (Figure 16.20). Cabbage for fresh market is selectively cut by hand and trimmed to a desired number of leaves. Heads intended for storage and/or processing are generally harvested all at once, either mechanically or by hand.

FIGURE 16.20 Proper stage of maturity to harvest cabbage.

Cabbage may be harvested earlier or later than typically preferred during the season depending on the market. A grower may harvest cabbage heads earlier and sacrifice some yield if there is an economical marketing advantage. Harvests can also be delayed, but an increased incidence of split heads may also result.

Postharvest Handling

Once cabbage is harvested it is removed immediately from exposure to the sun or it may "sun blister" and lose fresh weight. The optimum temperature for storage of cabbage is 32° to 36°F with a relative humidity of 90 to 98%. Under optimum conditions, cabbage can be stored for up to 5 months.

Market Preparation and Marketing

Cabbage for market is packed in wirebound crates, fiberboard cartons, nailed crates, or mesh bags.

Uses

Cabbage is used raw in salads, such as coleslaw, or fermented into sauerkraut. Other uses for cabbage include pickling, baking into casseroles, or combining with meats. Cabbage is also used in soups or may be combined with fruits in combination salads.

Nutritional Value

A serving of one medium-sized cabbage head is low in calories and is a good source of vitamin C (70% RDA) (Figure 16.21). This serving has no grams of fat and 2 g of

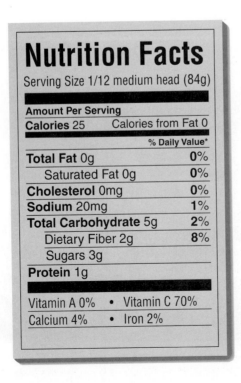

FIGURE 16.21 Nutritional value of cabbage. Percent daily values are based on a 2,000 calorie diet. Used with permission from Dole Food Co., Inc.

dietary fiber. Calcium and iron are available in cabbage in amounts as supplements to other sources. Folic acid (a B vitamin) is found in the green parts. Cabbage also provides an alkaline reaction in the body and is an ideal roughage that aids digestion.

Brussels Sprouts

Scientific Name

Brassica oleracea L. var. *gemmifera* Zenk.

Common Names in Different Languages

Brussels sprouts (English); bao zi gan lan (Chinese); choux de Bruxelles (French); Rosenkohl, Sprossenkohl, Brusseler kohl (German); cavolo di Bruxelles (Italian); me-kyabetsu, komochi-kanran, komochi-kyabetsu (Japanese); kapsta brjussel'skaja (Russian); col de Bruselas, repollo de Bruselas (Spanish)

Classification, History, and Importance

Brussels sprouts, *Brassica oleracea* var. *gemmifera*, are a type of nonheading cabbage. They are members of the Brassicaceae family and are a cool-season crop that is considered a delicacy by many people. Brussels sprouts are an herbaceous biennial that lack an apical head but have axillary heads or sprouts that are produced along an elongated stem (Figure 16.22). It is these sprouts that are formed in the axils

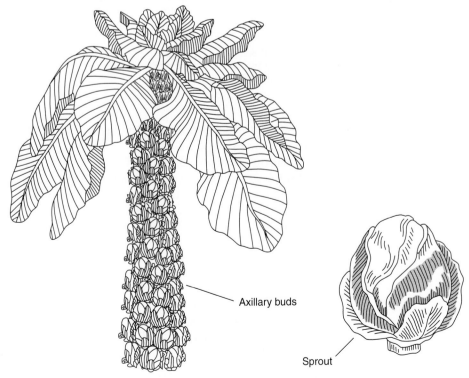

Axillary buds

Sprout

FIGURE 16.22   Diagram of Brussels sprouts plant. Source: Iowa State University.

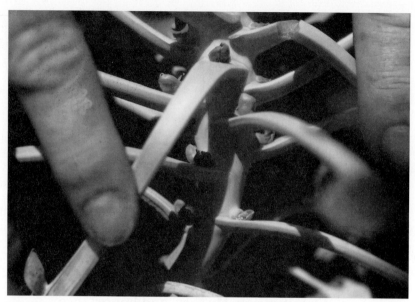

FIGURE 16.23   Young Brussels sprouts forming in the axil of the leaves. Courtesy of W. J. Lamont, Jr.

TABLE 16.9   *The top Brussels sprouts producing states in the United States*

Rank	State	Planted acres (000)	Harvested acres (000)	Yield (cwt/acre)	Production (000)
1	California	4.20	4.20	150.00	630
·	United States	4.20	4.20	150.00	630

Source: USDA, NASS, 1997 data.

of the leaves (Figure 16.23) and resemble small cabbagelike heads that are the edible portions of the plant.

Brussels sprouts probably originated in Europe and were grown there as early as the 1500s. It is probably one of the newest forms of cabbage. Brussels sprouts acquired their name partly because the plant is supposed to have grown since recorded time in the vicinity of Brussels, Belgium. The first description of the plant was made in 1587. By 1800 it was commonly grown in Belgium and France, and by 1850 it was becoming popular in England where it is considered a delicacy. Brussels sprouts have been reported in the United States since 1800 and were grown in California as early as 1909. Considerable expansion of Brussels sprouts production occurred after 1945 when the frozen food industry began processing Brussels sprouts.

Brussels sprouts are considered a minor vegetable crop in the United States. Most of the commercial acreage is in California with about 4,200 acres (Table 16.9). Brussels sprouts are grown in small amounts in other areas and are an important home garden crop.

Plant Characteristics

Height: 36 to 48 in.
Spread: 24 in.
Root Depth: 18 to 36 in.

Climatic Requirements

Brussels sprouts are a cool-season crop. They require a long, cool growing season and can withstand considerable frost. In many parts of the country they are best grown as a fall crop, preferring sunny days and light frosts at night during sprout formation. Hot weather results in soft, open sprouts, which are not desirable. Because Brussels sprouts prefer cool humid climates, commercial production is located in the California mountain valleys and around the Long Island, New York, area. About 85 to 95 days are required for Brussels sprouts to go from field seeding or transplanting stage to the harvesting stage of the sprouts.

Field Preparation

Brussels sprouts can be grown on a wide range of well-drained soils. However, they grow best on medium to heavy soils that are high in organic matter and fairly high in nitrogen. Soil pH for optimum Brussels sprouts production should be between 6.0 and 6.8. A complete fertilizer such as 5-10-10 applied at 2,000 lb/acre is often required. Additional boron and magnesium may also be incorporated into the fertilizer or supplied separately.

Selected Cultivars

Jade Cross is a widely adapted hybrid cultivar developed for shipping, processing, and home garden use. This cultivar is vigorous, uniform in size and appearance, with dark green leaves. It produces heavy yields of evenly spaced, very firm uniform sprouts that average 1 in. in diameter. Jade Cross E Hybrid, Royal Marvel, and Long Island Improved are three important commercial Brussels sprouts cultivars.

Planting

Brussels sprouts are usually established by transplanting. Plants are typically 6 weeks old when set in the field. These plants are set in rows 30 to 36 in. apart, with in-row spacings from 18 to 24 in. (Figure 16.24). The plants are usually watered at transplanting to prevent wilting. A starter fertilizer solution is also often applied at this time.

If Brussels sprouts are direct-seeded into the field, a Planet Jr. seeder or a precision seeder is typically used and final plant spacings are the same as for transplanted crops.

Irrigation

Irrigation is often necessary to supplement natural rainfall to ensure a quality crop. At least 1 to 1.5 in. of water per week is required for optimal growth.

FIGURE 16.24   An established planting of Brussels sprouts. Courtesy of W. J. Lamont, Jr.

Culture and Care

Brussels sprouts are an exacting crop that needs uninterrupted growth for maximum quality. Any delay in growth reduces yields and quality of the sprouts. For best development Brussels sprouts require large amounts of available nitrogen. This is accomplished by making at least three sidedressings with a nitrate fertilizer. The first application is made after the plants have been in the field for 3 weeks, with the next two applications 2 weeks apart.

Common Growing Problems and Solutions

Insects   Some of the more damaging insects on Brussels sprouts are Harlequin bugs, cabbage loopers, diamondback moths, imported cabbageworms, cutworms, cabbage maggots, and thrips.

Diseases   Powdery mildew and clubroot can be problems with Brussels sprouts. Clubroot is a disease that causes overgrowths or swellings of the underground stem and roots. The swellings or "clubs" on the roots interfere with the ability of the plant to take up food from the soil and as a result such plants never produce a crop. Once the organism responsible for this disease is introduced into a field, it will remain troublesome for 3 to 7 years even though no cole crops are grown in the field during this period.

Weeds   Shallow cultivations may be needed to control weeds.

Harvesting

Brussels sprouts are harvested when the sprouts attain 1 to 2 in. in diameter, and are firm and dark green. Harvesting usually occurs 3 months after the plants are set in the field.

FIGURE 16.25   Brussels sprouts maturing from the bottom of the plant upward. Courtesy of W. J. Lamont, Jr.

The sprouts begin to mature on the plant from the bottom upward (Figure 16.25). The sprouts can either be picked several times or the harvest can be delayed and the whole stalk taken at one time. To facilitate the harvest of the sprouts, the leaf below the sprout is typically broken away from the main stem. The sprout is removed by separating it from the stalk. Often the central growing point of the stalk is removed to hasten harvest. This is done when the sprouts are well formed.

As the lower leaves and sprouts are removed, the plant continues to grow upward producing more leaves and sprouts. One plant is capable of producing about 2.5 to 3.0 lb of sprouts, but commercial production is usually terminated at 2.0 lb.

The weather typically determines the frequency of harvest. During the earlier warm periods harvest may be every 7 to 14 days with about 2 to 6 sprouts being removed per harvest. As the weather begins to get cooler harvests may be delayed to every 3 to 4 weeks, with as many as 10 to 15 sprouts being removed from each plant at each harvest.

Postharvest Handling

Harvested Brussels sprouts are placed in hampers or baskets and taken out of the field as soon as possible for packaging and refrigeration. They are either vacuum cooled, then packaged and top iced, or held in refrigeration at 32°F. Brussels sprouts can be stored for periods as long as 30 days if kept at 32°F and 90 to 95% relative humidity.

FIGURE 16.26   Nutritional value of Brussels sprouts. Percent daily values are based on a 2,000 calorie diet. Used with permission from Dole Food Co., Inc.

Market Preparation and Marketing

Sprouts are cleaned, trimmed of loose leaves, and sorted to remove soft, damaged, or large-sized sprouts to prepare them for market. Brussels sprouts are packaged in either 2-lb drums, quart baskets, or 12-oz cellophane bags. Refrigeration is necessary to maintain the sprout's green color and eating quality.

Uses

Brussels sprouts are best prepared in a steamer to prevent overcooking. They may also be cooked by boiling or baking. Brussels sprouts are often served with a cheese or cream sauce or as a side dish with a meat. They can also be included in casseroles, salads, or soufflés.

Nutritional Value

Brussels sprouts (1 cup cooked) provide 160% of the RDA of vitamin C, one-tenth of thiamine, and more than one-tenth of iron (Figure 16.26). They are also a good source for fiber and are low in calories.

Review Questions

1. What is the effect of temperature on the development of broccoli heads?
2. What can a grower do to try to enhance his chances of producing a uniform field of broccoli?

3. What is bolting and what conditions may cause some cole crops to bolt?
4. How is harvested broccoli prepared for marketing?
5. What advantages does a rich, well-drained soil impart to cauliflower production?
6. What characteristics differentiate the various strains of cauliflower?
7. How and why is cauliflower blanched?
8. How does a self-blanching cauliflower cultivar differ from a more traditional cauliflower cultivar?
9. What factors contribute to "whiptailing" of cauliflower?
10. What are some of the characteristics of the various types of cabbage cultivars?
11. What factors determine when cabbage is ready for harvest?
12. What effect does removing the central growing point (terminal bud) have on the development of remaining sprouts on the Brussels sprouts plant?

Selected References

Ball, J. 1988. *Rodale's garden problem solver: Vegetables, fruits, and herbs.* Emmaus, PA: Rodale Press. 550 pp.

Camp, W. H., V. R. Boswell, and J. R. Magness. 1957. *The world in your garden.* National Geographic Society Press.

Davis, J. M. 1994. *Broccoli production for western North Carolina.* North Carolina State Univ. Coop. Ext. Serv. Leaflet 5-B.

Ehlert, G. R., and R. A. Seelig. 1966. Brussels sprouts. *Fruit & Vegetable Facts & Pointers.* United Fresh Fruit & Vegetable Association.

Ells, J. E., H. F. Schwartz, and W. S. Cranshaw. 1993a. *Commercial cabbage.* Colorado State Univ. Coop. Ext. Cir. 7.618.

Ells, J. E., H. F. Schwartz, and W. S. Cranshaw. 1993b. *Commercial cauliflower.* Colorado State Univ. Coop. Ext. Cir. 7.624.

Jett, J. W. 1996. *Growing Brussels sprouts.* West Virginia Univ. Coop. Ext. Serv. Misc. Publ. 372.

Johnson, J. R., J. W. Rushing, R. P. Griffin, and C. E. Drye. 1987. *Commercial cabbage production.* Clemson Univ. Coop. Ext. Serv. Hort. Leaflet 6.

Kays, S. J., and J. C. Silva Dias. 1995. Common names of commercially cultivated vegetables of the world in 15 languages. *Economic Botany* 49:115–152.

Nonnecke, I. L. 1989. *Vegetable production.* New York: Van Nostrand Reinhold. 656 pp.

Peirce, L. C. 1987. *Vegetables: Characteristics, production, and marketing.* New York: John Wiley & Sons. 433 pp.

Sanders, D. C. 1996a. *Broccoli production.* North Carolina State Univ. Coop. Ext. Serv. Leaflet 5.

Sanders, D. C. 1996b. *Brussels sprouts.* North Carolina State Univ. Coop. Ext. Serv. Leaflet 6.

Sanders, D. C. 1996c. *Cabbage production.* North Carolina State Univ. Coop. Ext. Serv. Leaflet 7.

Sanders, D. C. 1996d. *Cauliflower.* North Carolina State Univ. Coop. Ext. Serv. Leaflet 10.

Seelig, R. A. 1969. Cabbage. *Fruit & Vegetable Facts & Pointers.* United Fresh Fruit & Vegetable Association.

Seelig, R. A. 1971. Broccoli. *Fruit & Vegetable Facts & Pointers.* United Fresh Fruit & Vegetable Association.

Seelig, R. A., and P. F. Charney. 1967. Cauliflower. *Fruit & Vegetable Facts & Pointers.* United Fresh Fruit & Vegetable Association.

Yamaguchi, M. 1983. *World vegetables: Principles, production and nutritive values.* New York: Van Nostrand Reinhold. 415 pp.

Selected Internet Sites

USDA Agricultural Marketing Service Standards for Grades

Fresh
Broccoli   www.ams.usda.gov/standards/broccoli.pdf
Brussels sprouts   www.ams.usda.gov/standards/brussels.pdf
Cabbage   www.ams.usda.gov/standards/cabbage.pdf
Cauliflower   www.ams.usda.gov/standards/cauliflo.pdf

Processing
Broccoli   www.ams.usda.gov/standards/vpbroccl.pdf
Cabbage   www.ams.usda.gov/standards/vpcabba.pdf
Cauliflower   www.ams.usda.gov/standards/vpcauli.pdf

chapter 17

Greens

Greens are leafy crops that are grown for their foliage and are usually eaten after they have been cooked. As a group, they are generally easy and fast to grow and are nutritious. Most greens can withstand freezing temperatures and are usually grown during the cooler parts of the growing season. Crops in this grouping come from the Chenopodiaceae (spinach) and Brassicaceae (collards and kale) families.

Spinach

Scientific Name

Spinacia oleracea L.

Common Names in Different Languages

spinach (English); bo cai (Chinese); epinard (French); Spinat (German); spinacio, spinaci (Italian); hourensou, horenzo (Japanese); špinat ogorodnyj (Russian); espinaca (Spanish)

Classification, History, and Importance

Spinach, *Spinacia oleracea*, is a member of the Chenopodiaceae or goosefoot family. Other members of the family include Swiss chard and beets. *Spina* in Latin means spiny fruit and *oleracea* means herbaceous garden herb. Spinach is a low-growing fleshy-leafed annual that forms a heavy rosette of broad crinkly tender leaves (Figure 17.1). The leaves are the edible portion of the spinach plant.

Spinach has two distinct stages in its life cycle: the vegetative or rosette stage and the reproductive or bolting stage. During the vegetative stage, the seedling grows and develops leaves arranged in a rosette on a short stem near the ground's surface. The root system is relatively thick and shallow. During the reproductive stage, a branching seedstalk with pointed leaves emerges from the rosette center

FIGURE 17.1   Spinach growing in the field. Courtesy of W. J. Lamont, Jr.

and grows up to 2 to 3 ft high. Bolting (or premature flowering) can be stimulated by periods of long daylengths and warm temperatures and/or excessive crowding.

Spinach is thought to have originated in the region of Iran. Cultivation in Iran probably began during the Greco-Roman civilization. The earliest written record of spinach is Chinese, where it was called "herb of Persia." It is believed to have been introduced into China from Nepal in A.D. 647. The Arab Moors introduced spinach into Spain in A.D. 1100. Spinach soon spread throughout Europe. By the 14th century it was commonly grown in European monastery gardens.

A 1390 cookbook for the Court of Richard II contained recipes for "spynoches," presumed to be the early name for spinach. By 1538, spinach was well known in England and France. It is thought that European colonists introduced spinach to the United States and at least three cultivars were grown in U.S. gardens by 1806. The first savoyed (or crinkly) leaf cultivar was introduced in 1828.

Spinach for fresh market was a $58.7 million industry in 1997 (Table 17.1). This was up from $48 million in 1996. The spinach for processing market is considerably smaller than for the fresh market with a production value of $15.7 million in 1997. Per capita consumption of fresh spinach has increased from 0.3 lb in 1976 to 0.6 lb in 1996. The amount of spinach consumed, either canned or frozen, in 1996 was at similar levels to those recorded in 1976. Of the 16,800 acres of fresh market spinach grown in the United States, 7,100 are in California (Table 17.2). Other states producing fresh market spinach include New Jersey, Texas, Colorado, and Maryland. Texas is the major producer of spinach for processing with 5,000 acres grown.

Plant Characteristics

Height: 4 to 6 in.
Spread: 6 to 8 in.
Root Depth: Most of the plant's roots are usually limited to the upper 1 ft of soil, but the taproot measures up to 5 ft long.

TABLE 17.1   *The value of spinach production for fresh market and processing in the United States*

Crop	Value of production		
	1995	1996	1997
	(1,000 dollars)		
Fresh market	56,458	48,029	58,682
Processing[1]	15,970	17,105	15,729

Source: USDA, NASS.

[1]Value at processing plant door.

TABLE 17.2   *The top spinach producing states for fresh market and processing in the United States*

Rank	State	Planted acres (000)	Harvested acres (000)	Yield (cwt/acre)	Production (000)
Fresh Market					
1	California	7.10	7.10	175.00	1,243
2	New Jersey	2.40	2.30	125.00	288
3	Texas	3.10	2.40	65.00	156
4	Colorado	3.00	2.10	52.00	109
5	Maryland	1.20	1.20	89.00	107
Total	United States	16.80	15.10	126.00	1,903
Processing					
1	Texas	5.00	4.70	9.50	45

Source: USDA, NASS, 1997 data.

Climatic Requirements

Spinach is a very hardy cool-season crop that will withstand freezing temperatures better than most vegetable crops. Spinach grows best at a mean temperature of 50° to 60°F and it does not germinate well in hot weather. During extended periods of hot weather the spinach plant quickly develops a flower stalk, going to seed even after the development of only a few leaves. Spinach can also be grown during the winter and early spring months where the climate is mild.

Field Preparation

Spinach grows on a wide range of soils. Heavy soils tend to produce the highest yields of spinach; however light, sandy soils are desirable for winter and early spring crops since these soils are warmer, usually have good drainage, and permit earlier seeding and more rapid seed germination. All soils that are planted with spinach should be well drained and high in organic matter. Spinach is sensitive to acidic conditions and will not thrive on soils more acid than pH 5.5. The ideal pH range is 6.0 to 7.0.

Selected Cultivars

Spinach cultivars are classified according to leaf type. Leaf types include savoy (wrinkled), semisavoy, and flat leafed. The savoy types are preferred for fresh marketing, flat-leafed types are used in processing, and semisavoy for both. The savoy cultivars pack looser than the flat-leafed types. As a result, warming due to respiration in storage is less with the savoy types. Savoy types are less apt to wilt and turn yellow while in storage and transport and before purchase by consumer. Smooth-leafed cultivars are easier to clean and prepare for canning and freezing.

Spinach can also be classified according to seed type, being either prickly seeded or smooth seeded. Most commercial cultivars are smooth seeded, which makes them easier to handle and to plant accurately. Each leaf type can be further classified according to plant bolting. "Long-standing" means that the plant is slow in bolting and does not go to seed early. Some of the more common cultivars include the following:

- Seven R is a standard, semisavoyed cultivar that is often used for early spring and fall plantings. These plants are large and quick growing and the erect leaves are good for mechanical harvesting and processing.
- Marathon is a savoy leaf cultivar. It is considered better than Seven R for spring plantings since it is slower to bolt in warm weather. The leaves are large, dark, semierect, and long-standing and are used for both fresh market and processing.
- Melody F_1 is a semisavoyed type. The plants are large and quick growing with very deep-colored leaves that are thick and rounded.
- Vienna F_1 has large, savoyed leaves forming an erect plant type that tends to bolt in spring plantings.
- Grandstand has semisavoy leaves, is long-standing and erect. The leaves are medium large and light green and are used primarily for processing.
- Tyee F_1 is becoming a standard for savoyed spinach. The leaves are dark green with an upright growth habit that produces cleaner leaves. The plant does well in hot weather and is slow to bolt.
- Long-Standing Bloomsdale is a heavy savoy type with leaves that are dark green and medium large. The plants are medium and erect.

Planting

Spinach is planted and grown during the cool months of a region. It is a winter crop in much of the southern United States and in California. Spinach is typically planted in early spring and late summer in most other areas of the United States. Multiple plantings throughout the year are possible, but hot summer months are usually avoided. Spinach can be planted when soil temperatures are 50° to 60°F since warmer soil temperatures reduce seed germination.

Spinach seed is placed ½ in. deep in rows 1 to 2 ft apart. Plants are generally thinned to one plant every 3 to 4 in. (Figure 17.2). Spinach that has been seeded with a precision planter does not need to be thinned.

Irrigation

Irrigation is often necessary to grow good quality spinach since spinach has a relatively shallow root system and thrives best in a uniform moist soil. The first irrigation usually follows planting and a second irrigation is often needed 3 to 4 days after the

FIGURE 17.2   An established planting of spinach. Courtesy of W. J. Lamont, Jr.

plants germinate. One to three irrigations are normally required between emergence and harvest of spinach, depending on soil and climatic conditions.

Culture and Care

Spinach does not compete well with weeds, which remove water and nutrients from the soil and complicate the harvesting process. Cultivation is necessary to remove weeds and to prevent soil crusting. Shallow cultivation reduces the damage to a large proportion of spinach roots.

Common Growing Problems and Solutions

Insects   Insect pests of spinach include the green peach aphid, leaf miner, seed corn maggot, cabbage looper, cucumber beetles, and the spinach leaf miner. Aphids can be a major problem because they transmit virus diseases to spinach and are difficult to control, especially in savoyed leaves.

Diseases   Diseases on spinach tend to be those such as downy mildew and white rust that develop under cool, moist conditions. Symptoms characteristic of downy mildew are light yellow areas on the leaves. Infected young plants may be pale green and stunted with leaves heavily savoyed. During periods of high relative humidity or rainfall, sporulation will occur, appearing as a white mass that eventually turns purple. White rust is a serious problem in the southern Great Plains and Texas. White, blisterlike pustules appear, usually on the lower side of the leaf. Surrounding tissue browns and dies.

Physiological Disorders   The production of a premature flower stalk (bolting) in spinach and ensuing seed production can render the plant unmarketable. This can occur when spinach is grown under long days and warm conditions. Any deficits in soil moisture can intensify the effect of heat on spinach growth.

FIGURE 17.3   Mechanical harvesting of spinach. Courtesy of W. J. Lamont, Jr.

Harvesting

Most spinach cultivars mature in 40 to 50 days and may be harvested from the time the plants have five to six leaves until just before seedstalk develops. Spinach harvest for both fresh market and processing is usually mechanized (Figure 17.3). A small amount of fresh market spinach is hand-harvested and bunched. For fresh market, the entire rosette of the plant is harvested by cutting the taproot at the soil surface. Spinach harvested for processing is cut about an inch above the soil surface. This removes only the leaves, leaving the base of the plant and the growing point intact. Using this method, often more than one harvest can be made before the plant develops a seedstalk.

Postharvest Handling

For processing and the "poly bag" fresh pack, the spinach leaves are washed, trimmed as needed, and packed. Most fresh market spinach is shipped in wirebound crates or in bushel baskets for regrading, washing, and repacking in polyethylene bags at destination. Spinach has a high rate of respiration and must be cooled rapidly to prevent weight loss and decay. Cooling may be done by vacuum cooling by large operators or by hydrocooling by smaller operators. If hydrocooling is used, excess water is removed by centrifuging. Ice can also be placed in the bushel baskets of spinach soon after the plants are harvested to cool the leaf tissue. Harvested spinach from home gardens can be kept in a moisture-retentive container in the refrigerator for as long as 40 to 50 days. The shelf life of spinach is dependent on the quality of the plant at harvest.

Market Preparation and Marketing

Spinach is sold as bunches, loose, or more commonly, in consumer-sized polyethylene bags.

FIGURE 17.4   Nutritional value of spinach. Percent daily values are based on a 2,000 calorie diet. Used with permission from Dole Food Co., Inc.

Uses

The simplest and most nutritious way to eat spinach is raw substituted for or with lettuce in salads. Spinach is also cooked and eaten as a green or in omelets, quiches, and lasagna. If cooked and served as a green vegetable with a meal, spinach is generally better steamed than boiled so that the flavor and water-soluble nutrients are not destroyed.

Nutritional Value

One and one-half cup of shredded spinach provides 70% of the RDA for vitamin A, 25% for vitamin C, 20% of the iron, and 6% of the calcium (Figure 17.4). Compared to other vegetables, spinach is fairly high in protein and low in calories.

The value of spinach as a baby food has been debated. It has been claimed that eating spinach causes methemoglobinemia, a condition in which the oxygen-carrying chemical in red blood cells, hemoglobin, is attacked by chemicals called nitrites and prevented from carrying oxygen. Spinach contains nitrates, which can be converted chemically to nitrites by bacteria. This conversion takes place during storage of fresh spinach. This is a concern because newborn infants are particularly susceptible to methemoglobinemia.

Collards

Scientific Name

Brassica oleracea L. var. *acephala* DC.

FIGURE 17.5   Collard plant growing in the field.

Common Names in Different Languages

kales, tree kale, Scotch kale, borecole, collards (English); yu yi gan lan (Chinese); chou cavalier, c. fourrager, c. vert, c. frisé (French); Grünkohl, Blätterkohl, Federkohl, Futterkhol (German); cavolo riccio, c. da foraggio, c. cavaliere, c. a penna (Italian); keelu, ryokuyô-kanran (Japanese); kapusta listovaja kurčavaja (Russian); berza, b. col, col caballar, c. gallega, c. forragera (Spanish)

Classification, History, and Importance

Collards, *Brassica oleracea* var. *acephala,* are a nonheading member of the Brassicaceae. The term *acephala* means without a head. Collards are a cool-season, biennial crop of relatively minor importance in the United States. Collards are the standard winter greens grown in home gardens in the South for southern markets, and are close relatives to kale. The edible portion of the collard plant is the rosette of leaves (Figure 17.5), resembling cabbage leaves prior to heading.

Collards are one of the most primitive members of the cabbage group. These leafy nonheading cabbages originated in the eastern Mediterranean and have been used for food since prehistoric times. Collards were cultivated by the ancient Greeks and Romans. Either the Romans or Celts introduced them to Britain or France, reaching the British Isles in 400 B.C. The first mention of collards in America was in 1669.

Collards are a traditional southern U.S. crop and are grown mainly from December to April in Florida, South Carolina, Georgia, Virginia, and Alabama. Outside of these regions, collards are of minor importance for commercial production.

Plant Characteristics

Height: 24 to 30 in.
Spread: 24 in.
Root Depth: 18 to 24 in.

Climatic Requirements

Collards are a hardy, cool-season vegetable. They withstand summer heat and short periods of cold as low as 15°F. While collards are the most heat tolerant of the cole crops, they grow optimally under cool growing conditions. Fall frosts and mild winter temperatures appear to impart high sugar content and improved flavor to the leaves of the collard plant.

Field Preparation

Collards can be grown in a variety of soils, with loam soils often producing the greatest yields. Lighter, well-drained sandy soils are preferred for early spring crops. All soils used for growing collards should be well drained, rich in organic matter, and with a pH of 6.0 to 6.5. Collard growth is also satisfactory at a pH of 5.5 to 6.0.

For a medium fertility soil, a typical application is 1,200 to 1,500 lb per acre of a 5-10-10 fertilizer broadcast before planting. Usually not more than one-half of this amount is applied before planting and the remainder is applied as a side-dressing when the plants are established. Collards require ample nitrogen for good green color and tender growth.

Selected Cultivars

The Georgia cultivar (Figure 17.6) is the most widely grown for bunching. Vates (Figure 17.7) is a cultivar that is slightly harder to bunch, but has greater cold tolerance and bolts more slowly in the spring. Morris Heading is a slow-bolting, cold-tolerant cultivar that is used by home gardeners. The leaves of Morris Heading are more savoyed than Georgia and form a semihead, which makes bunching difficult. Blue Max, HeaviCrop, and Top Bunch are all hybrids that tend to be more cold tolerant, grow faster, and produce higher yields than the open-pollinated cultivars.

FIGURE 17.6   The Georgia cultivar of collards.

FIGURE 17.7   The Vates cultivar of collards.

FIGURE 17.8   An established planting of collards with wide rows.

Planting

Seeds of collards are sown ¼ to ½ in. deep in the field. Emerging seedlings are thinned to 6 to 12 in. apart to allow enough space for them to mature. The plants removed by thinning can also be eaten. Since collard plants can become large, at least 3 ft between rows is often needed (Figure 17.8).

Irrigation

Collards have a shallow root system between 18 to 24 in. deep. In general, collards will not need supplemental irrigation during the colder months of the year in many regions of the United States. One inch of water a week is needed for optimum growth.

Culture and Care

Frequent shallow cultivations are necessary to control weeds. The general close plant spacings and rapid growth of the collard plant also help to suppress weeds.

Common Growing Problems and Solutions

Insects   Curled or puckered foliage that turns yellow could be an indication that the plant is affected by aphids. Aphids have pear-shaped bodies about the size of a pinhead. Aphids are often found in clusters on the undersides of the leaves. Virus diseases are also spread by aphids.

Cabbage loopers eat ragged-shaped holes in leaves. The cabbage looper is a 1½ in.-long, light green caterpillar with long yellowish stripes on its back. Imported cabbageworms chew large ragged holes in collards, leaving trails of dark green frass or excrement. The cabbageworm measures 1¼ in. long and is light green with one yellow stripe. The adult is a common white butterfly with three or four black spots on its wings that is active during the day.

Diseases   Fusarium yellows can be a problem in summer collard crops when soils are warm and if fields have not been rotated. Infected plants and leaves will show one-sided yellowing and wilting. These leaves may die and drop, starting at the bottom of the plant.

Black rot begins as yellowed, V-shaped spots on the leaf edges. These spots spread to the centers of leaves and leaf veins turn black. Certified or tested seed and transplants must be bought by collard growers to eliminate black rot.

Downy mildew starts as small, irregular yellow or black spots on the top of leaves, followed by soft white to purplish mold growth directly under these spots. Alternaria leaf spot is recognized by round dark spots, sometimes with circles like a target or surrounded by a yellow halo.

Harvesting

All green parts of the collard plant are edible and may be harvested at any time during the growing season. Small plants can be harvested to thin the row and these plants can also be eaten. Plants grown 6 in. apart can be cut at ground level when they reach 6 to 10 in. in height. Large leaves can also be harvested when the plants are 10 to 12 in. tall. This allows the younger leaves to develop for later use.

Postharvest Handling

Collards should be stored as close to 32°F as possible. At this temperature they can be stored for 10 to 14 days. A relative humidity of 90 to 95% is required to prevent wilting.

Nutrition Facts

Serving Size 2 cups chopped (72grams)

Amount Per Serving

Calories 25   Calories from Fat 0

	% Daily Value*
Total Fat 0g	**0%**
Saturated Fat 0g	**0%**
Cholesterol 0mg	**0%**
Sodium 30mg	**1%**
Total Carbohydrate 5g	**2%**
Dietary Fiber 1g	**4%**
Sugars 2g	
Protein 1g	

Vitamin A 50% • Vitamin C 30%
Calcium 2% • Iron 0%

FIGURE 17.9   Nutritional value of collards. Percent daily values are based on a 2,000 calorie diet. Used with permission from Dole Food Co., Inc.

Market Preparation and Marketing

Collards are usually bunched with three to five stalks per bunch, depending on size. Stalks must be neatly clipped from the row and dirty, discolored, or ragged leaves removed when bunched.

Uses

Collards are generally boiled and eaten as a side dish. The taste of collards is similar to cabbage. A light frost near harvest time reportedly improves the flavor of collard greens.

Nutritional Value

Collards are extremely nutritious, low in calories, and contain substantial amounts of vitamins A and C. A serving of 2 cups of chopped collards has only 25 calories, with no grams of fat and 1 g of dietary fiber (Figure 17.9). This serving of collards also provides 50% of the RDA of vitamin A and 30% of vitamin C, and 2% of the RDA for calcium.

Kale

Scientific Name

Brassica oleracea L. var. *acephala* DC.

Common Names in Different Languages

kales, tree kale, Scotch kale, borecole, collards (English); yu yi gan lan (Chinese); chou cavalier, c. fourrager, c. vert, c. frisé (French); Grünkohl, Blätterkohl, Federkohl, Futterkhol (German); cavolo riccio, c. da foraggio, c. cavaliere, c. a penna (Italian); keelu, ryokuyô-kanran (Japanese); kapusta listovaja kurčavaja (Russian); berza, b. col, col caballar, c. gallega, c. forragera (Spanish)

Classification, History, and Importance

Kale, *Brassica oleracea* var. *acephala,* is a member of the Brassicaceae family. *Kale* is a Scottish word derived from *coles* or *caulis,* terms used by the Greeks and Romans in referring to the cabbagelike group of plants. The plant does not form a solid head and the highly curled, bluish green leaves are the edible portion.

Kale is one of the oldest forms of cabbage, originating in the eastern Mediterranean or in Asia Minor. Kale was known to the ancient Greeks and several cultivars were described by Cato around 200 B.C. Travelers and immigrants through the ages have introduced this green vegetable to many parts of the world and it was grown in the United States in the 17th century.

Since kale is a relatively low-priced commodity in greatest demand from December to April, extensive culture is restricted to areas with mild winters and markets available by truck. Kale production exists in Virginia, Maryland, and New York.

Plant Characteristics

Height: 8 to 12 in.
Spread: 8 to 12 in.
Root Depth: 18 to 24 in.

Climatic Requirements

Kale is a hardy, cool-season green of the cabbage family. Kale will tolerate summer heat, but grows best in the spring and fall in regions where it is grown. It will overwinter as far north as northern Maryland, southern Pennsylvania, and other areas that have similar winter climates. The quality of kale deteriorates when temperatures exceed 85°F.

Field Preparation

Most of kale production is in heavy, friable loam soils. These soils, with a pH of 6.5 to 6.8, produce the greatest yields. Kale is a heavy feeder and for a yield of 1,000 lbs/acre of greens, suggested fertilizer rates are 40, 12, and 40 lbs/acre of actual N, P, and K, banded under the seed followed by a side-dressing of 15 to 30 lbs/acre of N approximately 1 month after seeding.

Selected Cultivars

Cultivars of kale differ primarily in leaf color and texture. Scotch-type cultivars such as Dwarf Green Curled Scotch, Dwarf Blue Curled Scotch, and Tall Green Curled Scotch have extremely curled, wrinkled, and finely divided leaves with color ranging

FIGURE 17.10   A Scotch-type cultivar of kale.

FIGURE 17.11   A hybrid cultivar of kale.

from bright green to yellowish green (Figure 17.10). Hybrid kale cultivars such as Blue Knight, Blue Armor, and Winterbor tend to be more uniform in plant size, leaf texture, and the preferred blue-green color (Figure 17.11). The blue-green color is commonly associated with greater cold tolerance.

Cultivars sold as "flowering kale" are used for ornamental plants or for decoration. Although edible, flowering kale is not as palatable as regular kale. Smooth-leafed Siberian kale (Hanover salad) is not commonly grown in the United States and is classified as *Brassica napus,* var. *pabularia.*

FIGURE 17.12   An established planting of kale.

Planting

Kale is grown as a fall, winter, and spring crop in the southern United States and a fall and early spring crop in the northern United States. Kale is mostly direct-seeded into the field. Seeding is usually ¼ in. deep, with in-row spacing of 6-in. and up to 36 in. between rows (Figure 17.12). When seedlings are about 2 in. tall, they are thinned to 8 to 16 in. apart. Commercial crops are generally grown for a one time harvest that occurs about 40 to 55 days after planting.

Irrigation

Kale requires at least 1 in. of water every week from rain or from watering.

Culture and Care

Kale is relatively easy to grow, requiring only normal cultivation and watering, and is treated like collards or mustard greens. Cultivation is necessary to control weeds and to prevent surface crusting. The shallow-rooted character of kale prohibits deep cultivation, since it may cause severe root injury.

Common Growing Problems and Solutions

Disease and insect pests for kale are identical to those for collards and cabbage. Since kale is a short-season crop and often followed by other crops in the same year, weed control through cultivation is often preferred.

Harvesting

The lower leaves of kale can be individually harvested when they are small and tender (8 to 10 in. or shorter) or the entire kale plant may be cut. Another method of harvesting kale is by taking "stripped leaves." These leaves are stripped by removing the tender leaf parts from the coarse midrib of the plant. The main crown and stem of the plant are left in the field to produce additional leaves.

The quality of kale is improved by frost, and the plant can withstand a light freeze. Kale can be harvested in the field until a severe freeze in the winter.

Postharvest Handling

After harvest, the leaves of kale are washed, graded, and either bunched or packed. All long-distance shipping requires ice to preserve freshness, but ice may not be necessary for local distribution. Kale can be stored for 10 to 14 days at 32°F and 90 to 95% relative humidity.

Market Preparation and Marketing

If individual bunches of kale are packaged, perforated film bags are used. Kale can have a storage life of up to 4 to 5 weeks if optimum conditions of temperature and packaging are provided. Throughout storage and handling, extreme care is taken to prevent breakage of the leaves and the consequent infestation of rot organisms.

Uses

Kale is used as a green vegetable, steamed and served with butter or vinegar, or in soups. Much of the present production is used as decoration on salad bars because kale is less likely to wilt than lettuce or other greens.

Nutritional Value

Kale is among the most nutritious of the vegetables. One 3.5-oz serving of kale provides all the adult daily requirements of vitamins A and C and 13% of the calcium requirement.

Review Questions

1. What can stimulate bolting in spinach?
2. When might a light, sandy soil be advantageous for growing spinach?
3. Why is irrigation often necessary for growing quality spinach?
4. What characteristics of collards make it a more traditional southern United States crop?
5. What attributed effect does a light frost have on the taste of collards and kale?
6. Where are some of the commercial production areas of kale?
7. What are some of the ways that kale is harvested?

Selected References

Ball, J. 1988. *Rodale's garden problem solver: Vegetables, fruits, and herbs.* Emmaus, PA: Rodale Press. 550 pp.

Charney, P., and R. A. Seelig. 1966. Kale. *Fruit & Vegetable Facts & Pointers.* United Fresh Fruit & Vegetable Association. Washington, D.C.

Cook, W. P., R. P. Griffin, and A. P. Keinath. 1995. *Commercial collard production.* Clemson Univ. Coop. Ext. Serv. Hort. Leaflet 5.

Hodges, L. 1991. *Kale: The "new" old vegetable.* Univ. of Nebraska Coop. Ext. Serv. Publ. NF91-51.

Hodges, L. 1993. *Spinach and Swiss chard.* Univ. of Nebraska Coop. Ext. Serv. Publ. G92-1123-A.

Johnson, J. R., R. P. Griffin, and C. E. Drye. 1989. *Commercial spinach production.* Clemson Univ. Coop. Ext. Serv. Hort. Leaflet 61.

Kays, S. J., and J. C. Silva Dias. 1995. Common names of commercially cultivated vegetables of the world in 15 languages. *Economic Botany* 49:115–152.

Nonnecke, I. L. 1989. *Vegetable production.* New York: Van Nostrand Reinhold. 656 pp.

Peirce, L. C. 1987. *Vegetables: Characteristics, production, and marketing.* New York: John Wiley & Sons.

Sackett, C. 1975. Spinach. *Fruit & Vegetable Facts & Pointers.* United Fresh Fruit & Vegetable Association.

Sanders, D. C. 1996. *Spinach.* North Carolina State Univ. Coop. Ext. Serv. Leaflet 17.

Splittstoesser, W. E. 1990. *Vegetable growing handbook:Organic and traditional methods.* 3d ed. New York: Van Nostrand Reinhold.

Selected Internet Sites

USDA Agricultural Marketing Service Standards for Grades

Fresh
Collards   www.ams.usda.gov/standards/grnscolr.pdf
Kale   www.ams.usda.gov/standards/kale.pdf
Spinach (bunched)   www.ams.usda.gov/standards/spinbch.pdf
Spinach (leaves)   www.ams.usda.gov/standards/spinleav.pdf
Spinach (whole plants)   www.ams.usda.gov/standards/spinplnt.pdf

Processing
Spinach   www.ams.usda.gov/standards/vpspin.pdf

chapter
18

Leafy Salad Crops

I. Lettuce
II. Celery
III. Parsley
IV. Miscellaneous Leafy Salad and Greens Crops

Leafy salad crops are used mainly for their leaves and are eaten raw. These crops are often easy to grow and can serve as important sources of vitamins and minerals. Leafy salad crops with dark green leaves are generally good sources of vitamins A and C, iron, folic acid, and calcium.

While Iceberg lettuce is the most popular, there are many other lettuces and salad greens that can be used as salad crops. Major vegetables in this group come from the Asteraceae (lettuce) and Apiaceae (celery and parsley) families.

Lettuce

Scientific Name

Lactuca sativa L. *capitata* L. (head or butterhead lettuce)
Lactuca sativa L. var. *longifolia* Lam. (leaf or romaine lettuce)

Common Names in Different Languages

head or butterhead lettuce (English); wo ju (Chinese); latue pommée (French); Buttersalat, Krachsalat, Kopfsalat (German); lattuga a cappucio, l. cappuccia (Italian); lotasu (Japanese); salat kočannyj (Russian); lechuga acogollada, l. arrepollada (Spanish) leaf or Romaine lettuce (English); ye wo ju (Chinese); laitue romaine, l. lombard, l. à couper, chinchon (French); Blattsalat, römischer Salat, Bindersala (German); lattuga di taglio, l. romana (Italian); liifu letasu (Japanese); listovoj salat (Russian); lechuga romana, l. par cortar. l. de hojas, lechuguino (Spanish)

Classification, History, and Importance

Lettuce (*Lactuca sativa* L.) is a member of the Asteraceae (Compositae) family. Other members of the family include sunflower, dandelion, endive, and globe artichoke. Lettuce is named after the milky juice (latex) that it produces.

Lettuce is an herbaceous annual crop that has been used since ancient times for medicinal and food purposes. Earliest lettuce production was along the Mediterranean basin in Europe. As the popularity of lettuce quickly spread, more regionally adapted cultivars were developed for use in North America and other areas. The production and consumption of lettuce continued to increase so that it is now an important vegetable crop in many parts of the world.

Lettuce is a major salad crop in North America, Europe, Australia, New Zealand, and South America. Production of lettuce is increasing in the Middle East, Africa, and Japan. U.S. production of lettuce in 1989 was about 2.5 million tons. Over 70% of U.S. production is in California. European production in 1975 was 1.2 million tons.

Head lettuce was a $1.2 billion industry in 1997 (Table 18.1). This was up from $970 million in 1996. In 1997 the value of production of leaf lettuce and Romaine lettuce crops were $244.8 million and $171 million, respectively. Per capita consumption of lettuce continues to increase and is currently close to 30 lb (Table 18.2). Of the 202,900 acres of head lettuce grown in the United States, 141,000 are in California (Table 18.3). Other head lettuce producing states include Arizona,

TABLE 18.1   *The production value of leafy salad crops for fresh market in the United States*

Crop	Value of production		
	1995	1996	1997
		(1,000 dollars)	
Celery	306,828	199,398	273,445
Escarole/Endive	14,642	13,377	13,123
Lettuce, Head	1,463,348	970,798	1,187,830
Lettuce, Leaf	309,477	245,536	244,841
Lettuce, Romaine	215,026	163,132	170,954

Source: USDA, NASS.

TABLE 18.2   *U.S. per capita consumption of commercially produced fresh head lettuce and Romaine lettuce*

Use	1976	1986 (lb, farm weight)	1996
Head lettuce	24.2	21.9	23.3
Romaine lettuce	—	2.4	6.4
All	24.2	24.3	29.7

Source: USDA, Economic Research Service.

TABLE 18.3   *The top head lettuce-producing states in the United States*

Rank	State	Planted acres (000)	Harvested acres (000)	Yield (cwt/acre)	Production (000)
1	California	141.00	141.00	350.00	49,350
2	Arizona	54.20	53.90	319.00	17,192
3	Colorado	2.20	2.00	330.00	660
4	New Mexico	2.40	2.00	300.00	600
5	New Jersey	1.30	1.30	280.00	364
6	New York	0.80	0.70	280.00	196
7	Washington	1.00	0.90	200.00	180
.	United States	202.90	201.80	340.00	68,542

Source: USDA, NASS, 1997 data.

TABLE 18.4   *The top leaf lettuce-producing states in the United States*

Rank	State	Planted acres (000)	Harvested acres (000)	Yield (cwt/acre)	Production (000)
1	California	35.70	35.70	210.00	7,497
2	Arizona	5.80	5.70	275.00	1,568
3	Ohio	0.65	0.60	205.00	123
4	Florida	0.35	0.33	160.00	53
.	United States	42.50	42.33	218.00	9,241

Source: USDA, NASS, 1997 data.

TABLE 18.5   *The top Romaine lettuce-producing states in the United States*

Rank	State	Planted acres (000)	Harvested acres (000)	Yield (cwt/acre)	Production (000)
1	California	24.00	24.00	265.00	6,360
2	Arizona	8.30	8.20	320.00	2,624
3	Florida	0.90	0.85	235.00	200
4	Ohio	0.55	0.50	190.00	95
.	United States	33.75	33.55	277.00	9,279

Source: USDA, NASS, 1997 data.

Colorado, New Mexico, New Jersey, Washington, and New York. Leaf lettuce is grown on 42,500 acres, with 35,700 of that acreage in California (Table 18.4). Other leaf lettuce-growing states include Arizona, Ohio, and Florida. Romaine lettuce is grown on 33,750 acres, with major production in California and Arizona (Table 18.5).

Plant Characteristics

Height: 4 to 8 in.
Spread: 6 to 12 in.
Root Depth: 18 to 36 in., but the taproot can grow to 5 ft long

Climatic Requirements

Lettuce thrives best in fairly cool growing environments and can be grown in most temperate regions by selecting proper cultivars for the particular climate. The climatic requirements for crisphead types are more critical than for the other lettuce types. Temperatures above 75°F inhibit lettuce seed germination and optimal head formation requires a mean maximum day temperature of about 73°F and a mean minimum night temperature of approximately 45°F. These cool nights are essential for quality lettuce production. High temperatures induce the plant to go from a vegetative state to an irreversible flowering state. While high temperatures stimulate seedstalk formation, lower temperatures inhibit growth. In addition, high temperatures may result in a loose leaf structure with the leaves acquiring a bitter taste. Lettuce grown in high temperatures also has physiological disorders such as tipburn (tips of the inner leaves in the head exhibit necrosis).

Field Preparation

The lettuce plant root system is somewhat weak and is concentrated in the upper layer of soil. Lettuce prefers soils that are high in organic matter with adequate moisture and nutrients in the upper 10 to 12 in. A well-drained, fertile soil is optimal for lettuce production.

In much of the large commercial acreage, lettuce is grown in muck, sandy loam, and silt loam soils. Although lighter soils tend to warm up faster, these soils also tend to dry out more readily. In addition, the planting time of lettuce can dictate the soil type for maximum performance. For instance, early plantings may produce better in a sandy loam, while late plantings may grow better in mucks and clay loams.

Lettuce is grown in a variety of soil types and climatic conditions. If a hardpan soil layer is suspected in the field, the field may need to be tilled to a deeper depth using a deep plow or subsoiler. A disk, plow, or soil rotavator is used for primary tillage. Raised beds are essential with furrow irrigation or when excess rainfall occurs. The raised beds encourage wet soils to dry quicker and prevent waterlogging of plant roots.

Lettuce does not grow optimally in highly acidic soils and lime is generally applied if the soil pH is below 5.5. Lettuce is somewhat tolerant of alkaline soil, but it may still be beneficial to reduce the soil pH to avoid mineral deficiencies and toxicities.

Fertilization

Approximately 80% of lettuce growth occurs during the 3 to 4 weeks before harvest. It is critical that the plant's nutritional requirements are fulfilled during this time. When compared to other crops, lettuce does not normally demand a high uptake of nutrients. A summer crop of lettuce in Salinas Valley, California, contains or withdraws 47, 7, 117 lb/acre of N, P, and K, respectively.

The lettuce plant has a weak, shallow root system that requires a well-fertilized soil for high yields. Since lettuce is not normally afflicted with micronutrient disorders, an average application in Salinas Valley, California, usually includes nitrogen (150 lb/acre), phosphoric acid (100 lb/acre), and potash (150 lb/acre).

Many growers apply 1,000 lb/acre of a complete fertilizer such as 5-10-10 to sandy-silt loam soils or a 3-9-18 to muck soils. Fertilizers should be applied early, well before rapid plant growth. One-third to one-half of the N and P and all the K is typically applied at or before seeding. The remaining N and P is applied at the time of thinning. These recommendations may need to be modified based on soil type (e.g., sandy soils may leach some nutrients faster than heavier soils and more frequent applications of less fertilizer may be needed).

Lettuce growth generally shows a favorable response to animal manures. Disadvantages with the use of manures may include the presence of excess amounts of salt, introduction of weed seeds, and a poor cost-to-benefit ratio. Also, some manures may need to be properly fermented before applying to the fields. In high temperatures, manures may also encourage bolting of the lettuce.

Selected Types and Cultivars

General Types of Lettuce   Lettuce can be broken down into various types based on appearance (Figure 18.1). The general types of lettuce include the following:

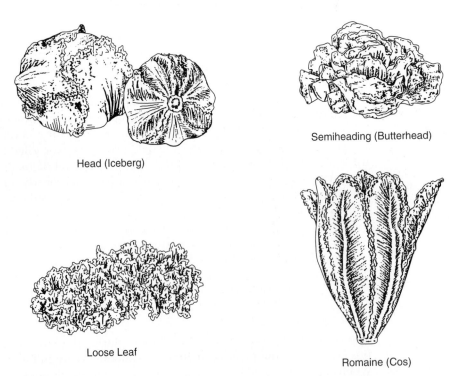

Head (Iceberg)

Semiheading (Butterhead)

Loose Leaf

Romaine (Cos)

FIGURE 18.1   Diagram of four lettuce types. Source: Iowa State University.

Crisphead lettuce   Crisphead lettuce, also called Iceberg lettuce, has brittle-textured leaves that are tightly folded (Figure 18.2). The outer leaves are green whereas the interior of the head is white or yellow (Figure 18.3). Crispheads, if handled properly, can withstand long-distance shipping.

FIGURE 18.2   A crisphead or Iceberg head of lettuce.

FIGURE 18.3   A crisphead lettuce head with green outer leaves and yellow or white inner leaves.

FIGURE 18.4   A butterhead lettuce cultivar type.

FIGURE 18.5   A romaine lettuce type.

Butterhead lettuce   Butterhead lettuce, also called Bibb or Boston lettuce, has crumpled
   leaves with a soft buttery texture that form loose heads (Figure 18.4). The veins
   and midribs of butterheads are less prominent than in the crisphead types.

Romaine, or Cos, lettuce   Romaine lettuce, or Cos lettuce, has long leaves with heavy
   midribs that form elongated heads (Figure 18.5). The outer leaves are fine
   textured and light green. Although this lettuce has a tough appearance due to
   heavy rib formation, the inner leaves are tender and the eating quality is
   excellent.

FIGURE 18.6   Leaf lettuce types of various colors in the field.

Leaf lettuce   Leaf lettuce is primarily a local marketing and home garden lettuce.
   The leaves are variable in shape, margin, and color, and form a compact
   rosette-shaped head (Figure 18.6). Most leaf-type cultivars withstand greater
   environmental variation than the heading types. The quality of the leaf types,
   particularly nutritionally, is superior to the other types.

Stem lettuce   Stem lettuce, also sold as "celtuce," is a popular crop in the Orient.
   The stems of this type enlarge during growth and are peeled and used as a
   cooked vegetable.

Latin lettuce   Latin lettuce leaves are elongated and more leathery than Romaine.
   Latin lettuce is primarily grown in the Mediterranean region and South
   America.

Crisphead Lettuce Types   Horticultural improvement in crisphead lettuce has been
concerned with type and quality. Several characteristics such as color, leaf texture,
shape, and butt appearance determine lettuce type. In the history of lettuce breeding in the United States, there have been five types of crisphead lettuce.

1. *Imperial type*   Lettuce cultivars of the Imperial type form symmetrical, partly
   exposed heads with many wrapper leaves. They have light or medium green
   leaves with a relative soft texture. The leaf margins are serrated or wavy.
   Imperial cultivars vary in butt color and ribbiness. The quality ranges from
   good to excellent.

2. *Great Lakes type*   Cultivars of the Great Lakes type form extremely firm, mostly
   exposed heads. They have bright green leaves in various shapes. The leaves are
   very crisp and serrated. The heads are characterized by white insides and the
   butt is whitish with prominent ribs. Great Lakes cultivars were bred for
   tolerance to summer heat (warm season), but they are generally not good for
   shipping. Great Lakes types are moderately slow bolting and resistant to
   tipburn.

3. *Vanguard type*   Vanguard-type cultivars have erratic heading with firmness ranging from normal size and quite firm to heads that are large and puffy. The leaves are dark green with a softer texture than the Great Lakes type. The leaf margins are wavy and the butt is green with flat ribs. Vanguard cultivars are tolerant of cold weather.
4. *Empire type*   Cultivars of the Empire type have heads that are conical in shape. They have light green leaves that are deeply serrated and very crisp. Empire types are warm-season cultivars (derivative of Great Lakes) and are often grown in deserts.
5. *Salinas type*   Salinas-type cultivars have thick dark gray-green leaves with scalloped margins and a creamy interior. Salinas cultivars have a soft texture and are resistant to tipburn. They are grown in mountain valleys and are good for shipping.

Planting

Lettuce seed is an achene or dry fruit. The seeds can be sized for uniform growth, pelleted for precision planting, or greenhouse germinated for later transplanting. One pound of lettuce seed contains 200,000 to 450,000 seeds. The small-sized, irregular-shaped lettuce seeds are often difficult to separate and handle during planting.

Precision seeding requires proper spacing and depth of seeds. Commercial lettuce is grown primarily from seed sown directly in the field at a rate of 1 to 2 lb/acre. This rate, however, can be decreased as seed quality and performance are improved.

The distance between the rows varies with respect to cultivation, irrigation, fertilization, and cultivar. For example, the butterhead cultivars are smaller and can be grown closer than the crisphead types.

Western U.S. growers usually plant lettuce in two-row beds (40 to 42 in. from center to center and 18 to 27 in. wide) (Figure 18.7). The plants are located 3 to 4 in. from the shoulder and generally 12 to 14 in. apart between rows on the same bed.

FIGURE 18.7   An established planting of crisphead lettuce with two rows planted per bed.

Within-row spacing varies from 10 to 14 in. depending on location, cultivar, and season. Bed height is an additional factor that varies with soil type, drainage, and season.

A seeding depth of 0.25 in. is usually satisfactory unless high temperatures persist. Under high-temperature conditions, seeds are usually planted closer to the ground surface. Sprinkler irrigation is often needed for successful seedling emergence and plant establishment.

A disadvantage of direct seeding is the requirement of the laborious and expensive operation of thinning. To reduce plant competition, thinning is done when the seedlings are established. Direct plantings are often thinned 10 to 16 in. apart. When the initial leaves are formed, the clusters are thinned to one per plant. Late thinning often causes root damage and prolongs root recovery.

Lettuce transplants can be produced in greenhouses or in the field in protected areas using plastic or fabric materials for coverings. The optimum germination temperature for lettuce in a greenhouse is 70°F. At this temperature, lettuce seed will usually germinate within 7 days. After germination, temperatures in the greenhouse should be maintained at 55° to 58°F. It will take 6 to 8 weeks for the lettuce transplants to develop properly for planting to the field.

Some growers practice direct seeding into a field in close spacing to produce transplants. When the emerging seedlings are of suitable size they are "pulled" from the plant bed and transplanted to the field. These transplants are called bare-root transplants. Bare-root transplants are not as uniform or as quick-growing as transplants produced using flats.

Using transplants reduces the amount of time the plant is in the field and may increase uniformity of the crop and harvest. Transplants can be planted into a field using manual labor or with a mechanical transplanter.

Irrigation

Adequate, available moisture and cool temperatures are necessary at the time of heading for crisphead types. Low moisture and high temperatures may increase the incidence of tipburn. Excessive rains or irrigation may leach nutrients from the soil and also increase the incidence of diseases and insects.

Culture and Care

The vegetative growth phase of lettuce is characterized by increased root, stem, and leaf mass. A rosette of leaves on a short stem forms as light intensity and daylength increase. Leaf width during this time also increases. Heading varieties produce leaves that cup and curl inward in an overlapping manner. Nonheading varieties produce a rosette of leaves that curve away from the center of the plant.

Flowering and seed production characterize the later, reproductive phase. "Bolting" is the stress-induced transition the plant makes by quickly shifting from a vegetative state to a reproductive state. A seedstalk from which flowers and seeds develop forms within the head. The entire seed-to-seed cycle requires 6 to 8 months.

Weeds can be a major nuisance in lettuce fields. Lettuce is not a good competitor against many annual grasses and broad-leaved weeds. This suggests that weed removal or prevention is an important cultural practice for growing lettuce. Weeds also harbor insects and diseases that could spread to the lettuce crop. Successful land preparation is an effective method of weed control in lettuce fields. Early shallow cultivation reduces weed damage. Herbicides may not be necessary if the soil is properly managed during the short life cycle of lettuce.

Common Growing Problems and Solutions

Insects   Aphids transmit diseases to lettuce and need to be controlled. Other insects of lettuce include cutworm, armyworm, cabbage looper, corn earworm, leafhopper, and spider mites.

Diseases   Mosaic is the most serious disease of lettuce and is seedborne. Other diseases include spotted wilt, aster yellows, big vein, downy mildew, powdery mildew, sclerotinia, anthractnose, bottom rot, and botrytis.

Physiological Disorders   Physiological problems with lettuce include tipburn, russet spotting, and rib discoloration.

Harvesting

The time from seedling emergence to harvest of a mature lettuce head (crisphead type) under normal daylength and warm temperature conditions ranges from 55 to 70 days. It may take up to 140 days for lettuce heads to mature when grown under cold temperatures and short daylength because of a slower-growing response. Using transplants can reduce from 1 to 4 weeks the amount of time the lettuce is in the field before harvest.

Lettuce is typically hand harvested (Figure 18.8). A sufficient number of early maturing heads that are firm to hard indicate that a field is ready to harvest. Two to four harvests of a field may be necessary if maturity in a field is irregular. In uniformly maturing fields, up to 90% of the heads may be harvested in a single harvest.

A knife is used to cut the lettuce heads at or slightly below ground level (Figure 18.9). All but three to five wrapper leaves are removed and the heads placed in boxes. Water is sometimes sprayed on the cut ends of the heads prior to placing heads in boxes to remove the latex and soil particles that cling to the cut surface.

FIGURE 18.8   Hand harvesting of crisphead lettuce.

In many areas, lettuce is packed into 11 × 12.5 × 21.5 in. boxes designed to hold 24 heads. This size box may be packed with 18 large or 30 small heads. Packed boxes weigh from 45 to 55 lb or more.

In some areas, lettuce is packed in plastic crates or in wooden boxes. If these containers are reused, good sanitation (cleaning) practices must be followed so that harvested lettuce is not infected with bacteria, fungus, or foreign matter from previous harvests.

Some lettuce may be wrapped with porous or perforated clear plastic after harvest (Figure 18.10). This is usually done in the field. This lettuce is cut in the conventional way but trimmed to one wrapper leaf. Lettuce grown for processing is bulk packed. Harvested heads are placed in large bulk bins and transported to the vacuum cooler or to the processing plant. Growers often supply the bulk bins. After arrival at the processing plant, the heads are removed from the bins for processing. The empty bulk bins are immediately returned to the grower for use with future harvests. Inadequate storage area in the processing plant for storing the empty bins is a common problem.

Postharvest Handling

Proper postharvest handling is essential to maintain the quality of harvested lettuce. The lettuce head, separated from the root system, can no longer bring water and nutrients into the plant. From the moment of harvest all effort is directed toward reducing the demand for the limited reserves of water and nutrients in the plant. This prolongs the viability of the harvested head.

The estimated loss of lettuce at the wholesale, retail, and consumer levels is 21.3%. Lettuce production in tropical areas can result in 22 to 78% of postharvest wastage. General problems occur at each handling step during harvesting, packing, transporting, grading, storage, and retail.

FIGURE 18.9   A laborer cutting the lettuce head with a knife.

FIGURE 18.10 Wrapping of lettuce heads with clear plastic in the field prior to placing in boxes.

Temperature is perhaps the single most important factor in prolonging postharvest life. The higher the temperature, the faster the metabolic rate and depletion of reserves in the lettuce leaves. Ideally, lettuce is placed in auxiliary cooling to reach 33° to 36°F in the harvested lettuce for maximum shelf life. Auxiliary cooling is often not available at the field, but proper handling procedures in the field (removal from exposure to the sun and harvesting during cool parts of the day) can reduce high-temperature stress of lettuce.

Market Preparation and Marketing

Lettuce is one of the most perishable of the commercially shipped vegetables. Therefore, prompt cooling of the lettuce before shipping is desirable and the preferable temperature for shipping is 34°F. A thermostat setting in the truck of 32°F might cause freezing damage before the thermostat shuts off the cold airflow.

Uses

Lettuce is the major salad vegetable in North America and in much of the world. Lettuce is also used with all kinds of sandwich ingredients as well as being used to make hot or cold dishes more attractive.

Nutritional Value

Head lettuce is low in calories and is a source of vitamin A and vitamin C. One medium head has only 15 calories, with no grams of fat and 1 g of dietary fiber (Figure 18.11). This serving of lettuce provides 6% of the RDA of vitamin C, 4% of the RDA of vitamin A, and 2% of the RDA for both calcium and iron. The loose leaf (Figure 18.12) and Romaine (Figure 18.13) types have higher nutritional values (vitamin C, vitamin A, and calcium) than those of crisphead types. Every lettuce type is low in calories and aids in the digestion of other foods eaten with it.

Nutrition Facts

Serving Size 1/6 medium head (89g)

Amount Per Serving

Calories 15   Calories from Fat 0

	% Daily Value*
Total Fat 0g	0%
Saturated Fat 0g	0%
Cholesterol 0mg	0%
Sodium 10mg	0%
Total Carbohydrate 3g	1%
Dietary Fiber 1g	4%
Sugars 2g	
Protein 1g	

Vitamin A 4%	•	Vitamin C 6%
Calcium 2%	•	Iron 2%

FIGURE 18.11   Nutritional value of head lettuce. Percent daily values are based on a 2,000 calorie diet. Used with permission from Dole Food Co., Inc.

Nutrition Facts

Serving Size 1 1/2 cups shredded (85g)

Amount Per Serving

Calories 15   Calories from Fat 0

	% Daily Value*
Total Fat 0g	0%
Saturated Fat 0g	0%
Cholesterol 0mg	0%
Sodium 30mg	1%
Total Carbohydrate 4g	1%
Dietary Fiber 2g	8%
Sugars 2g	
Protein 1g	

Vitamin A 40%	•	Vitamin C 6%
Calcium 4%	•	Iron 0%

FIGURE 18.12   Nutritional value of loose leaf lettuce. Percent daily values are based on a 2,000 calorie diet. Used with permission from Dole Food Co., Inc.

Nutrition Facts

Serving Size 6 leaves (85g)

Amount Per Serving	
Calories 20	Calories from Fat 0

	% Daily Value*
Total Fat 0.5g	1%
Saturated Fat 0g	0%
Cholesterol 0mg	0%
Sodium 0mg	0%
Total Carbohydrate 3g	1%
Dietary Fiber 1g	4%
Sugars 2g	
Protein 1g	

Vitamin A 20%	•	Vitamin C 4%
Calcium 2%	•	Iron 2%

FIGURE 18.13　Nutritional value of Romaine lettuce. Percent daily values are based on a 2,000 calorie diet. Used with permission from Dole Food Co., Inc.

Celery

Scientific Name

Apium graveolens L. var. *dulce* (Mill.) Pers.

Common Names in Different Languages

celery, blanched celery (English); qin cai (Chinese) céleri à côte ś, c. à branchir (French); Bleichsellerie, Stangensellerie (German); sedano, s. da coste (Italian); serurii (Japanese); selderej čere škovyj (Russian); rungia (Spanish)

Classification, History, and Importance

Celery, *Apium graveolens* var. *dulce*, is a member of the Apiaceae family. *Graveolens* means strong scented and *dulce* means sweet. The characteristic flavor and odor of celery is due to volatile oils in the stems, leaves, and seeds. Other members of the family include carrots, parsnips, and parsley. Celery is a cool-season biennial.

Celery probably originated in the Mediterranean region. It grows wild in marshy places over a large area covering parts of Sweden south through Europe, Algeria, Egypt, Asia Minor, and the Himalayas. It also has been found growing wild in the southern tip of South America and New Zealand.

Celery was known to the Greeks and Romans as a medicine and as a flavoring. It was mentioned as "selinon" in Homer's *Odyssey* in 850 B.C. The Chinese also used celery as a medicine and it was mentioned in Chinese writings from the 5th century A.D. Cultivation of celery began in the 16th century in Italy and northern Europe

TABLE 18.6   *The top celery producing states in the United States*

Rank	State	Planted acres (000)	Harvested acres (000)	Yield (cwt/acre)	Production (000)
1	California	23.90	23.90	690.00	16,491
2	Michigan	2.30	2.10	490.00	1,029
3	Texas	0.90	0.90	590.00	531
4	Ohio	0.05	0.03	375.00	11
.	United States	27.15	26.93	671.00	18,062

Source: USDA, NASS, 1997 data.

and in 1623 celery was used as a food in France. In 1726, celery seed was sold in England for planting for use in soups and broths. In 1806 celery was listed as being grown in the United States.

Celery for fresh market was a $273.4 million industry in 1997 (see Table 18.1). This was up from $199.4 million in 1996. Of the 27,150 acres of celery grown in the United States, 23,900 acres are in California (Table 18.6). Other celery growing states include Michigan, Texas, and Ohio.

Plant Characteristics

Height: 16 in.
Spread: 8 to 12 in.
Root Depth: Roots are generally shallow and fibrous, limited to the upper 6 in. of soil, with some going as deep as 18 in.

Climatic Requirements

Celery requires a long, cool growing season, especially with cool nights. The optimum temperature range is 60° to 65°F. Although it is a cool-season crop, exposure of juvenile celery plants to temperatures below 40° to 50°F for more than 5 to 10 days can cause premature bolting (flowering). A steady water supply and proper amounts of nutrients are necessary for constant growth.

Field Preparation

Celery can be grown on most fertile, medium-textured mineral soils with irrigation. The best soils for celery have high levels of organic matter such as a fertile muck soil. Optimum pH for celery production on mineral soils is 6.0 to 6.5. On organic soils lime must be applied if soil pH is below 5.5.

Fields planted to celery are generally deep plowed to increase water-holding capacity of the soil. Celery uses large quantities of fertilizer and, in general, 200 to 250 lb of P_2O_5 and 600 lb of potassium per acre are broadcasted and incorporated before planting. Nitrogen is typically applied at 75 to 100 lb per acre. Approximately two to three side-dress applications of 50 lb N per acre are usually required during the growing season and growers often side-dress 4 to 6 weeks after transplanting and 3 to 4 weeks before harvest.

Selected Cultivars

Celery cultivars can be grouped into two distinct groupings: the yellow, or golden, and the green. The golden and green stalks are typically covered with soil to the tip of the petiole to blanch the stalks for tenderness and reduce puffiness. The darker green cultivars are more tender, fleshier, crisper, and sweeter and are the dominant ones grown for commercial production and home gardens.

An important breeding improvement of celery was the development of slow-bolting cultivar types that resist the temperature-triggered response to premature bolting. Green celery, termed pascal, may be divided into three types: Summer pascal—less susceptible to bolting, but smaller with a less compact head; Utah—the preferred standard, with large dense heads, susceptible to bolting; and Slow bolting—less susceptible to bolting than Utah but with fewer petioles and less attractive.

Planting

There are three methods of planting celery: (1) direct seeding, (2) growing seedlings in greenhouses or cold frames and transplanting to the field, and (3) growing seedlings in outdoor seedbeds and transplanting to the field. Direct seeding is only feasible when irrigation is available since it is necessary to keep the soil surface moist during germination and early growth.

About 56 to 70 days are required to produce seedlings for transplanting and plants are ready to transplant when they are 4 to 6 in. high and about 1/4 in. in diameter. Trimming the tops of the transplants may be practiced in hot weather. This trimming or topping allows more light to penetrate to the base and thus toughens the plant tissue.

Celery plants are spaced about 7 in. apart in single rows with 24-in. centers. Double-row plantings are planted into rows spaced 14 in. apart on beds with centers 40 in. apart and plants spaced 7 to 10 in. within row (Figure 18.14).

FIGURE 18.14 An established planting of celery.

Irrigation

Successful production of celery requires continuous growth. Irrigation immediately after transplanting is often done. Frequent irrigation of 1 to 1.5 in. of water per week during the growing season is needed. Irrigation during the 6 weeks prior to harvest is especially important due to rapid plant growth.

Culture and Care

Cultivation is required to control weeds and also to loosen soils.

Common Growing Problems and Solutions

Insects   Major insects of celery include aphids, leafhoppers, carrot weevils, flea beetles, armyworms, and loopers.

Diseases   Major diseases of celery include damping-off, root rot, pink rot, basal stalk lesions, early blight, late blight, bacterial blight, western celery mosaic virus, cucumber mosaic virus, aster yellows, fusarium yellows, and nematodes.

Physiological Disorders   Adequate irrigation can prevent black heart, a physiological disorder of celery. This disorder occurs when young leaves in the center of the plant do not get adequate water and calcium for proper growth. The resulting dead tissue turns black and often decays.

Harvesting

Celery is harvested when the petioles (stalks) from the soil line to the first node are at least 6 in. long (Figure 18.15). Plants must be compact and tight without excessive open space in the center of the stalk. Celery is harvested by cutting the taproot

FIGURE 18.15   Celery at the proper stage of maturity for harvesting.

FIGURE 18.16   Hand-harvested celery cut at ground level with a knife.

FIGURE 18.17   Mechanical harvest of a celery field.

below the ground. The crop is typically ready to harvest 85 to 120 days after transplanting, depending on the cultivar and the weather. Overmature plants are not harvested because they often contain cracked and pithy petioles.

Commercially grown celery can be either hand harvested or mechanically harvested. When celery is harvested by hand, the stalks are cut at ground level with a knife (Figure 18.16). The outer unmarketable petioles are removed and the stalks are placed on the conveyer of a mule train. They are then washed, graded, and packed into crates. Mechanical harvesters may cut off 10 or 12 rows of celery below the soil surface (Figure 18.17) and operate ahead of a mule train.

Postharvest Handling

Celery should be precooled to remove field heat as soon as possible after harvesting. Precooling of celery can be done by refrigeration, forced air, hydrocooling, or vacuum cooling. Proper storage conditions of celery are between 31° to 32°F and near 95% humidity. If stored at these conditions, the shelf life for celery is 2 weeks.

Market Preparation and Marketing

Celery is shipped with the butt upward to prevent water accumulation and butt discoloration.

Uses

Raw celery is used as a salad ingredient or as an appetizer. For salads, it is trimmed of leaves, washed, and chopped. As an appetizer, celery sticks are prepared as for salad and can be served alone or on a platter with any number of dips. When cooked, celery can be served as a main vegetable or, more often, as part of a vegetable dish. Celery is used extensively in Chinese cooking and to add crispness to vegetable sautés and other dishes. Celery leaves can be dried and used as an herb, as can celery seed, which is used in soups and pickles.

The caloric value of celery is very low, suggesting that celery is an ideal food for people that are dieting. Eating raw celery is also an excellent dental detergent.

Nutritional Value

Celery is low in calories and is a source of vitamin C, vitamin A, calcium, and iron. A serving size of two medium stalks has only 20 calories, with no grams of fat and 2 g of dietary fiber (Figure 18.18). This serving of celery also provides 15% of the RDA of vitamin C, 4% of the RDA of calcium, and 2% of the RDA for both vitamin A and iron.

Nutrition Facts

Serving Size 2 medium stalks (110g)

Amount Per Serving

Calories 20   Calories from Fat 0

	% Daily Value*
Total Fat 0g	0%
Saturated Fat 0g	0%
Cholesterol 0mg	0%
Sodium 100mg	4%
Total Carbohydrate 5g	2%
Dietary Fiber 2g	8%
Sugars 0g	
Protein 1g	

| Vitamin A 2% | • | Vitamin C 15% |
| Calcium 4% | • | Iron 2% |

FIGURE 18.18   Nutritional value of celery. Percent daily values are based on a 2,000 calorie diet. Used with permission from Dole Food Co., Inc.

Parsley

Scientific Name

Petroselinum crispum (Mill.) Nym. var. *crispum*

Common Names in Different Languages

parsley (English); yang yan sui (Chinese); persil, p. commun, p. friśe (French); Petersilie (German); prezzemolo (Italian); paserii (Japanese); petruška (Russian); perejil (Spanish)

Classification, History, and Importance

Parsley, *Petroselinum crispum,* is a member of the Apiaceae family. Other members of the family include carrots, parsnips, and celery. Parsley is a popular culinary biennial herb, commercially cultivated as an annual in many parts of the world. Parsley is grown for its attractive and aromatic leaves, which are used for flavoring and for garnishing.

Parsley is native to the Mediterranean region where it still can be found in the wild. It has been cultivated for thousands of years. Ancient belief was that the pleasing aroma of parsley absorbed the inebriating fumes of wine, thereby preventing intoxication. Parsley was said to be sacred to Pluto and, in the ancient days of Greece and Rome, a wreath of parsley was presented to the winners of many games.

In A.D. 79 parsley was widely used in salads. Parsley was introduced into England about 1548 where it was largely used for making parsley pies. Two kinds of parsley, the plain and common or curled-leaf, were mentioned as being grown in American gardens in 1806.

Plant Characteristics

Height: 9 to 18 in.
Spread: 6 to 9 in.

Climatic Requirements

Parsley is a cool-season crop that grows well in a temperature range of 45° to 60°F. Parsley is planted in many areas of the United States in the spring as soon as the ground can be worked, or planted in the fall for a winter crop in the southern United States.

Field Preparation

A rich moist soil with good drainage and a pH of 5.3 to 7.3 is preferred for growing parsley. The rate of fertilizer application depends on soil type and prior cropping history of the field site. On well-drained, lightly textured soils and on muck soils, higher fertilizer rates are frequently used. In these cases, generally one-third is broadcast fertilizer. Regardless of soil type a fine seedbed is required for parsley production. A rototiller and bedshaper are often used after plowing and disk harrowing to smooth the field and prepare it for seeding or transplanting.

Selected Cultivars

There are a number of cultivars available for commercial parsley cultivation. Common, or curled-leaf, parsley (var. *crispum*) (Figure 18.19) is used both fresh (primarily as a garnish) and dried (primarily in food products). Types of common parsley include Banquet, Dark Moss Colored, Decorator, Deep Green, Forest Green, Improved Market Gardener, Moss Curled, and Sherwood. Plain, flat-leaf or Italian, parsley (var. *neapolitanum* Danert) (Figure 18.20) is commonly used as a flavoring in sauces, soups, and stews. Italian types include the Plain type and Plain Italian Dark Green type. Hamburg or turnip-rooted parsley (var. *tuberosum* Crov.) is grown for its enlarged, edible root and is popular in specialized markets.

Planting

Parsley seeds are planted no deeper than 1/4 in. in raised beds. Parsley can be seeded with three or four rows, or in double rows on a 36- to 42-in. bed. Transplants are typically spaced 4 to 8 in. apart on 36-in. rows.

Irrigation

Irrigation needs of parsley are similar to those of other leafy green vegetables. Overhead sprinklers and drip irrigation are both successfully used in growing parsley.

Common Growing Problems and Solutions

Parsley is susceptible to the same growing problems as celery.

FIGURE 18.19   Common, or curled-leaf, parsley in the field.

FIGURE 18.20   Plain, or flat-leaf, parsley in the field.

Harvesting

Hand labor is the preferred method of harvesting parsley. This is done to obtain the lowest amount of crop damage acceptable for fresh market use. Bunches of parsley plants are grouped by hand and the stalks are cut with a knife. A rubber band is often slipped around the stalks to maintain bunch integrity. Multiple harvests either by hand or by machine are possible, depending on crop quality.

Postharvest Handling

Parsley is washed after harvest and any faded or yellowing leaves discarded. Parsley may be packed and shipped hydrocooled or with packaged ice to maintain crispness and fresh appearance. Optimum storage and handling temperatures for parsley are 32° to 36°F at 95% relative humidity. In these conditions, parsley can be held for up to 2 months.

Market Preparation and Marketing

Parsley is marketed as a fresh-market culinary green with important quality considerations including a healthy dark green color, a favorable aroma, freedom from cosmetic defects, a trueness to type, and long stalks for bunching.

The parsley leaves are washed and packed into crates or baskets, usually bunched or packed loose. While a percentage of the crop is dehydrated, most of the parsley grown in the United States is used fresh at restaurants.

Uses

Parsley foliage is used for garnishing and for flavoring soups, stews, gravy, and poultry stuffing. Parsley sprigs are probably the most popular of garnishes and are used largely with fish, meat, and poultry.

FIGURE 18.21   Nutritional value of parsley. Percent daily values are based on a 2,000 calorie diet. Used with permission from Dole Food Co., Inc.

Nutritional Value

While parsley is a source of vitamin C, vitamin A, and iron (Figure 18.21), it is not an important factor contributing to human nutrition since only small amounts of parsley are actually eaten. One tablespoon of chopped parsley has 0 calories and no grams of fat. This serving of parsley provides 6% of the RDA of vitamin C, 4% of the RDA of vitamin A, and 2% of the RDA for iron.

Miscellaneous Leafy Salad and Greens Crops

Specific Leafy Salad and Greens Crops

Cress   Cress (*Lepidium sativum* L., Brassicaceae family) comes in four types—common, curled, broadleaf, and golden. Curly cress, or peppergrass, germinates quickly and is ready to eat in about 10 days. Cress prefers cool weather or a slightly shady spot and forms little stems of frilly green leaves. The leaves are used for garnishing, adding pungency to salads and resembling a radish in flavor. Cress is harvested by cutting the stems while the plants are young and tender. Mature cress plants tend to become pungent.

Corn Salad   Corn salad (*Valerianella locusta* (L.) Laterrade em. Betcke, Valerianaceae family), also known as mâche, lamb's lettuce, or fetticus, has a mild, nutty flavor. Corn salad is a salad plant, but it may also be used as a cooking green. Since it does have a distinct flavor, it is often mixed with other greens such as mustard. Corn salad has rounded leaves that grow slowly into small rosettes of soft, buttery leaves. It needs a long, cool growing season and may do best as a fall crop. Since it is quite cold tolerant, it can overwinter in many areas providing early spring salads. Whole rosettes of the plants can be harvested if plants need thinning or individual leaves may be harvested for repeated harvests.

Endive and Escarole   Endive and escarole (*Cichorium endivia* L., Asteraceae family) are two popular salad vegetables in Europe. Endive has curly, deeply cut, lacy leaves with creamy inner leaves. Escarole has broad, coarse, crumpled leaves that blanch the inner leaves so they are crunchy yet tender. Cultural practices are the same as those used for lettuce. Endive and escarole are often planted early in the spring or in the fall since they may become bitter-tasting in hot weather.

Arugula   Arugula or rocket salad (*Eruca sativa* Miller, Brassicaceae family) leaves have a peppery/sweet tangy flavor. This green is easy to grow, but the young leaves are harvested frequently for best flavor. Arugula becomes bitter in hot weather. If the plants bloom, the flowers can be harvested and added to salads for color.

Mesclun   Mesclun is a mixture of many different salad greens grown and harvested together for an instant mixed salad (Figure 18.22). Commercial seed mixes are available or the grower can mix his own seed lot. Mesclun is harvested when individual leaves are bite-sized, with immature lettuce usually being the main ingredient. Cultivars in a particular seed mix are chosen for their color, texture, flavor, leaf size, and shape. Other ingredients, which vary by seed company, include endive, arugula, oriental greens, radicchio, mâche, mustard, and cress. Beet and chard thinnings may be included. Herbs such as fennel, sorrel, tarragon, basil, and mints are sometimes included in very small quantities. Some companies blend their mesclun mixtures for different flavors, such as tangy or mild. Generally the ingredients need cool growing temperatures and should be planted in early spring or for a fall garden.

FIGURE 18.22   A field planting of a mixture of mustard and turnip plants grown for greens. Courtesy of W. J. Lamont, Jr.

Chicory   Chicory terminology can be confusing because the same plant is used in three different ways and cultivars have specific uses. Chicory (*Cichorium intybus* L., Asteraceae family) is native to Europe but was brought to the United States in the 18th century. It has naturalized over much of the country. Young plants resemble dandelions but later the flower stem becomes bristly and bears rigid branches reaching 18 to 24 in. tall. The flowers are dandelion-like, sky blue, 1 to 1½ in. wide, and they close by midday. The roots can be dug and roasted and substituted for or mixed with coffee. Commercially, chicory grown for dried roots is planted in deep, rich soil in rows 18 in. apart. The parsniplike roots are dug in the fall, roasted, and ground.

Witloof Chicory   Witloof chicory, root chicory, or Belgian endive is the same plant treated differently. There are improved cultivars of Belgian endive/witloof chicory available from seed companies. The plants are grown in well-drained, deep, rich soil. In the fall, the tops are removed and the roots dug and put into cold storage for at least 3 months. After this dormancy period, the roots are planted in containers 18 in. deep at 55°F. For high-quality white buds or chicons, the newly planted roots must be kept in complete darkness. In about 3 weeks the pale yellow leaves will form cone-shaped buds about 6 to 8 in. long. The chicons are removed and the spent roots are dicarded. Since the chicons are grown in the dark, they, like blanched asparagus, have few vitamins and are not considered a good source of nutrition.

Radicchio   Radicchio (also *Cichorium intybus* L., Asteraceae family) is also known as red chicory. Radicchio is a popular salad vegetable and garnish in Europe. It has a distinctively bitter flavor. The beautiful red-and-white heads can grow from orange to grapefruit size. The leaves are slightly bitter like endive, but also sweeten slightly with cooler day temperatures. In the past, only a portion of radicchio plants made tight heads, but new cultivars, including some hybrids, have improved uniformity in color and heading. Radicchio is grown much like head lettuce and needs a long, cool growing period. Spring crops are started very early and transplanted to the garden when 3 to 4 weeks old. A fall crop is likely to be of higher quality than a spring crop. The fall crop is often direct-seeded about 85 days before the first average fall frost. The row is covered with a board or very light mulch to keep soil temperatures cooler while the seeds germinate. Long days and/or high temperatures can cause radicchio to bolt, increase in bitterness, and develop tipburn. Radicchio is frost tolerant, but growth will be slow in cold weather. Radicchio is used raw or lightly grilled or roasted and added to salads. Radicchio can also serve as a colorful garnish.

Beet Greens   Beet (*Beta vulgaris* L., Chenopodiaceae family) greens are leaves harvested from beet plants. These leaves are cooked and used as a side dish. The leaves may be harvested as thinnings from a young crop or as individual leaves from young plants. Leaves are of best quality when fully expanded or slightly earlier. Beet greens must be washed, graded, and bunched to be sold commercially. Storage can be only for short periods.

New Zealand Spinach   New Zealand spinach (*Tetragonia tetragonioides*, Tetragoniaceae family) has a flavor similar to but milder than common spinach. It is a heat-resistant, warm weather plant that is sensitive to frosts. The plant reaches a height of 1 to 2 ft and is amply branched. The 2 to 3 in. of new growth at the end of each branch, including tender shoots, tips, and leaves, are harvested and cooked like spinach. New growth will arise along cut branches. Whole plants can be harvested when small.

Edible Common Weeds Used for Salads or Greens

Purslane   Purslane (*Portulaca oleracea* L., Chenopodiaceae family) is a common edible weed that is high in omega-3 fatty acids and vitamin E. Some seed catalogs sell cultivated purslane seed that is reported to have better flavor than the common weed form found in gardens. Purslane is a succulent plant with fleshy, drought-tolerant leaves and stems. Whole young plants, and especially leaves and tender stem tips, can be used in salads. Purslane has a cool, citrusy flavor. When mixed with other greens, purslane adds crunch and texture to an otherwise routine salad. When planting cultivated purslane, the soils must be warm and the danger of frost passed. It does best in hot weather with full sun. Once sown in the field, the plant produces flowers containing thousands of tiny black seeds, thus reseeding itself for future crops.

Dandelions   Dandelions (*Taraxacum officinale* Wiggers, Asteraceae family) are a perennial weed that can be a good source of vitamins. Young leaves can be used raw in salads while older leaves are usually steamed or braised. Commercially, the intact plant is cut below leaf whorl. The leaves are an excellent source of vitamin A, vitamin C, calcium, and several other minerals. The yellow flower petals can also be added to salads, butters, or sauces to add color and interest. In some parts of the world, dandelion roots are roasted and used as a coffee substitute. The roots have also been used as a medicinal. Dandelion flowers are also used in wine making. Improved dandelion seed can be purchased through some seed catalogs. One French cultivar, Montmagny, has 8- to 9-in. leaves that taste like mild chicory. Commercial cultivars are selected for their large, tender leaves.

Lamb's-quarters   Lamb's-quarters (*Chenopodium album,* Chenopodiaceae family) can grow more than 4 ft tall. If allowed to grow large it can be difficult to pull or hoe. Lamb's-quarters are harvested when the plants are young and tender. The leaves have a taste similar to spinach and the plant is related to common cultivated spinach. The undersides of the leaves have a slightly rough texture and are best eaten cooked rather than raw in a salad.

Review Questions

1. What effect does temperature have on lettuce growth and development?
2. At what stage of plant development does most of the growth of lettuce occur?
3. What are some of the general types of lettuce?
4. What is tipburn in lettuce, and what environmental conditions favor tipburn formation?
5. What soils are best for growing celery?
6. What are the two general groupings of celery cultivars?
7. How is parsley harvested and marketed?
8. What are some of the uses of parsley?
9. What effect does hot weather have on the taste of arugula?
10. What is mesclun?
11. What common weeds may be used for greens or salads?

Selected References

Ball, J. 1988. *Rodale's garden problem solver: Vegetables, fruits, and herbs.* Emmaus, PA: Rodale Press. 550 pp.

Decoteau, D. R., D. Ranwala, M. J. McMahon, and S. B. Wilson. 1995. *The lettuce growing handbook.* Oak Brook, IL: McDonald's International.

Ells, J. E., H. F. Schwartz, and W. S. Cranshaw. 1993. *Commercial lettuce.* Colorado State Univ. Coop. Ext. Cir. 7.623.

Kays, S. J., and J. C. Silva Dias. 1995. Common names of commercially cultivated vegetables of the world in 15 languages. *Economic Botany* 49: 115–152.

McCarthy, W. H., and D. C. Sanders. 1994. *Commercial celery production in eastern North Carolina.* North Carolina Coop. Ext. Serv. Leaflet 27.

Meyers, C. 1991. *Specialty and minor crops handbook.* The Small Farm Center, Univ. of California, Division of Agriculture and Natural Resources Publ. 3346.

Neild, R. E., and R. D. Uhlinger. 1990. *Lettuce.* Univ. of Nebraska Coop. Ext. Serv. NebGuide G73-71-A.

Nonnecke, I. L. 1989. *Vegetable production.* New York: Van Nostrand Reinhold. 656 pp.

Peirce, L. C. 1987. *Vegetables: Characteristics, production, and marketing.* New York: John Wiley & Sons.

Roberts, E. W. 1960. Parsley. *Fruit & Vegetable Facts & Pointers.* United Fresh Fruit & Vegetable Association. Washington, DC.

Sackett, C., and J. Martin. 1977. Celery. *Fruit & Vegetable Facts & Pointers.* United Fresh Fruit & Vegetable Association. Washington, DC.

Sanders, D. C. 1996. *Lettuce production.* North Carolina Coop. Ext. Serv. Leaflet 11.

Schoneweis, S. D. 1995. *Lettuce and other salad crops.* Univ. Nebraska Coop. Ext. Inst. of Ag. and Nat. Resources Leaflet G95-1268-A.

Seelig, R. A. 1967. Lettuce. *Fruit & Vegetable Facts & Pointers.* United Fresh Fruit & Vegetable Association. Washington, DC.

Simon, J. E., J. Rabin, and L. Clavio. 1991. *Parsley: A production guide.* Purdue Univ. Coop. Ext. Serv.

Stephens, J. M. 1988. *Manual of minor vegetables.* Florida Coop. Ext. Serv. Bull. SP-40.

Yamaguchi, M. 1983. *World vegetables: Principles, production and nutritive values.* New York: Van Nostrand Reinhold. 415 pp.

Selected Internet Sites

USDA Agricultural Marketing Service Standards for Grades

Fresh
Beet greens   www.ams.usda.gov/standards/grnsbeet.pdf
Dandelion greens   www.ams.usda.gov/standards/grnsdand.pdf
Endive, escarole, or chicory   www.ams.usda.gov/standards/endive.pdf
Mustard and turnip greens   www.ams.usda.gov/standards/grnsmust.pdf
Celery   www.ams.usda.gov/standards/celery.pdf
Lettuce   www.ams.usda.gov/standards/lettuce.pdf
Parsley   www.ams.usda.gov/standards/parsley.pdf
Romaine lettuce   www.ams.usda.gov/standards/romaine.pdf

chapter
19

Perennial Crops

   I. Asparagus
  II. Rhubarb
 III. Globe Artichoke

Perennial crops are in the field for more than 2 years. Many perennial crops that are of commercial importance as vegetables have aboveground parts that are killed each year in the temperate regions but whose roots remain alive to send up shoots in the spring. Plants in this grouping include members from the Liliaceae (asparagus), Asteraceae (globe artichoke), and Polygonaceae (rhubarb) families.

Asparagus

Scientific Name

Asparagus officinalis L.

Common Names in Different Languages

asparagus (English); lu sun (Chinese); asperge (French); Spargel (German); asparago (Italian); asuparagasu (Japanese); sparža aptečnaja (Russian); esparrago (Spanish)

Classification, History, and Importance

Asparagus, *Asparagus officinalis,* is a member of the Liliaceae or lily family. The word *asparagus* is a Latinized form of the Greek *aspargos* or *asparagos,* meaning shoot or spear. *Officinalis* means medicinal. Other members of the family include onions, garlic, leeks, hyacinths, and gladiolas. Asparagus is a native of temperate Europe and Western Asia and has been known and prized by gourmets since Roman times. In the 17th century, King Louis XIV of France was so fond of asparagus that he made his gardeners build greenhouses so that he could enjoy asparagus year-round. Northern Europeans and

Britons have been eating asparagus for centuries. Asparagus was introduced into America in the 1600s.

Asparagus is a long-lived monocotyledonous, herbaceous perennial that produces spears each year. A well-planned row or bed can last for 15 years or more. The plant is composed of ferns, a crown, and the root system (Figure 19.1). The fern is a photosynthetically active modified stem called a cladophyll (Figure 19.2). The true leaves are the scalelike structures that form at the tip of the spear and down the stems. The crown is a series of rhizomes (underground rootlike stems) attached to the base of the main plant (Figure 19.3). New crown buds (Figure 19.4), from which spears (immature ferns) arise, are formed from the previous year. Larger buds generally result in larger spears. Bud size is influenced mainly by the plant's overall vigor the previous year.

Because asparagus production is seasonal and commercial production areas are limited, it is often priced as a gourmet item. Asparagus for fresh market was a $133.6 million industry in 1997 (Table 19.1). This was up from $103 million in

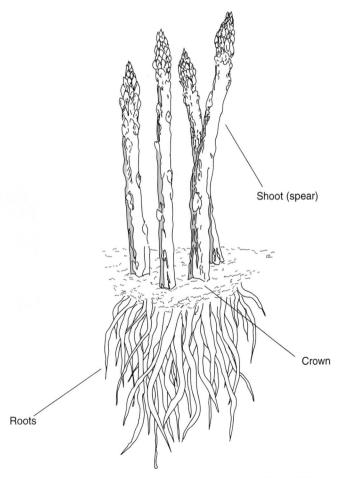

FIGURE 19.1 Diagram of an asparagus plant. Source: Iowa State University.

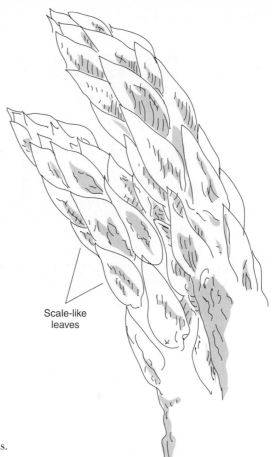

Scale-like
leaves

FIGURE 19.2 Diagram of
asparagus cladophylls, which are
photosynthetically active food stems.
Source: Iowa State University.

FIGURE 19.3 An asparagus plant
with the crown and roots exposed.
Courtesy of R. J. Dufault.

FIGURE 19.4   Spears arise from the buds of the crown. Courtesy of R. J. Dufault.

TABLE 19.1   *The production value of asparagus and globe artichoke in the United States*

	Value of production		
Crop	**1995**	**1996**	**1997**
	(1,000 dollars)		
Fresh market			
Artichoke[1]	61,965	65,416	67,620
Asparagus	124,171	103,480	133,653
Processing[2]			
Asparagus	52,999	53,143	47,571

Source: USDA, NASS.

[1] Includes processed and fresh market.

[2] Value at processing plant door.

1996. The asparagus for processing market is considerably smaller than for fresh market with a production value of $47.6 million in 1997. Per capita consumption of canning and frozen asparagus has decreased in the last couple of decades, while fresh asparagus per capita consumption increased from 0.4 lb. in 1976 to 0.6 lb. in both 1986 and 1996 (Table 19.2). Of the 77,600 acres of asparagus grown in the United States, 32,700 is in California, 24,000 in Washington, and 18,000 in Michigan (Table 19.3).

TABLE 19.2   *U.S. per capita consumption of commercially produced fresh and processing asparagus*

Use	1976 (lb, farm weight)	1986	1996
Fresh	0.4	0.6	0.6
Canning	0.5	0.3	0.2
Freezing	0.3	0.1	0.1
All	1.2	1.0	0.9

Source: USDA, Economic Research Service.

TABLE 19.3   *The top asparagus producing states in the United States*

Rank	State	Planted acres (000)	Harvested acres (000)	Yield (cwt/acre)	Production (000)
1	Washington	24.00	23.00	36.00	828
2	California	32.70	30.10	27.00	813
3	Michigan	18.00	17.50	15.00	263
4	New Jersey	1.00	1.00	23.00	23
5	Illinois	0.20	0.20	11.00	2
·	United States	77.64	73.54	27.00	1,979

Source: USDA, NASS, 1997 data.

Plant Characteristics

Height: 5 to 6 ft
Spread: 3 ft
Root Depth: 4 to 10 ft

Climatic Requirements

Asparagus does well in areas where winters are cool and the soil occasionally freezes at least a few inches deep. It is considered a hardy plant and grows best when the growing conditions include high-light intensity, warm days, cool nights, low relative humidity, and adequate soil moisture. Spear initiation and root growth begin when the soil temperature is above 50°F, and optimum productivity occurs at 78° to 85°F in the day and 55° to 66°F at night. High daytime temperatures during harvest will loosen the spear tip and develop fiber in the stem, both of which reduce marketable and eating quality.

Field Preparation

Asparagus can be grown in many soil types but deep loam or sandy soils with good surface water and air drainage provide the best production. Asparagus roots can develop to a 10-ft depth in well-drained soils. Asparagus will thrive in soils having a salt content too high for many other crops, but will not tolerate extreme acidity. Al-

FIGURE 19.5   A successful planting of asparagus requires good drainage and full sun.

though asparagus will tolerate less than optimum soil conditions, yields are likely to be reduced and the life of the planting will be shortened in these soils.

Asparagus occupies the land for many years so it is very desirable that the soil is fertile and free of troublesome weeds. Because it is more difficult to improve soil after crowns are planted, soil improving practices are usually started at least a year before planting. If soil pH is below 6.0 it is generally raised to 6.5 by applying lime. Asparagus thrives in soils well supplied with organic matter. Applications of animal manure or turning under (plowing into the soil) a green crop is often practiced to increase the organic matter in the soil.

For planting asparagus, a site is usually chosen that has good drainage and full sun (Figure 19.5). The plant bed is typically prepared as soon as possible prior to planting and enriched with additions of manure, compost, bone, or blood meal, leaf mold, wood ashes, or a combination of these. Asparagus is a medium-heavy feeder requiring high P, K, and organic matter at planting. Nitrogen is applied annually in late winter or early spring and is also side-dressed after harvest.

Depending on the soil fertility, the amount of N, P, and K application can vary. If the fertility in the soil is low, a typical recommendation may include 60 to 80 lb/acre, 30 lb/acre, and 30 lb/acre.

Once the plants are established, maintaining plant vigor becomes the primary objective of successful asparagus production. Asparagus has a very fleshy root system that is capable of storing large quantities of nutrients. It has been estimated that roots store 150 lb N/acre, 3 lb P/acre, and 20 lb K/acre. These stored nutrients are used for the development of spears in the spring. Asparagus is often not fertilized until after harvest and then it is top-dressed to encourage fern growth.

Selected Cultivars

Asparagus is dioecious, which means that is has both male and female plants. Females generally produce larger spears than males, but the males produce greater numbers of smaller-diameter spears. After the first year small red berries form on female plants in late summer. Breeding programs have been directed toward producing all-male lines, since this would reduce seed production by the plants. Unwanted seed production can result in a weed problem because some of the seeds germinate around the parent plant.

There are very few cultivars of asparagus. Cultivars in the past have been various strains of Mary and Martha Washington, mainly because of their rust-resistant characteristics. Hybrids and F_2 cultivars, such as UC157-F_2, are available with good resistance to rust and fusarium rot. Also, several all-male cultivars such as Jersey Giant and Jersey Knight have been developed that produce large uniform spears and high yields.

Planting

Asparagus is typically planted in early spring as soon as the ground can be worked. One-year-old crowns or plants are generally preferred. These young plants have compact buds in the center (crown) with dangling pencil-sized roots.

The crowns are placed in a trench that is 12 to 18 in. wide and 6 to 8 in. deep (Figure 19.6). The crowns are uniformly spaced 18 to 24 in. apart in an upright centered position and covered with 2 in. of soil. As the spears develop, soil is pulled over to cover a portion of the spear. This is done until the trench is filled. Gradual filling of the trench allows the bud to grow to the surface and the process of gradually filling the trench is usually completed by midsummer.

In addition to crowns, two other methods are available for establishing asparagus: direct seeding and transplanting. Direct seeding is generally not recommended and practiced. Using greenhouse-grown 10-week-old transplants is increasing in popularity, but a disadvantage of using transplants compared to crowns is the low yield 1 year from planting.

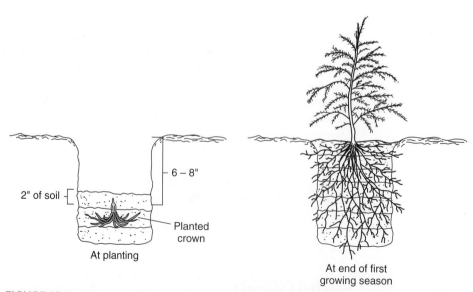

At planting

At end of first
growing season

FIGURE 19.6   Diagram of the trench used for planting asparagus. The crown is covered with 2 in. of soil after planting. As the spears develop the soil is gradually pulled over the plant until, by the end of the first year, the trench is filled in. Source: Iowa State University.

Irrigation

Irrigation is important to relieve stress of asparagus, particularly during the first two seasons after crown planting. An extended dry period early in the development of the fern after the cutting season is undesirable and may reduce yield the following year. Dry weather during the cutting season or late in the fern-growing season generally has little effect on yield.

Asparagus is very deep rooted and draws water from a large volume of soil. This allows the crop to withstand considerably dry weather. Deep rooting also allows longer intervals between irrigation applications than in other irrigated crops.

Asparagus roots must be kept free of too much moisture or else damage to the roots could occur. For this reason it is of great importance to plant asparagus on level land or, if the land is rolling, asparagus must not be planted in the low-lying areas.

Culture and Care

As the asparagus plants grow, they produce a mat of storage roots that orientate horizontally rather than vertically. During the first year, the top growth is generally spindly in appearance and, as the plant becomes older, the stems become larger in diameter.

Asparagus plants need stored nutrients and time to recover from harvests. They also need a weed-free environment, moderate soil fertility, and adequate soil moisture to build up food reserves in their crowns. Often neglecting asparagus fields after harvest contributes more to poor yields in subsequent years than insect or disease damage.

Common Growing Problems and Solutions

Weeds can be a major problem for asparagus. They compete with developing spears, often decreasing the yield and quality.

Insects   Asparagus beetles can rapidly ruin an asparagus planting if they are not controlled (Figure 19.7). The adult beetles are recognized by their bright, shiny coloration. They have bluish black wing covers with square yellow spots and red outer margins. When the adults lay eggs (dark brown to black and oblong-shaped) on spears the spears may become unmarketable because it is almost impossible to wash the eggs off. Both larvae and adults eat shoots and leaves but are devastating when they chew the tips of the spears, causing them to scar and turn brown. In addition to their feeding damage, larvae also secrete a dark fluid that stains the plant.

Cutworms can cause sporadic and occasionally severe damage to asparagus plantings. The adults are robust moths with dark-colored forewings and light-colored hind wings. Larvae are greasy-looking dark-colored caterpillars that feed at night. Young cutworms climb plants and feed on the spears and ferns of asparagus. Older larvae are active at night, feeding on spears at the soil surface and cutting them at ground level. Damage by cutworms is characterized by spears being cut partially or completely at the soil surface.

Aphids can cause damage to asparagus by extracting plant juices and causing the foliage of the plants to turn yellow. Another indicator of aphids on asparagus is the ants that are attracted to the sticky honeydew secreted by the aphids.

FIGURE 19.7   Asparagus beetle damage to a planting.

Diseases   Rust is a serious disease problem of asparagus. Soon after the cutting season, the spores of the rust disease create reddish-brown masses on the ferns of the plant. When these areas are touched, they give off a dusty cloud. In late summer, black masses of spores may be produced. The disease retards fern growth, causes early maturity, and reduces food storage in the crown of the plant. Diseased tops should be removed and burned.

Crown rot is one of the most serious diseases of asparagus. The crowns of diseased plants are discolored and have a low number of shoot buds. Feeder roots are frequently rotted and become hollow and limp. Spindly spears are produced in the spring by plants infected with crown rot while shoot growth during the summer is limited because shoots tend to wilt or turn yellow during hot weather. Crown rot is often associated with already-weakened plants. Factors that favor crown rot include overharvesting, planting on poor soils, repeated defoliation by asparagus rust, and infection by asparagus viruses.

Cercospora needle blight on asparagus is caused by a fungus, *Cercospora asparagi*. Affected ferns develop elliptical tan- to gray-colored spots. The lesions are encircled by a dark purple margin. Heavy infection causes premature defoliation in late summer or early fall and can result in decreased yields the following year. The disease development is favored by periods of high relative humidity or abundant rainfall and relatively warm temperatures.

Physiological Disorders   Weak, spindly plants and/or too few spears produced during the harvest may be indicators that the plants were harvested too soon after planting and before the crowns were mature enough to support a harvest, or that the crowns were affected by damage from rot or from inadequately prepared heavy soil. Asparagus spear tips that are curved may be due to damage from high winds and abrasion caused by windborne particles. If curving is excessive, the spear becomes unmarketable. Windbreaks and grain cover crops can reduce wind damage, particularly on sandy sites.

FIGURE 19.8   Asparagus spears at the proper maturity for harvest.

FIGURE 19.9   Using a knife designed for asparagus harvesting to cut the spear under the ground.

Harvesting

While asparagus can be harvested the second year after planting the crowns, it is usually not harvested for more than 1 month during the second year. During this time the plant is still expanding its feeder roots and storage root system. Excessive removal of spears will weaken the plant. Starting the third year, spears may be harvested for 8 to 10 weeks with harvesting ending when the diameter of the spear is less than that of a pencil.

Spears are harvested when they are 5 to 8 in. (Figure 19.8) and are usually cut or snapped from the plant. To cut a spear, a special pronged knife is run under the ground where the spear is emerging (Figure 19.9). The cutting is done carefully to

FIGURE 19.10　An asparagus field late in the summer.

avoid damaging developing spears and the crown below the soil surface. To snap a spear, it is bent from the top toward the ground. The spear will break at the point where it is free of fiber, generally just above the soil surface.

For high-quality asparagus, the spear must be cut or snapped while the head of the spear is tight and the buds under the scale below the head of the spear have not elongated. If the head of the spear is loose (ferned-out) it may indicate poor quality and that excessive fiber (toughness) has developed in the base of the spear.

When the harvest is over, the remaining spears are allowed to grow into an attractive fern (Figure 19.10). If asparagus has been overcut (too many spears harvested) in any given year the result is that usually smaller and fewer spears will be produced the following year. To correct this situation, growers will shorten the cutting season for at least one season to permit a longer, fuller fern production season to replenish the crown shoots.

Postharvest Handling

Asparagus deteriorates rapidly after harvest, giving off heat from respiration and losing moisture and quality if not handled properly. Loss of quality begins at harvest and exposure to temperatures from 90° to 100°F for even a few hours results in tremendous quality loss. Cold water baths in the packing sheds remove some of the field heat. Once packed, asparagus is rapidly cooled to 40°F by hydrocooling and maintained at that temperature or cooler (Figure 19.11). Asparagus that is held for less than 10 days is kept at 34° to 36°F. If held for more than 10 days and up to 3 weeks, the storage temperature should be 36°F to avoid chilling injury due to prolonged low-temperature exposure. Relative humidity should be 90 to 95% and cold rooms require good air circulation.

Following harvest, asparagus spears are very susceptible to market diseases such as soft rot (*Erwinia carotovora*), fusarium stem rot, and phytophthora rot. The latter two diseases can be effectively controlled by holding asparagus at 32° to 36°F during this period.

FIGURE 19.11   Harvested asparagus in pyramid-shaped boxes in a cooler prior to shipping.

Market Preparation and Marketing

Asparagus is kept near 32°F when displayed for retail sale. In addition, spears are also occasionally sprinkled with cold water to preserve quality for a longer period of time. Displaying the asparagus with their butts (or cut ends) standing in trays of ice water also helps maintain quality.

Blanched or white asparagus is a variation of the traditional asparagus. White asparagus refers to the white color of the harvested spears. This whiteness is achieved through a blanching process that is accomplished by mounding the soil over the plants in the early spring. The soil covering allows the asparagus plant to grow without exposure to light thereby preventing the development of chlorophyll, which does not occur in the dark. The soil that is mounded around the plant is leveled when the harvest is over and the ferns are allowed to develop. The asparagus produced by this technique is almost perfectly smooth and rounded at the tip, with no visible bract development. These spears also tend to be thicker and more tender and have a more subtle flavor than asparagus spears grown in the traditional manner.

Uses

Asparagus is served in salads, soups, hot dishes, and in combination with many sauces. Fresh steamed asparagus is most often served with butter or a cream sauce. About 40% of asparagus is consumed fresh, 50% canned, and the rest frozen.

Nutritional Value

Asparagus is low in calories and is a good source of vitamin A and vitamin C. One serving of five spears has only 25 calories, with no grams of fat and 2 g of dietary fiber (Figure 19.12). This serving of asparagus also provides 15% of the RDA of vitamin C and 10% of the RDA of vitamin A.

Nutrition Facts

Serving Size 5 spears (93g)

Amount Per Serving

Calories 25   Calories from Fat 0

% Daily Value*

Total Fat 0g	**0%**
Saturated Fat 0g	**0%**
Cholesterol 0mg	**0%**
Sodium 0mg	**0%**
Total Carbohydrate 4g	**1%**
Dietary Fiber 2g	**8%**
Sugars 2g	
Protein 2g	

Vitamin A 10% • Vitamin C 15%

Calcium 2% • Iron 2%

FIGURE 19.12   Nutritional value of asparagus. Percent daily values are based on a 2,000 calorie diet. Used with permission from Dole Food Co., Inc.

Rhubarb

Scientific Name

Rheum rhabarbarum L.

Common Names in Different Languages

rhubarb, pie plant (English); tang da huang (Chinese); rhubarbe, rhapontic (French); Rhabarber, Rhapontic (German); rabarbaro (Italian); rubaabu, kara daaio (Japanese); reven' ogorodnyj (Russian); ruibarbo, rapóntico (Spanish)

Classification, History, and Importance

Rhubarb, *Rheum rhabarbarum*, belongs to the Polygonaceae family. The Polygonaceae family is also called the buckwheat family, with rhubarb and buckwheat the only members of the family that are of commercial importance. The sourness of the juice is a characteristic of the family.

Rhubarb is a cool-season, herbaceous perennial grown for its large tender leaf stalks. The leaves of rhubarb contain a high concentration of oxalic acid salts that can be very toxic. The leaf stalk has less oxalates and those present are insoluble and therefore not dangerous for human consumption. During World War I, rhubarb leaves were recommended as a substitute for green vegetables. Many instances of acute poisoning resulted and death occurred in some cases. Rhubarb contains a highly irritant anthraquinone, a glycoside. This compound rather than oxalate is the suspect.

Rhubarb is a native of the cooler areas of Asia, probably Siberia. The earliest records of rhubarb date back to about 2700 B.C. in China or Mongolia. The root of rhubarb was used as a medicine. Dioscorides, who was the physician to Anthony and Cleopatra, recommended rhubarb for diseases of the liver and weaknesses of the stomach. He used it to cure ringworm by applications of a combination of roots in vinegar.

Rhubarb as we know it today was introduced into Europe from the East about the 1600s. By the 1700s, rhubarb was reportedly grown in Europe and England and used as a filling for tarts and pies. A Maine grower who obtained seeds or rootstalks from Europe grew rhubarb in the United States as early as 1800. He supposedly introduced rhubarb to gardeners in Massachusetts where its growth spread, and by 1822 it was being sold in produce markets in Massachusetts.

Plant Characteristics

Height: 4 to 6 ft
Spread: 3 to 5 ft

Climatic Requirements

Rhubarb is a cool-season, herbaceous, perennial crop that grows best when mean summer temperatures are below 75°F or when winter mean temperatures are below 40°F. It is typically grown in the northern tier of the United States in a region extending from Maine south to Illinois and west to Washington. The plant requires temperatures below 50°F to break dormancy. Rhubarb is not easily grown in the southern United States because high temperatures disrupt growth. Cultivars that exhibit good pink or red petiole color during the cool season usually become very green when temperatures remain above 85°F. The vegetative parts of the plant are killed at 26° or 27°F.

Field Preparation

Rhubarb will grow in almost any type of soil as long as it is well drained. Rhubarb prefers deep, fertile loams and does best in slightly to moderately acid soil. When new plantings are made, the soil is deeply plowed and worked down to a fine seedbed. Liberal amounts of fertilizer are typically incorporated into the soil before planting. Typical amounts of fertilizer include 1,500 lb of 10-10-10 fertilizer per acre broadcast before planting. Annual applications of fertilizer are usually needed. Before spring growth, usually 400 lb of ammonium nitrate and 200 lb of muriate nitrate are worked into the soil.

Selected Cultivars

Rhubarb can be grown as an outdoor plant and as a forcing plant. Cultivars of rhubarb may also be chosen according to the color of their petioles. For outdoor production, the cultivar MacDonald has large red-stalks. It is vigorous and upright growing. Valentine is another red-stalked cultivar that is popular for growing in home gardens. For forcing, Timperley, Early, Victoria, Crimson Red, and Sutton are used. Victoria, a green-stalked cultivar, is a heavy producer that is excellent for commercial purposes. Sutton is a pink-stalked cultivar.

Planting

Rhubarb is usually propagated by dividing the crowns formed during the previous season. Crowns are divided in late fall or early spring. Plants must be divided and re-set every 4 years to keep the bed in vigorous condition. Plants not divided may become large and stalks may become more numerous than desired. Seedstalks should be removed as they appear.

Typical spacing for a commercial rhubarb planting is rows 4 ft apart with plants 4 ft apart in row. Trenches are dug prior to planting, generally 6 in. deep. Crown pieces are placed 2 in. below the surface of the soil in the trench. Soil is pulled over the soil and firmed. Good drainage is essential for growing rhubarb. Home garden–grown rhubarb may be planted on raised beds to ensure good drainage and guard against rotting of the crowns.

Irrigation

The rhubarb plant responds well to moisture, although reliable yields can be obtained with minimal watering.

Culture and Care

Rhubarb grown for forcing is planted in a field for 2 years in which the stalks and petioles are not harvested. After 2 years of growth the crowns are ready to be dug for forcing. In order to get the best response to forcing, the rhubarb plant must be exposed to a period of low temperatures to break dormancy once the forcing conditions are started under cover. The crowns and roots must be dug before the soil freezes, taken to a sheltered storage, and held where dehydration and water loss are at a minimum until the forcing operation begins (usually early the next year). When ready for forcing, the crowns are packed together on 4- to 5-ft wide beds with soil and water added to initiate growth, and then placed in the dark. The optimum forcing temperature is around 50°F. Regular watering and near constant temperature in the forcing shed are the basic requirements. Because of the dark conditions, the stalks are quite long, bright pink, and very tender. The first harvest occurs after about a month. Harvest lasts 6 to 8 weeks with one or two harvests per week. Once used the roots are discarded in order to prevent spread of diseases. A new batch of rhubarb plants is used each year for forcing.

In outdoor production, rhubarb is not harvested the first year and only for a short time during the second year to permit the roots to build up food reserves. Beginning with the third year, a harvest period of 8 to 10 weeks is typical under ideal growing conditions. Seedstalks are removed whenever they appear. Rhubarb can also be grown in greenhouses under a full light-forcing regime.

Common Growing Problems and Solutions

Grass weeds can be controlled with glyphosate or Round-up, or with Grammaxone or Paraquat.

Insects   The rhubarb cucurlio is a serious pest. This cucurlio is a rusty snout beetle about ¾ in. long that bores into stalks, crowns, and roots. Destruction of wild dock plants in the area of rhubarb plantings helps in control. Other insects are potato borers and nematodes.

Diseases   Crown rot and virus diseases are ever-present diseases of rhubarb. The removal of diseased plants from any planting is a major control method of both these types of diseases. Phytophthora crown rot is probably the most serious disease affecting rhubarb. Slightly sunken lesions develop at the base of the stalk and enlarge rapidly to cause collapse of the whole leafstalk. Rhubarb stalks may continue to collapse in moist, warm weather until the plant is killed.

Harvesting

The harvest season for field production of rhubarb is short, rarely extending for more than 2 months. The stalks are usually picked twice a week when they reach 12 to 20 in. long. Harvest of the petioles of rhubarb should be delayed until the second year of growth, and only a few should be removed during the second year. This allows the plant to produce food for good crowns and roots. Only the best quality and largest stalks are harvested.

The leafstalks separate readily from the crown and are easily harvested by grasping them near the base and pulling slightly to one side. The stalks can also be cut at ground level. No more than two-thirds of the developed stalks should be removed at any one time. As stalks are removed for commercial harvest, the leaves are trimmed to within 1 in. of the top of the stalk and left in the field. Some rhubarb is mechanically harvested (Figure 19.13).

Forced rhubarb is a very delicate, perishable commodity requiring great care in handling. Any rough handling detracts from appearance.

Postharvest Handling

Rhubarb can be stored for 2 to 4 weeks at 32°F and high relative humidity.

FIGURE 19.13   Mechanical harvesting of rhubarb. Courtesy of M. D. Orzolek.

Market Preparation and Marketing

Harvested rhubarb stalks are washed and cut to a length of 18 in. Three to five stalks weighing about 1 lb are often bunched and marketed. Rhubarb displayed for retail sale is kept near 32°F and displayed with the base end of the stalk standing in trays of ice water or on shaved ice.

Uses

Rhubarb has long been used in pies and is commonly referred to as the pie plant. The petioles are chopped and used in pies and in mixes with other fruits to make preserves. Rhubarb is also used in tarts, sauces, puddings, punch, jams, and jellies. It is easily prepared and preserved and readily adapts to freezing.

Nutritional Value

Rhubarb is low in calories and is a source of vitamin C. One ½-cup serving of diced rhubarb has only 15 calories, with no grams of fat and 1 g of dietary fiber (Figure 19.14). This serving of rhubarb also provides 8% of the RDA of vitamin C and 4% of the RDA of calcium.

Globe Artichoke

Scientific Name

Cynara scolymus L.

Nutrition Facts

Serving Size 1/2 cup diced (61grams)

Amount Per Serving	
Calories 15	Calories from Fat 0
	% Daily Value*
Total Fat 0g	**0%**
Saturated Fat 0g	**0%**
Cholesterol 0mg	**0%**
Sodium 5mg	**0%**
Total Carbohydrate 2g	**1%**
Dietary Fiber 1g	**4%**
Sugars 1g	
Protein 1g	
Vitamin A 0%	• Vitamin C 8%
Calcium 4%	• Iron 0%

FIGURE 19.14 Nutritional value of rhubarb. Percent daily values are based on a 2,000 calorie diet. Used with permission from Dole Food Co., Inc.

Common Names in Different Languages

globe artichoke (English); chao xian ji (Chinese); artchaut (French); Artischoke (German); carciofo, articiocche (Italian); aachi chohku (Japanese); artišok (Russian); alcachofero, alcachofa (Spanish)

Classification, History, and Importance

Globe artichoke, *Cynara scolymus,* is a member of the Asteraceae family. This family also includes lettuce, endive, aster, sunflower, thistles, and other cultivated and weedy species. The artichoke is a thistlelike herbaceous perennial that can be grown from seed (achene) or from vegetative material (stumps or crown pieces). Artichokes are grown for the flower heads, which are harvested before they bloom (Figure 19.15). If the artichoke flower is allowed to bloom, large, attractive heads of violet, blue, or white flowers are the result (Figure 19.16).

The globe artichoke probably originated in the Mediterranean area. It is native to the central and western Mediterranean area and was introduced into Egypt and farther eastward at least 2,000 years ago. The artichoke was used as a cooked vegetable and as a salad. It was very popular in 2nd century Rome, often considered an expensive luxury food. Little of artichoke was mentioned during the Dark Ages. It reappeared in Italy in the 15th century. It was during this time that the artichoke was grown and harvested much the way that it is now. Artichokes were introduced into France in 1548 and were first mentioned in the United States in 1806. Introduction into the United States was primarily through French immigrants in Louisiana and Italian immigrants in California. Castroville, California is now considered the primary area of production in the United States.

Artichokes are grown on all the continents of the world. Countries bordering the Mediterranean Sea grow nearly 95% of the world's artichokes, with Italy, Spain,

FIGURE 19.15   The marketable part of the globe artichoke plant is the immature flower (bud), which is harvested before the flower blooms.

FIGURE 19.16   Globe artichoke flower at full bloom.

TABLE 19.4   *Artichoke production in the United States*

Rank	State	Planted acres (000)	Harvested acres (000)	Yield (cwt/acre)	Production (000)
1	Washington	9.10	9.10	95.00	865
.	United States	9.10	9.10	95.10	865

Source: USDA, NASS, 1997 data.

TABLE 19.5   *U.S. per capita consumption of commercially produced fresh artichokes*

Use	1976 (lb, farm weight)	1986	1996
Fresh	0.4	0.6	0.5

Source: USDA, Economic Research Service.

and France producing over 80% of the crop. Italy is the largest producer and consumer of artichokes.

Production in the United States is restricted along the central California coast and is about a $67.6 million industry (Table 19.1) that is produced on 9,100 acres (Table 19.4). About 66% of U.S. production is for fresh consumption. The rest is processed primarily as marinated hearts and bottoms. Per capita consumption of fresh artichokes increased from 1976 to 1986 (Table 19.5).

Plant Characteristics

Height: 4 to 5 ft
Spread: 5 to 6 ft

Climatic Requirements

Artichokes require a frost-free area with cool, foggy summers. Under optimum growing conditions, the plant produces compact tender buds. Periods of hot weather may cause artichoke buds to open quickly and ruin the tenderness of the edible parts. Cold weather also damages the artichoke. While commercial production is mainly restricted to California, artichokes are grown in gardens in many other areas of the United States.

Field Preparation

Artichokes can be grown in a wide range of soils. Preferred soils in California are deep, fertile, well-drained sandy loams to clay loams. Fertilizer is typically applied in the range of 150 to 300 lb/acre of N, 20 to 40 lb/acre of P, and 25 to 85 lb/acre of K. All of the P and K and most of the N are applied after plants have been cut back at the end of the previous cutting period, or when a new planting is established. Two to three side-dress applications are put in furrows or through sprinkler irrigation water.

Selected Cultivars

There are very few cultivars of globe artichoke grown around the world. In the United States there is only one commercially important cultivar, Green Globe, that is responsible for almost all of the artichokes grown. Other available but less important cultivars include Desert Globe, Big Heart, and Imperial Star.

Planting

Propagation of globe artichoke is done primarily by using basal stem pieces with attached root sections called *stumps* or *crown pieces*. Buds at the base of these stumps develop roots and shoots. Most fields in the United States and around the world are planted by hand. In California, planting is usually done from June through September. Planting densities are relatively low (approx. 6,000 plants/acre), with spacings of 8 × 9 ft and 4.5 × 10 ft.

The plant grows to a height of 3 to 4 ft or more and spreads to cover an area about 6 ft in diameter (Figure 19.17). The plant sends up seasonal shoots from a permanent crown. The number of shoots will vary with age and health of the plant.

Irrigation

Globe artichokes require a relatively constant supply of water. In commercial production, frequent irrigations are often necessary. Furrow irrigation is commonly practiced when fields are flat. Although artichokes require constant water, they are not tolerant of flooding conditions.

FIGURE 19.17   An established planting of globe artichoke plants. The plants spread to cover a large area.

Culture and Care

The artichoke plant will grow for 15 years or longer in the same location, but commercial growers often replant with stumps every third year so that they can obtain a more vigorous plant and a finer marketable head. After the last harvest of the season, the plants are cut back below the ground and are not watered for several weeks so that they are dormant during part of the summer. Fertilizer, particularly nitrogen, is added after the dormancy period and the plant is irrigated. Rapid growth occurs, and new stems with new buds develop for fall production. New shoots then come up from the underground stump or crown. Several buds develop on each flower stalk. The largest flower bud is found on the tip of the main stem, with two medium buds lower on the stem and several smaller buds farther down.

Common Growing Problems and Solutions

Various types of aphids attack artichokes. Slugs and snails may also be a problem. Plume moth larvae attack the bracts destroying the plant base. There are no serious diseases of artichoke except the viral diseases spread by aphids.

Harvesting

Globe artichokes are harvested by cutting the stem below the base of the bud (Figure 19.18). Harvesting usually occurs every 4 to 10 days in California. After harvesting, the buds are separated into sizes usually determined by bud-count-per-shipping-package.

FIGURE 19.18   Globe artichokes are harvested by cutting the stem below the base of the bud.

Postharvest Handling

Harvested buds are typically cooled from below 40°F to close to 32°F by hydrocooling in commercial packing facilities. Damaged or injured buds are removed prior to cooling.

Market Preparation and Marketing

Harvested artichoke buds are packed in boxes lined with perforated film, in waxed cartons, polyethylene bags, and other methods to reduce water loss. Since the bracts are subject to rapid moisture loss, it is essential to maintain high humidity either by constant top icing or by refrigerated shelves with fine sprinkling capability, to keep the artichokes at peak level. Store shelves are typically refrigerated at about 40°F to preserve high-quality artichokes.

Uses

Artichokes, which have a delicate nutty flavor, can be eaten entirely or by pulling each leaf off and dipping it into a sauce. Artichokes can also be served as a hot vegetable with butter or special sauce, or served cold.

Nutritional Value

Artichokes are fat free, low in sodium, cholesterol free, a good source of fiber, and a source of vitamin C. One artichoke has only 25 calories and 3 g of dietary fiber (Figure 19.19). This serving of artichoke also provides 10% of the RDA of vitamin C, and 2% of the RDA for vitamin A, calcium, and iron.

FIGURE 19.19   Nutritional value of globe artichokes. Percent daily values are based on a 2,000 calorie diet. Used with permission from Dole Food Co., Inc.

Review Questions

1. Why is it important to plant asparagus in fields that are fertile and free of troublesome weeds?
2. What effect does the sex of the asparagus plant have on spear production?
3. What are some of the methods by which asparagus is planted?
4. What indicator can be used to determine when the asparagus harvest should be terminated during the season for a plant?
5. Why is rhubarb not grown in the southern United States?
6. How is rhubarb grown for forcing?
7. Where are most of the world's globe artichokes produced?
8. Where is the principal region in the United States for the production of globe artichokes and why?
9. How are globe artichokes propagated?

Selected References

Ball, J. 1988. *Rodale's garden problem solver: Vegetables, fruits, and herbs.* Emmaus, PA: Rodale Press. 550 pp.

Camp, W. H., V. R. Boswell, and J. R. Magness. 1957. *The world in your garden.* National Geographic Society Press. Washington, DC.

Hutchinson, W. D., F. L. Pfleger, C. J. Rosen, L. B. Hertz, V. A. Fritz, J. A. Wright, and R. L. Burrows. 1991. *Growing asparagus in Minnesota.* Minn. Coop. Ext. Serv. Publ. FO-1861-GO.

Kays, S. J. and J. C.Silva Dias. 1995. Common names of commercially cultivated vegetables of the world in 15 languages. *Economic Botany* 49:115–152.

Marr, C. W., N. Tisserat, and K. McReynolds. 1993. *Asparagus.* Kansas Coop. Ext. Serv. Publ. MF-1093.

Motes, J., B. Cartwright, and J. Damicone. 1996. *Asparagus production.* Oklahoma Coop. Ext. Serv. OSU Extension Facts F-6018.

Nonnecke, I. L. 1989. *Vegetable production.* New York: Van Nostrand Reinhold. 656 pp.

Pennel, J. T. 1976. *Rhubarb production.* USDA, ARS Leaflet 555.

Rowland, W. A. 1969. Rhubarb. *Fruit & Vegetable Facts & Pointers,*. United Fresh Fruit & Vegetable Association. Washington, DC.

Ryder, E. J., N. E. De Vos, and M. A. Bari. 1983. The globe artichoke (*Cynara scolymus* L.). *HortScience*18:646–653.

Seelig, R. A. and P. F. Charney. 1967. Artichokes. *Fruit & Vegetable Facts & Pointers.* United Fresh Fruit & Vegetable Association. Washington, DC.

Wolf, J. A. 1977. *Rhubarb.* Purdue Univ. Hort. Leaflet HO-97.

Yamaguchi, M. 1983. *World vegetables: Principles, production, and nutritive values.* New York: Van Nostrand Reinhold. 415 pp.

Selected Internet Sites

USDA Agricultural Marketing Service Standards for Grades

Fresh

Artichokes   www.ams.usda.gov/standards/artichoke.pdf
Asparagus   www.ams.usda.gov/standards/asparagu.pdf
Rhubarb   www.ams.usda.gov/standards/rhubarb.pdf

Processing

Asparagus   www.ams.usda.gov/standards/vpar.pdf

chapter

20

Root Crops

I. Carrot
II. Beet
III. Radish
IV. Turnip and Rutabaga
V. Sweet Potato

Root crops have a prominent, fleshy underground structure. The underground structure (depending on the crop) may be a root, tuberous root, or a hypocotyl with a taproot forming below the hypocotyl. The marketable underground structure of most root crops generally has a long storage life. Major root crops come from the Apiaceae (carrot), Chenopodiaceae (beet), Brassicaceae (turnip and rutabaga), and Convolvulaceae (sweet potato) families.

Carrot

Scientific Name

Daucus carota L. ssp. *sativus* (Hoffm.) Arcang.

Common Names in Different Languages

carrots (English); hu luo bo (Chinese); carotte (French); Mohre (German); carota (Italian); ninjin (Japanese); moskov' kul'turnaja (Russian); zanahoria (Spanish)

Classification, History, and Importance

The carrot, *Daucus carota*, is a biennial member of the Apiaceae family. Other members of the family include dill, caraway, fennel, anise, parsley, parsnip, and celery. The carrot is a biennial plant grown for its taproot. The carrot root is the hypocotyl and crown (Figure 20.1).

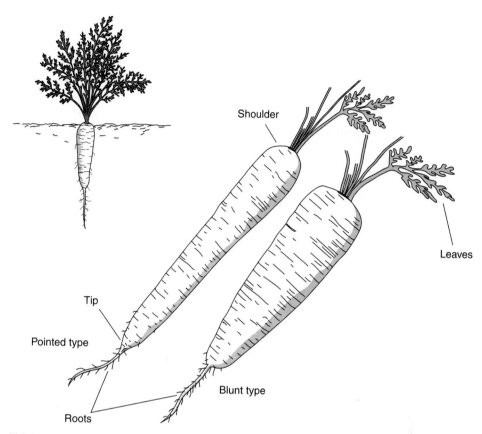

FIGURE 20.1   Diagram of a carrot plant. Source: Iowa State University.

Carrots arose as a natural variation of the Eurasian wildflower Queen Anne's lace. Centuries ago, a wild variety of Queen Anne's lace was found in Afghanistan with a thickened sweet-tasting root. It is thought that this plant evolved to become the carrot. When garden carrots were brought to North America, it is believed that a few plants reverted to the wild type and escaped to establish Queen Anne's lace as an introduced wildflower across much of the continent.

Carrot roots were purple for the first few hundred years of cultivation. Orange roots appeared as a chance mutation mentioned in the mid-1700s and quickly became the favorite in Germany, the Netherlands, and England.

Carrots for fresh market was a $441 million industry in 1997 (Table 20.1). This was up from $355.8 million in 1996. The carrot for processing market is considerably smaller than for fresh market with a production value of $37.5 million in 1997. 1996 per capita consumption of fresh and frozen carrots has increased compared to 1976 amounts, while the amount of canned carrots has decreased (Table 20.2). Of the 99,000 acres of carrots grown in the United States for fresh market sales, 70,900 acres are in California (Table 20.3). Of the 23,800 acres of carrots grown in the United States for processing, Washington has 6,600 acres, Texas 4,800 acres, and Wisconsin 3,600 acres.

TABLE 20.1 *The production value of selected root crops for fresh market and processing in the United States*

Crop	Value of production		
	1995	1996	1997
	(1,000 dollars)		
Fresh market			
Carrots	394,356	355,829	441,193
Processing[1]			
Beets	8,814	8,092	8,136
Carrots	46,443	38,098	37,447

Source: USDA, NASS.

[1]Value at processing plant door.

TABLE 20.2 *U.S. per capita consumption of commercially produced fresh and processing carrots*

Use	1976	1986	1996
		(lb, farm weight)	
Fresh	6.4	6.5	10.2
Canning	1.9	1.0	1.6
Freezing	1.7	1.8	2.8
All	10.0	9.3	14.6

Source: USDA, Economic Research Service.

Plant Characteristics

Height: 12 to 18 in.
Spread: 12 to 24 in.
Root Depth: Most of the plant's roots are usually limited to the upper 2 ft of soil, but the taproot measures up to 5 ft long.

Climatic Requirements

Carrots are a hardy, cool-season crop that can be planted as soon as the soil can be worked in the spring. Carrot seed germinates at soil temperatures of 40°F or higher. Young carrot seedlings can withstand light frosts, but hard frosts heave the soil and break the taproots, which can result in stubby and forked carrots.

Carrots require about 90 to 120 days to mature, depending on location, cultivar, and climate. While they are growing, carrots require relatively large amounts of moisture and are not drought tolerant. Prolonged hot weather, particularly in later stages of development, not only retards growth but also induces undesirable strong flavors and coarseness in the roots. Exposure to temperatures below 55°F tends to make the roots longer, more slender, and paler in color than is typical. The best temperature range for highest quality roots is between 60° and 70°F.

TABLE 20.3   *The top carrot-producing states for fresh market and processing in the United States*

Rank	State	Planted acres (000)	Harvested acres (000)	Yield (cwt/acre)	Production (000)
Fresh Market					
1	California	70.90	70.90	360.00	25,524
2	Colorado	5.50	4.90	500.00	2,450
3	Michigan	5.50	5.30	250.00	1,325
4	Florida	6.80	6.50	185.00	1,203
5	Washington	2.20	2.10	430.00	903
6	Arizona	2.40	2.40	265.00	636
7	Texas	3.70	3.50	165.00	578
8	New York	0.65	0.60	280.00	168
9	Minnesota	0.50	0.40	400.00	160
.	United States	99.03	97.46	345.00	33,599
Processing				Tons/Acre	Tons
1	Washington	6.60	6.40	30.00	192
2	California	2.80	2.70	32.04	87
3	Texas	4.80	4.20	20.00	84
4	Wisconsin	3.60	3.10	22.30	69
5	Michigan	1.60	1.50	25.00	38
6	New York	1.50	1.50	17.00	26
7	Minnesota	1.10	1.00	20.43	20
8	Oregon	0.35	0.35	25.66	9
.	United States	23.85	22.20	24.84	551

Source: USDA, NASS, 1997 data.

Field Preparation

Carrots prefer a deep, loose, well-drained sandy loam soil. On clay soils, carrots produce more leaves and forked roots. Stony, cloddy, or trash-laden soils increase the incidence of root defects. Soil preparation is particularly important with carrots since the seeds are small and early seedling growth is slow. As a result, the soil must be worked to produce a firm, well-pulverized seedbed. Raised beds are helpful when more drainage is needed. Soil pH should be 5.8 to 6.8 and fertilizer should be applied according to soil-test recommendations.

Selected Cultivars

There are four classes of carrots. These are based on root shape and date of maturity. The classes are Danvers, Imperator, Nantes, and Chatenay.

Danvers   Roots of the Danvers class are medium to long with broad shoulders that taper toward the tip (Figure 20.2). This type was developed in Danvers, Massachusetts in the 1870s.

FIGURE 20.2   A carrot cultivar of the Danvers class.

FIGURE 20.3   A carrot cultivar of the Imperator class.

Imperator   Roots of the Imperator class are more slender at the shoulder than the Danvers types and usually slightly longer (Figure 20.3). They taper smoothly from top to bottom and are widely grown for winter market consumption in which a long smoothly tapered root is desired by consumers. Imperator carrots are late maturing and good for storing.

Nantes   Roots of the Nantes class have a nearly cylindrical shape and almost no taper with the shoulder and tip having rounded ends (Figure 20.4). They are medium to long, early maturing, and usually eaten fresh in summer. They originated near Nantes, France.

FIGURE 20.4   A carrot cultivar of the Nantes class.

FIGURE 20.5   An established planting of carrots.

Chatenay   Chatenay types, named for the region in France where they originated, tend to be medium to short and tapered with a blunt end. They also mature by mid-summer.

Planting

Carrots are direct-seeded into the fields. Seeds of carrots are placed ⅛ to ⅜ in. deep. A special plant shoe called a "spreader shoe" is often used for direct-seeding carrots. This shoe plants a band of carrots 2 in. wide.

Single- or double-row beds of carrots are usually 36 to 42 in. on center (Figure 20.5). Maximum yields per acre have been obtained with three rows spaced

18 in. apart on a 48-in.-wide bed (5 to 6 ft on center). Crusting of the soil around the seedbed can contribute to poor seed germination and seedling establishment. In areas where this may be a problem, covering the soil with vermiculite or a fine compost and keeping the soil evenly moist until the seedlings have emerged can enhance carrot seedling establishment. Seeded carrots in the field are thinned to 2 to 3 in. apart. Pelletized seed is also used for carrots, making them easier to plant.

A general recommendation for growing carrots in medium fertility soil would be to broadcast 50 lb of ammonium nitrate and 40 lb of 8-24-24 per acre. Another 75 to 100 lb of ammonium nitrate may be needed if the soil becomes heavily leached.

Irrigation

Carrots require an evenly distributed and plentiful soil moisture supply throughout the growing season. Irrigation is used to prevent soil drying and crusting. In general, ½ to ¾ in. of water is applied every 4 to 7 days until carrots emerge. After emergence, the seedlings are irrigated weekly for the first 3 weeks. After the crop is established, irrigation is used as necessary for optimum plant growth and root development. However, too much moisture toward the end of the season will cause roots to crack.

Culture and Care

Weed control is particularly important when carrot seedlings are young since they are slow growing and cannot compete with weeds. Frequent shallow cultivations are used to control the weeds and keep the soil surface loose.

Common Growing Problems and Solutions

Insects   Carrot root flies, flea beetles, leafhoppers, armyworms, and carrotworms feed on carrots. Carrot root flies are maggots that feed on and destroy the roots. Flea beetles, which chew small round holes in leaves, spread rapidly and can destroy the entire crop if not checked.

Diseases   Aster yellows is a common disease of carrots. A carrot infected with this disease will have multiple tops, yellowish leaves, and a large number of very small hairlike roots emerging from the shrunken taproot. This disease is caused by a viruslike organism vectored to the plant by a leafhopper. Aster yellows seldom affects more than one or two carrots in an area of the field. Infected plants are removed and destroyed to prevent spreading of the disease.

Physiological Disorders   Forked carrot roots (Figure 20.6) may be due to rocky or stony soil, or heavy soil. Carrot plants with full-foliage top growth and small or limited roots can result from seeding too close in the field and not thinning. Excessive nitrogen fertilization can also contribute to extensive top growth at the expense of root growth. Poor-tasting carrots (tasteless, woody, or bitter) often result when carrots are grown under inappropriate environmental conditions while the root is maturing. Pale yellow coloring of the root also can be an indication of poor environmental conditions during growth.

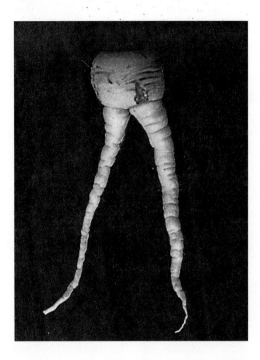

FIGURE 20.6   Forked carrot roots due to rocky or heavy soil.

Harvesting

Carrots for processing are usually a minimum of 1½ in. at the shoulder and 4 or more inches long. Fresh market carrots are harvested when they reach ¾ to 1½ in. in diameter. Generally, the smaller carrots are juicier and more tender.

Postharvest Handling

Carrots will keep for 4 to 5 months in 90 to 95% relative humidity at 32°F. If a carrot loses as little as 5% of its moisture, its texture often becomes rubbery.

Market Preparation and Marketing

Carrots are often topped and all misshapen or injured specimens sorted out before being placed in storage. Under good conditions topped carrots can be stored for 4 to 5 months. Bunch carrots may be stored at 32°F for 10 days to 2 weeks, and those shipped to distant markets are often iced.

Packages of "baby-cut" or "mini-cut" carrots are available in most supermarkets. Many of these are simply mature carrots that were cut into short pieces and mechanically whittled down to give them the appearance of "baby" vegetables. This process chips off some of the outer flesh, which is often sold for use in dried soup mixes.

Uses

Carrots are a versatile vegetable that can be served alone, raw or cooked, or in combination with meats or other vegetables. They may be boiled as a main vegetable dish, roasted with meats, baked, fried, sautéed, and pickled. Carrots are generally used for lending body and seasoning to soups, sliced raw for hors d'oeuvres, grated or sliced in salads, and sliced into thin sticks for snacks.

FIGURE 20.7   Nutritional value of carrots. Percent daily values are based on a 2,000 calorie diet. Used with permission from Dole Food Co., Inc.

Nutritional Value

Carrots are low in calories and are an excellent source of vitamin A. One 7-in.-long carrot has only 35 calories, with no grams of fat and 2 g of dietary fiber (Figure 20.7). This serving of carrot also provides 270% of the RDA of vitamin A, 10% of the RDA of vitamin C, and 2% of the RDA for calcium.

The reputation that eating carrots is good for one's eyesight is based on the high carotene content of carrots. Each carotene molecule can be cleaved in the human digestive system to form two molecules of vitamin A, which is important for human vision. Since vitamin A deficiency also results in night blindness in some people, carrots were found to be the ideal vegetable for alleviating this condition long before medical reasons were understood.

Because vitamin A is a fat-soluble vitamin, a surplus of vitamin A is stored in the liver. Fish also store vitamin A in their liver; thus generations of children were often given a daily dose of cod liver oil to compensate for the low natural carotene levels in wintertime diets, which were partially contributed to the relatively low availability and consumption of carrots. Modern marketing practices for many vegetables and fruits have made crops such as carrots available year-round, and fewer doses of cod liver oil are used as a supplement to winter diets.

Beet

Scientific Name

Beta vulgaris L. Crassa group

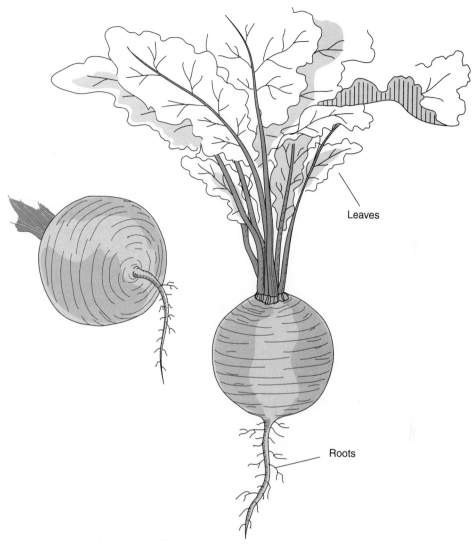

FIGURE 20.8   Diagram of a beet plant. Source: Iowa State University.

Common Names in Different Languages

garden, table or red beet (English); gen tian cai (Chinese); betterave rouge, b. potagére (French); Rote Rübe, Rote Beete (German); bietola da orta, barba bietola, d'insalata (Italian); biito, shokuyōbīto, kaensai (Japanese); svekla stolovaja (Russian); remolacha hortelana, r. roja, r. de mesa (Spanish)

Classification, History, and Importance

The garden beet or red beet, *Beta vulgaris,* is a biennial member of the Chenopodiaceae family. The root is the marketable product and is composed of hypocotyl and crown (Figure 20.8). The garden beet is closely related to Swiss chard, sugar beets, and mangels. Mangels are also known as stock beets and are considered too coarse for human consumption, and instead are grown for animal feed.

Beets are thought to have originated in Europe and North America and spread eastward from the Mediterranean area in prehistoric times. Wild beets are found around the Mediterranean, Asia Minor, and the Near East. Originally only the leaves of beets were used as a potherb or green.

The first accounts of an improved beet were recorded in Germany about 1558 and in England about 1576. Prior to this time, beet roots were used in cooking, since recipes for cooking beet roots exist from the 3rd century. All through the 17th and 18th centuries very few kinds of garden beets were known. In the United States in 1806 only one cultivar was listed.

The beet is one of the more popular home garden vegetable crops. Most of the commercial production of beets is for processing. Beets for processing was an $8 million industry in 1997 (see Table 20.1). Per capita consumption of canned beets has declined steadily from 1.8 lb in 1976 to under 1.0 lb in 1996 (Table 20.4). Of the 7,800 acres of beets grown in the United States for processing, almost half are in Wisconsin (Table 20.5).

Plant Characteristics

Height: 6 to 12 in.
Spread: 4 to 8 in.
Root Depth: Most roots to 2 ft, some reaching down to 5 ft

Climatic Requirements

Beets prefer a cool climate, withstanding cold weather short of severe freezing temperatures, but are not harmed by spring or fall frosts. Temperatures of 60° to 65°F and bright sunny days are ideal for beet plant growth and development. They are also somewhat tolerant of short periods of warm temperatures, but their roots may become tough during hot weather. Beets are biennials requiring a period of cold temperature of 40° to 50°F for 2 weeks or longer for flower initiation.

TABLE 20.4   *U.S. per capita consumption of commercially produced processing beets*

Use	1976	1986 (lb, farm weight)	1996
Canning	1.8	1.2	0.9

Source: USDA, Economic Research Service.

TABLE 20.5   *The top beet-producing states for processing in the United States*

Rank	State	Planted acres (000)	Harvested acres (000)	Yield (tons/acre)	Production (000)
1	Wisconsin	3.60	3.30	15.17	50
2	New York	2.70	2.70	15.00	41
.	United States	7.78	7.46	16.38	122

Source: USDA, NASS, 1997 data.

Field Preparation

Beets can tolerate a wide range of soil types and grow best in loose, well-drained soils. Beets are extremely sensitive to soil acidity. A low soil pH results in stunted growth. Preferred pH is between 6.2 to 6.8, but they can tolerate a pH between 6.0 to 7.5.

Selected Cultivars

Beets may be red, white, or yellow and vary in shape from oblate to long and cylindrical. The oblate to globe-shaped, red-rooted types are most popular in the home garden.

Planting

Beets are fairly frost-hardy and can be planted 30 days before the frost-free date. Although beets grow well during warm weather, the seedlings are established more easily under cool, moist conditions.

Beet seed is actually a dried fruit or seed ball containing several tiny seeds. Heat, drought, or soil crusting can interfere with seed germination and emergence. Beets are seeded ½ in. deep and in rows 12 to 18 in. apart in a well-prepared seedbed. Seeds are spaced 1 in. apart in rows and when the emerging beet seedlings are 2 in. tall, they are thinned to about one plant per inch. In an established planting, the seedlings are often thinned to about 3 to 4 in. between plants (Figure 20.9). These thinned beets can be used as greens.

Irrigation

Beets are very responsive to surrounding soil moisture conditions. Beet root quality is reduced if the plants become stressed by lack of water or from waterlogging due to poor drainage.

FIGURE 20.9   An established planting of beets.

Culture and Care

Frequent shallow cultivation is important in the growing of good-quality beets since beets compete poorly with weeds. Because beets are extremely shallow-rooted, hand weeding and early, frequent, and shallow cultivations are the most effective methods of controlling weeds in a row.

Common Growing Problems and Solutions

Insects   The most common insect pests of beets are leaf miners, aphids, and leafhoppers. Extensive damage can also be caused by flea beetles and webworms.

Diseases   Diseases are relatively minor problems in beet production, and the most common disease problems are leaf spot and curly top virus. Occasional infestations of downy mildew and root rot may also develop.

Physiological Disorders   Poor or no root formation in beets may be due to overcrowding. Woody root and poor coloring of the root are most often due to inappropriate environmental conditions during growth.

Harvesting

Beets are harvested as soon as the roots are large enough to use (Figure 20.10). Beet roots are usually ready to harvest 8 to 9 weeks after the seeds are sown. Roots are most tender when less than 2 in. in diameter. As beets get larger they tend to get more fibrous. When harvested, topped beets typically are left with at least 1 in. of foliage on the root to avoid bleeding during cooking. Bunched beets are graded into similar size and appearance. Dead or damaged leaves are removed. Beet greens are best when 4 to 6 in. tall.

The diameter of beets for processing ranges from 1 to 4 in. Smaller sizes are used for whole-packed products and larger sizes are used for sliced or diced products. Accordingly, the payment to a grower differs according to size as well as condition, with smaller, high-quality sizes generally commanding the higher prices.

Postharvest Handling

Beets store best at 32°F with 95% humidity. Under these conditions, beets with tops can be stored for 10 to 15 days. Topped beets store considerably longer.

Market Preparation and Marketing

Fresh beets can be bunched with or without tops. Processed beets packed whole, sliced, or cubed probably represent the largest marketable uses of beets. Also, considerable quantities of beets are pickled.

Uses

Beet tops may be cooked or served fresh as greens. The roots may be pickled for salads or cooked whole, sliced, or diced. Beets are also added to soups and beet juice is used for coloring in many different products. Beets are the primary ingredient of borscht (a Russian stock soup).

FIGURE 20.10   Beet roots harvested and placed on a wagon in the field.

Nutrition Facts

Serving Size 1 medium beet (100g)

Amount Per Serving

Calories 50   Calories from Fat 5

	% Daily Value*
Total Fat 0.5g	1%
Saturated Fat 0g	0%
Cholesterol 0mg	0%
Sodium 150mg	6%
Total Carbohydrate 11g	4%
Dietary Fiber 2g	9%
Sugars 6g	
Protein 1g	

Vitamin A 0%	•	Vitamin C 4%
Calcium 0%	•	Iron 0%

FIGURE 20.11   Nutritional value of beets. Percent daily values are based on a 2,000 calorie diet. Used with permission from Dole Food Co., Inc.

Nutritional Value

Beets are low in calories and are a source of vitamin C. One medium-sized beet has only 50 calories, with no grams of saturated fat and 2 g of dietary fiber (Figure 20.11). This serving of beet also provides 4% of the RDA of vitamin C. Beet tops (greens) are a good source of vitamin A and a better source of most minerals and vitamins than the roots.

Radish

Scientific Name

Raphanus sativus L. Radicula group

Common Names in Different Languages

radish (English); ying tao luo bo (Chinese); radis, ravonet (French); Garten-Rettich, Radies, Radieschen (German); rafano, ravanello, radice (Italian); hatsuka daikon (Japanese); rabanito, rabanete, rabanillo (Spanish)

Classification, History, and Importance

The radish, *Raphanus sativus,* is a member of the Brassicaceae family. Other members of the family include cabbage, cauliflower, kale, Brussels sprouts, turnips, and horseradish.

The name of the genus, *Raphanus*, is a Latinized form of the Greek *raphanos*, which means "easily reared." Radishes are cool-season, fast-maturing, easy-to-grow annual or biennial herbaceous plants that are grown for their roots (Figure 20.12).

China is believed to be the center of origin for the radish. Radishes were common food in Egypt before the building of the pyramids. The ancient Greeks prized radishes so much that they made small replicas of them in gold. Greeks in the third century B.C. wrote of radishes and one Greek physician devoted an entire book to the plant. Roman writers at the beginning of the Christian era described small, mild, early, round, and large forms of radishes. Other types were described that weighed several pounds each. In 1544 a German botanist reported seeing radishes that weighed 100 lb. Radishes were common in England in 1586 and were among the first European crops introduced into America by the Spaniards and grown by the early colonists. By 1629 they were being cultivated in Massachusetts.

Garden radishes are extremely popular in home gardens because they can be grown almost anywhere there is sun.

Plant Characteristics

Height: 6 to 8 in.
Spread: 6 in.

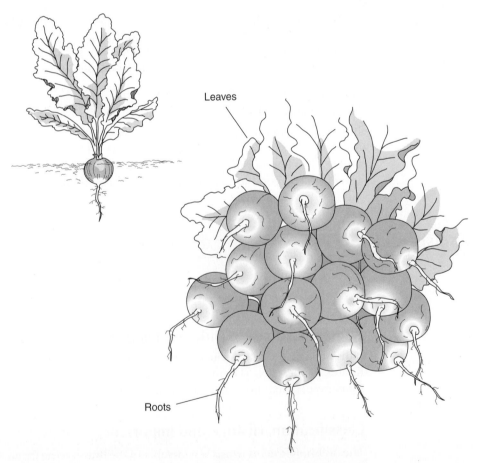

FIGURE 20.12   Diagram of radish plant. Source: Iowa State University.

Climatic Requirements

Radish is a cool-season vegetable that does not grow well in hot weather. Radishes are frost tolerant and emerge quickly when soil temperatures exceed 60°F. Radishes generally mature 3 to 5 weeks after planting.

Field Preparation

Since radishes grow so rapidly, a rich fertile soil is essential. For an early crop, sandy or sandy loam soils are preferred. Organic or moist soils generally give better results during the warmer, drier seasons of the year. Radishes are intolerant of poorly drained soils and soils used for growing radishes should be free of stones, clods, lumps, and undecayed matter. Thorough seedbed preparation is essential to ensure uniform depth of planting. Soil pH should be at least 6.0.

Spring radishes are sown in 6- to 10-in. rows, frequently 3 to 5 rows per bed. In-row spacing is 1 to 1½ in. apart. Daikon and winter radishes are sown 2 to 6 in. in-row, depending on root size. Seeds are placed ½ in. deep and uniform seed depth is important for uniform maturity.

Selected Cultivars

Radishes are available in three distinct types: spring (or summer) types are round, elongated French breakfast forms that require 25 to 35 days; daikon types are white elongated radishes that require 40 to 50 days; and winter radish types require 50 to 60 days and come in several shapes (often larger and longer shaped than the spring types) and colors and can be stored exactly the same as carrots.

The most popular spring cultivar types are those that have bright red or red-and-white round roots such as Cherry Belle, Early Scarlet Globe, Scarlet Prince, Champion, Comet, Cherry Beauty, and Red Boy. White Icicle is the most common and popular long-rooted spring type (Figure 20.13) The most popular winter cultivars such as China Rose, White Chinese, Long Black Spanish, and Round Black Spanish are usually grown as fall crops in many regions of the United States for winter storage.

Planting

Radishes can withstand cold weather and can be planted early in the spring in many areas of the United States. The seeds are drilled ½ in. deep in rows 8 to 12 in. apart. Often there are 3 to 5 rows per bed. Seeds that have been graded to uniform size can assist in ensuring uniform emergence and growth. A general fertilizer recommendation for radish is 50 lb of nitrogen, 100 lb of P_2O_5, and 100 lb of K_2O per acre before planting.

Irrigation

Radishes require 1 in. of water per week and anything that restricts the rapid growth of radishes can result in pithy, cracked, harsh-tasting roots.

Culture and Care

Radishes must be grown in environmental conditions that favor rapid growth for the roots to be mild, tender, and attractive. Slow growth or checked growth often results in roots that are tough, woody, pithy, and pungent.

FIGURE 20.13   White Icicle cultivar of radish is a long-rooted spring type.

Generally, no chemical weed control is needed due to the short growing season. Mechanical cultivation is kept shallow to avoid root damage.

Common Growing Problems and Solutions

Insects   The major insects of radish are root maggots, aphids, and flea beetles, although other insects found on cole crops may be found on radish occasionally.

Diseases   Radishes share many diseases with other members of the Brassicaceae family such as clubroot, black leg, rhizoctonia root rot, and phytophthora. Wire stem, pythium, powdery and downy mildews, black spot, black rot, yellows, and white rust are the more serious of the fungal and bacterial diseases. Radishes are also susceptible to nematode damage.

Physiological Disorders   Radishes that are too old will crack and split making them unmarketable. If roots fail to develop the radishes may have been planted too thick and not thinned, or the weather may be too hot. Radishes will not bulb properly when forced to mature during temperatures above 80°F. Off-flavored radishes are caused by planting at the wrong time or poor cultural practices such as low fertility or low moisture resulting in slow growth.

Harvesting

Most of the commercial radishes are harvested by machines. The radish lifter lifts the entire plant, removes the top, and drops the root in a container. This means that there is only one harvest per planting.

Some growers bunch radishes for sale. These radishes are pulled and tied in the field. About 8 to 12 roots are put in one bunch. The bunch is firmly tied with tape, string, twist-ties, or rubber bands. The ideal size for radish roots is ¾ to 1¼ in. in diameter. The quick-maturing spring varieties become pithy and pungent if they are not harvested as they reach edible size.

Postharvest Handling

Radishes must be cooled as soon as possible to 33°F with high humidity. Radishes remain in edible condition for only a short time (up to 2 to 3 weeks in refrigeration) before they become pithy and spongy and bitter tasting.

Market Preparation and Marketing

Commercial radishes harvested by machines are frequently marketed in transparent bags. Bunched radishes are packed in baskets, crates, or hampers, and iced for market.

Uses

Radishes are eaten fresh either whole or sliced in salads or as a side item.

Nutritional Value

Radish is low in calories and is a good source of vitamin C. One serving size of 7 radishes has only 15 calories, with no grams of fat (Figure 20.14). This serving of radish provides 30% of the RDA of vitamin C, and 2% of the RDA for calcium.

Nutrition Facts

Serving Size 7 radishes (85g)

Amount Per Serving

Calories 15	Calories from Fat 0

	% Daily Value*
Total Fat 0g	**0%**
Saturated Fat 0g	**0%**
Cholesterol 0mg	**0%**
Sodium 25mg	**1%**
Total Carbohydrate 3g	**1%**
Dietary Fiber 0g	**0%**
Sugars 2g	
Protein 1g	

Vitamin A 0%	•	Vitamin C 30%
Calcium 2%	•	Iron 0%

FIGURE 20.14   Nutritional value of radish. Percent daily values are based on a 2,000 calorie diet. Used with permission from Dole Food Co., Inc.

Turnip and Rutabaga

Scientific Name

Brassica rapa L. var. *rapa* (DC.) Metzg. (Turnip)
Brassica napus L. *napobrassica* (L.) Reichb. (Rutabaga)

Common Names in Different Languages

turnip (English); wu jing (Chinese), navet, n. potager (French); Mairube, Speis-erübe (German); rapa, r. coltivata (Italian); kabu (Japanese); turneps (Russian); nabo (Spanish)

rutabaga, swede, swede turnip (English); rui dian wu jing (Chinese); rutabaga, navet de Suède (French); Steckrübe (German); rutabaga, navone (Italian); rutabaya (Japanese); brjukva (Russian); rutabaga, nabo succo, n. de Suecia (Spanish)

Classification, History, and Importance

Turnips (*Brassica rapa*) and rutabagas (*Brassica napobrassica*) are members of the Brassicaceae family. Other members of the Brassicaceae include broccoli, Brussels sprouts, cabbage, cauliflower, cabbage, and mustards. Rutabagas are also known in Europe as Swedes or Swedes turnip.

Turnips and rutabagas are among the most commonly grown and widely adapted root crops. They are similar in plant size and general characteristics. Turnips are usually light green, thin and hairy, while rutabagas are bluish green, thick and smooth. The roots of turnips generally have little or no neck and a distinct taproot, while rutabagas are often more elongated and have a thick, leafy neck and roots originating from the underside of the edible root as well as the taproot. Rutabagas are higher in total dry matter and total digestible nutrients than turnips. Neither turnip nor rutabaga is as high in sugar as table beets.

Turnips and rutabagas have been grown for nearly 4,000 years and have spread all over the world from their original home in temperate Europe. They reached Mexico in 1586, Virginia in 1610, and New England in 1628.

Plant Characteristics

Turnip
Height: 10 to 15 in.
Spread: 6 to 8 in.

Rutabaga
Height: 15 in.
Spread: 15 in.

Climatic Requirements

Turnips and rutabagas are cool-season, frost-hardy crops that have their best root growth during relatively low-temperature (40° to 60°F) conditions. Both crops are biennials, and if an extended period of cool weather occurs after turnips or rutabagas are planted, a seedstalk may form (bolt) rendering the root as unsalable.

Turnips are easy to grow and will mature in about 2 months. Rutabagas grow less rapidly than turnips, requiring an additional 4 weeks longer to mature.

Field Preparation

A moderately deep, highly fertile soil with a pH of 6.0 to 6.5 is best for growing turnips and rutabagas. A general recommendation for turnips and rutabagas is 40 to 60 lb of N, 40 to 60 lb of P_2O_5, and 60 to 100 lb of K_2O per acre.

Selected Cultivars

Turnip and rutabaga cultivars differ mainly in the color and shape of their roots. There are white- and yellow-fleshed cultivars of both crops, although most turnip cultivars are white fleshed and most rutabaga cultivars are yellow fleshed.

Turnip

- Purple-Top, White Globe—bright purple crown, white below the crown, 5- to 6-in.-diameter globe, leaves dark green and cut (Figure 20.15). It can be grown for both greens (Figure 20.16) and roots.
- Just Right F_1—white root, 7- to 8-in.-diameter flattened globe, light green leaves that are deeply cut.
- Seven-Top—a winter-hardy green used mostly for greens with dark green foliage.

FIGURE 20.15   Root of a Purple-Top, White Globe cultivar of turnip.

FIGURE 20.16   Turnips in a field grown for greens. Courtesy of W. J. Lamont, Jr.

Rutabaga

- American Purple-Top—deep purple color, yellow below the crown, globe-shaped roots 5 to 6 in. in diameter with yellow flesh, medium blue-green cut leaves (Figure 20.17).
- Laurentian—purple crown, light yellow below crown, globe-shaped roots 5 to 5½ in. in diameter with yellow flesh, medium blue-green cut leaves.

Planting

Turnips and rutabagas are planted in multiple rows on a raised seedbed. Seedbeds can range from 3 to 5 ft wide. Seeds are typically drilled ½ in. deep, 4 in. apart in-row, with rows 12 to 15 in. apart. After emergence, plants are thinned to 3 or 4 in. apart in-row.

If turnips are planted for greens, the tops are harvested when they are 4 to 6 in. tall.

Irrigation

Turnips and rutabagas require an abundant supply of moisture and most soils require 1.5 inches of water every 7 to 10 days for optimum plant growth and production.

Culture and Care

Cultivation is often needed to control weeds during the early growth of turnips and rutabagas. If cultivation is used to control weeds that emerge, it is typically shallow (less than 2 inches deep) so as to not disrupt the root growth of the planted crop.

FIGURE 20.17   American Purple-Top rutabaga in the field.

Common Growing Problems and Solutions

Insects   Turnip aphids, flea beetles, root maggots, and wireworms are serious pests of turnips and rutabagas.

Diseases   Clubroot, root knot, leaf spot, white rust, anthracnose, and alternaria are common disease problems of turnips and rutabagas.

Physiological Disorders   Conditions that result in slow growth or stress can cause the leaves of turnips and rutabagas to acquire bitter, off-flavors.

Harvesting

Turnips and rutabagas are of best quality (mild and tender) when they are medium sized. Turnip roots are harvested for bunching when they are 2 in. in diameter. Turnip roots that are topped are harvested when 3 in. in diameter. Rutabagas are generally harvested when roots are 4 or 5 in. in diameter. Turnips with tops are washed and tied in bunches of about four to six plants. Topped turnips and rutabagas for the general market are sold either by volume or weight. "Topping" is the removal of the leaves from the fleshy roots. The roots are commonly packed in transparent film bags for individual consumers. Topping is recommended for sales in most wholesale and retail outlets.

Postharvest Handling

Turnips and rutabagas require storage temperatures of 32° to 35°F and relative humidity of 90 to 95%. Turnips and rutabagas may be dipped in warm wax to prevent moisture loss.

Market Preparation and Marketing

Prior to shipping, rutabagas are trimmed, washed, graded, and waxed. Rutabagas maintain a longer supermarket shelf life if hot wax-treated. Turnip and rutabaga roots in storage will shrivel less and maintain their appearance for several weeks longer if they are blemish-free when stored.

Uses

Turnips are used in roasts, stews, soups, casseroles, and as a boiled vegetable or sliced in salads. They may be eaten in wholes or halves, or mashed and served with a variety of sauces. When young and tender, turnip greens are cooked and eaten. Brief cooking in a small amount of water keeps the leaves crisp yet tender.

Because rutabagas are large they are usually cut up before cooking. The flavor of rutabaga makes it suitable not only as a cooked vegetable but also in soups, salads, and baked goods. Rutabagas may be boiled or roasted with meat and eaten in pieces, or boiled and mashed. Fresh uncooked rutabaga can be shredded into salads.

Nutritional Value

A cup of turnip greens provides 165% of the RDA of vitamin C, 182% of vitamin A, one-sixth of the iron and thiamin, and one-fifth of the riboflavin.

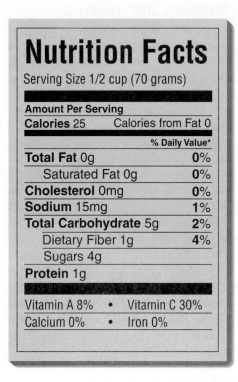

FIGURE 20.18   Nutritional value of rutabagas. Percent daily values are based on a 2,000 calorie diet. Used with permission from Dole Food Co., Inc.

Rutabagas have much more vitamin A than turnips, considerably more ascorbic acid, and more of the other vitamins, but less iron (Figure 20.18). This comparison refers only to the roots. Rutabagas are also much lower in sodium than turnips.

Sweet Potato

Scientific Name

Ipomoea batatas (L.) Lam

Common Names in Different Languages

sweet potato (English); gan shu (Chinese); patate douce (French); Batate Süsskartoffel (German); pata americana, pata dolce, batata (Italian); satsuma-imo (Japanese); batat (Russian); batata, boniato, camote, papa dulce, moiato (Spanish)

Classification, History, and Importance

The sweet potato, *Ipomoea batatas,* is a member of the Convolvulaceae (morning glory) family. Sweet potato is a tuberous-rooted perennial plant (Figure 20.19). It is an important food crop in many of the warmer regions of the world and is often considered a tropical crop.

Sweet potatoes are a native-American crop that was noted by Columbus in the records of his fourth voyage to the New World. DeSoto, in 1540, found sweet potatoes growing in Indian gardens in what is now Louisiana. It is believed that Spanish

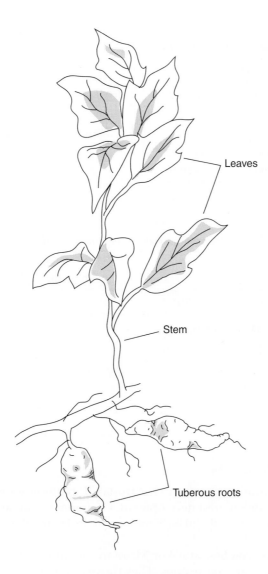

Leaves

Stem

Tuberous roots

FIGURE 20.19   Diagram of a sweet potato plant.

explorers brought the sweet potato to the Philippines and the East Indies. Portuguese voyagers are credited with bringing sweet potatoes to India, China, and Malaya. Sweet potatoes were cultivated in Virginia in 1648 and possibly taken to New England by 1764. Native Americans were growing sweet potatoes in what is now Georgia by the time the first English settlers arrived.

Production of sweet potato was a $204.6 million industry in 1996 (Table 20.6). This was up from $134.4 million in 1986. Per capita consumption of sweet potato has decreased from 5.4 lb per person in 1976 to 4.6 lb per person in 1996 (Table 20.7). Of the 13.5 million cwt (hundredweight) of sweet potatoes produced in the United States, 4.3 million cwt was grown in North Carolina, 3.4 million cwt in Louisiana, 2.2 million cwt in California, and 1.3 million cwt in Mississippi. Other states that produce sweet potatoes include Alabama, Georgia, New Jersey, South Carolina, Texas, and Virginia.

TABLE 20.6   *U.S. acreage, yield, production, and value of sweet potatoes for the years 1976, 1986, and 1996*

Year	Acreage Planted	Acreage Harvested	Yield (cwt/acre)	Production (1,000 cwt)	Farm value ($1,000)
	(acres)				
1976	119,800	114,800	116	13,273	99,054
1986	94,500	90,800	136	12,368	134,436
1996	89,100	84,800	159	13,456	204,658

Source: USDA, NASS, 1997 data.

TABLE 20.7   *U.S. per capita consumption of commercially produced sweet potatoes*

Use	1976	1986 (lb, farm weight)	1996
Fresh and processing	5.4	4.4	4.6

Source: USDA, Economic Research Service.

Plant Characteristics

Height: 10 to 12 in.
Spread: 4 to 8 sq ft

Climatic Requirements

Sweet potato is a root crop native to the tropics. It requires warm days and nights for optimum growth and root development. Optimum temperatures for sweet potato growth are 70°F for soil and 84°F for air with little growth below 59°F soil temperature.

Sweet potatoes grow best with 4 or 5 frost-free months, but will produce smaller roots under shorter growing seasons. They thrive best in the hot conditions of the southern United States, but can be grown as far north as southern Michigan and in the mild climates of the Pacific Northwest.

Field Preparation

Sweet potatoes yield more roots and better-quality roots on a well-drained, sandy loam or silt loam soil. Rich, heavy soils produce high yields of low-quality roots. Extremely poor, sandy soils generally produce low yields of high-quality roots. Poor surface drainage of the field may cause wet spots and reduce yields, which cause sweet potatoes to be large, misshapen, cracked, and rough skinned.

Sweet potatoes are fairly tolerant of variations in soil pH between 5.2 and 6.7. The optimum pH for high yields of quality sweet potatoes is 5.8 to 6.0.

A moderately high supply of available nutrients is required for high yields and quality. Only moderate amounts of nitrogen are required, while potassium is used in relatively high amounts. High phosphorus may be necessary if the crop is grown on heavy clay soils. Half of the fertilizer is typically applied before planting and the rest is applied as a side-dressing or topdressing 2 to 3 weeks after planting.

On heavy soils where drainage is slow, ridges 12 to 15 in. high are often used. The ridges are typically spaced 42 to 48 in. apart. On light, sandy soils, ridges 8 to 10 in. high spaced 30 in. apart are usually adequate. The tops of the ridges are flattened prior to planting.

Selected Cultivars

Sweet potato cultivars are grouped into two general types—those with deep yellow-orange color that are soft, moist, and sweet when cooked; and those with firm, dry, light-colored mealy flesh. The moist-fleshed types are often called "yams." However, the true yam is an entirely different genus, *Dioscorea*, that is grown only in tropical climates. Soft-fleshed cultivars are becoming more popular and include Centennial, Nemagold, and Goldrush. Yellow Jersey is a prominent firm-fleshed cultivar.

Planting

Sweet potatoes are grown from plants or sprouts (slips) produced from roots of the previous crop and from vine cuttings. Slips are produced from seed roots that are typically 1½ to 3 in. in diameter (Figure 20.20). Slips can be grown in cold frames (Figure 20.21) or heated beds, generally taking 5 to 6 weeks for slip production in hotbeds and 7 to 8 weeks in cold frames. Preheating of seed roots is useful for rapid slip production and is accomplished by raising the seed storage temperature for 2 weeks to 85°F with 85 to 95% relative humidity. Under these conditions, 1- to 2-in. sprouts will be formed by time of bedding. Fungicide-treated roots are placed in the plant bed with the sprouts upright. The roots are separated and placed in plant beds according to size in order to get an even depth of covering and uniform sprouting. Tar paper or plastic is then often placed directly over the bed surface. When the slips push the covering about 2 in., the covering material is removed. Air temperature in the beds should be kept under 90°F. Plants are pulled when they are about 8 in. tall and have at least five leaves, stocky stems, and a healthy root system (Figure 20.22).

Sweet potato storage root development depends on good soil aeration that is achieved by quality field selection and by bedding the field prior to transplanting. Beds should remain 8 to 10 in. high after setting and transplanting. Field transplanting should be as soon as possible after slip pulling and one- or two-row transplanters are commonly used commercially.

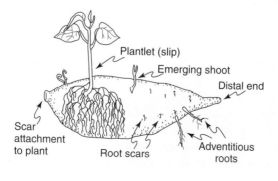

FIGURE 20.20   Diagram of sweet potato slips or plantlets produced from a seed root.

FIGURE 20.21 Sweet potato slips produced in a cold frame are removed by pulling them away from the seed root.

FIGURE 20.22 Sweet potato plants at the proper size for planting.

A common spacing is 12 in. between sweet potato plants and 36 to 42 in. between rows. Plant spacing depends on soil fertility and availability of irrigation water. Wide spacing on fertile soils results in excessive jumbo roots and rougher potatoes. Close spacing on very sandy soils results in undersized roots.

Irrigation

Irrigation is commonly needed immediately following transplanting to reduce transplant shock. Inadequate soil moisture is a consistent limiting factor in sweet potato production and supplemental irrigation should be available to supply up to 1½ in. of water each 7 to 10 days. Moisture should be withheld toward the end of the growing season to condition the soil and roots for harvesting and to discourage the development of cracks and jumbo roots.

Culture and Care

Sweet potatoes are poor competitors against weeds. Control measures for weeds are shallow cultivation and some hilling around the plants to smother weeds until the exposed areas between the rows are covered by the crop. After 4 to 6 weeks in the field, sweet potatoes compete better against weeds.

Common Growing Problems and Solutions

Insects   The major insect pests of sweet potatoes are soil insects such as wireworms, flea beetles, cucumber beetles, and grubworms. Wireworms and leaf beetles cause small, shallow holes that enlarge or deepen as the potato grows and grub injury is recognized as the grubs gorge broad, shallow areas in the roots.

Diseases   Several diseases affect sweet potato production including scurf, pox (soil rot), stem rot (fusarium wilt), root-knot nematode, and black rot.

Physiological Disorders   While sweet potatoes respond to nitrogen fertilizer, too much nitrogen can cause excessive vine growth with little root growth.

Harvesting

The time to harvest sweet potatoes is determined from regular field inspection. Sweet potatoes may be harvested any time after a sufficient number of roots have reached marketable size (Figure 20.23). If soil temperatures fall below 55°F, some damage to the quality, storability, and slip production of the roots will result.

Most harvesters require vines to be cut with a flail or rotary mower (Figure 20.24) so that they do not interfere with digging. For large plantings a chain-type digger is often used (Figure 20.25).

Postharvest Handling

Sweet potatoes stored for later marketing or seed stock are generally cured for 7 to 10 days immediately after harvest to minimize storage losses. Curing is a wound-healing process that occurs most rapidly at 80° to 90°F and relative humidity of 85

FIGURE 20.23   A sweet potato field is harvested when a certain number of roots reach marketable size and quality.

FIGURE 20.24   Mowing of a sweet potato field prior to harvest.

to 95%. Good ventilation is also required to remove carbon dioxide from the cutting area. Wounds and bruises heal and a protective cork layer develops over the entire root surface. In addition, suberin, a waxy material, is deposited. The cork layer and suberin act as a barrier to decay-causing organisms and to moisture loss during storage.

FIGURE 20.25   A mechanical harvester for sweet potatoes.

Sweet potatoes are stored between 55° and 60°F. Storage temperatures below 55°F often result in chilling injury to the sweet potato roots. Relative humidity is maintained between 75 to 80% to prevent excessive water loss from roots.

If sweet potatoes are stored slightly above 55°F at high humidity, they can be stored for 6 months.

Market Preparation and Marketing

Sweet potatoes are usually washed (Figure 20.26), graded, and often waxed before marketing. Most of the crop is packed in 50-lb corrugated boxes and sold as bulk potatoes (Figure 20.27).

Uses

Sweet potatoes may be prepared for eating by boiling, baking, or frying. They are also used to make biscuits, bread, muffins, pies, custards, cookies, and cakes.

Nutritional Value

Sweet potato has a moderate amount of calories and is an excellent source of vitamin A and a good source of vitamin C. One medium-sized sweet potato has 130 calories, with no grams of fat and 4 g of dietary fiber (Figure 20.28). This serving of sweet potato provides 440% of the RDA of vitamin A, 30% of the RDA of vitamin C, and 2% of the RDA for both calcium and iron.

Yellow-fleshed sweet potatoes are high in carbohydrates and beta-carotene (provitamin A). Roots with orange-colored flesh are higher in carotene than the white or pale yellow varieties. The tender leaves and shoots are used as greens in Southeast Asia and are higher than the roots in vitamins A and C and protein.

FIGURE 20.26 Harvested sweet potatoes going through a washing line.

FIGURE 20.27 Different cultivars of sweet potatoes in boxes ready for marketing.

FIGURE 20.28   Nutritional value of sweet potatoes. Percent daily values are based on a 2,000 calorie diet. Used with permission from Dole Food Co., Inc.

Review Questions

1. What was the role of the carrot in the introduction of the wildflower Queen Anne's lace to North America?
2. Why is proper soil preparation important for growing quality carrots?
3. What are the four carrot cultivar classes?
4. How are "baby" carrots produced?
5. What environmental factors can interfere with beet seed germination and emergence?
6. What influence does removing the leaves (topping) have on the storage life of beets?
7. What factors contribute to pithy, cracked, and/or harsh-tasting radish roots?
8. Why are turnips more often grown than rutabagas in the southern United States?
9. What is often done to the rutabaga after harvest to enhance its storage life?
10. What effect does soil type have on sweet potato production?
11. How are sweet potato slips produced and what are they used for?
12. When is a sweet potato field ready for harvest?
13. What is curing of sweet potatoes and how is it done?

Selected References

Anderson, C. A. 1998a. *Beets.* Univ. of Arkansas Coop. Ext. Serv. Publ.

Anderson, C. A. 1998b. *Turnips-rutabagas.* Univ. of Arkansas Coop. Ext. Serv. Publ.

Ball, J. 1988. *Rodale's garden problem solver: Vegetables, fruits, and herbs.* Emmaus, PA: Rodale Press. 550 pp.

Charney, P., and R. A. Seelig. 1967. Sweetpotatoes. *Fruit & Vegetable Facts & Pointers.* United Fresh Fruit & Vegetable Association. Washington, DC.

Fritz, V. A. 1996. *Growing carrots and other root vegetables in the home garden.* Minnesota Coop. Ext. Serv. Publ. FS-0435-GO.

Kays, S. J., and J. C. Silva Dias. 1995. Common names of commercially cultivated vegetables of the world in 15 languages. *Economic Botany* 49: 115–152.

McCraw, D., J. Motes, B. Cartwright, and J. Damicone. 1998. *Sweet potato production.* Oklahoma Coop. Ext. Serv. Publ. F-6022.

Murray, J. 1977. Radishes. *Fruit & Vegetable Facts & Pointers.* United Fresh Fruit & Vegetable Association. Washington, DC.

Nagel, D. 1997a. *Commercial production of carrots in Mississippi.* Mississippi Coop. Ext. Serv. Information Sheet 1502.

Nagel, D. 1997b. *Commercial production of radishes in Mississippi.* Mississippi Coop. Ext. Serv. Information Sheet 1504.

Neild, R. E., and L. Hodges. 1992. *Sweet potatoes.* Univ. of Nebraska Coop. Ext. Serv. Publ. G73-73-A.

Nonnecke, I. L. 1989. *Vegetable production.* New York: Van Nostrand Reinhold. 656 pp.

Peirce, L. C. 1987. *Vegetables: Characteristics, production, and marketing.* New York: John Wiley & Sons.

Sanders, D. C. 1996. *Turnips and rutabagas.* North Carolina State Univ. Coop. Ext. Serv. Leaflet 26.

Schultheis, J. R. 1993. *Radish.* North Carolina State Univ. Coop. Ext. Serv. Leaflet 25.

Seelig, R. A. 1966. Beets. *Fruit & Vegetable Facts & Pointers.* United Fresh Fruit & Vegetable Association. Washington, DC.

Seelig, R. A. 1970. Rutabagas. *Fruit & Vegetable Facts & Pointers.* United Fresh Fruit & Vegetable Association. Washington, DC.

Seelig, R. A. 1973. Turnips. *Fruit & Vegetable Facts & Pointers.* United Fresh Fruit & Vegetable Association. Washington, DC.

Splittstoesser, W. E. 1990. *Vegetable growing handbook: Organic and traditional methods.* 3d ed. New York: Van Nostrand Reinhold.

Wott, J. A., and A. E. Boger. 1977. *The sweet potato.* Purdue Univ. Coop. Ext. Serv. Publ. HO-136.

Yamaguchi, M. 1983. *World vegetables.* New York: Van Nostrand Reinhold.

Zandstra, B. H., and D. D. Warncke. 1986. *Carrots.* Michigan State Univ. Coop. Ext. Serv. Bull. E-1437.

Selected Internet Sites

USDA Agricultural Marketing Service Standards for Grades

Fresh
Beets   www.ams.usda.gov/standards/beets.pdf
Carrots (bunched)   www.ams.usda.gov/standards/carotbch.pdf
Carrots (topped)   www.ams.usda.gov/standards/carottop.pdf
Radishes   www.ams.usda.gov/standards/radishes.pdf
Sweet potatoes   www.ams.usda.gov/standards/sweetpot.pdf
Turnips or rutabagas   www.ams.usda.gov/standards/turnips.pdf

Processing
Beets   www.ams.usda.gov/standards/vpbeets.pdf
Carrots   www.ams.usda.gov/standards/vpcarro.pdf
Sweet potatoes   www.ams.usda.gov/standards/vpswtpot.pdf

chapter
21

Bulb Crops

  I. Onion
 II. Leek
III. Garlic

Bulb crops are all species of *Allium* and are members of the Alliaceae family. These plants are native to the temperate regions of South America, South Africa, and the Mediterranean. The plants have bulbs or corms and are valued for their pungency. Crops of importance include onion, leek, and garlic.

Onion

Scientific Name

Allium cepa L. Cepa group

Common Names in Different Languages

onion (English); yang cong (Chinese); oignon (French); Speisezwiebel, Küchenzwiebel (German); cipolla (Italian); tam an egi (Japanese); luk repčatyj (Russian); cebolla (Spanish)

Classification, History, and Importance

Onions are biennial herbaceous members of the Alliaceae family that are typically grown as annuals. Other members of the family include garlic and leeks. The common onion (*Allium cepa*) is grown from either seed, transplants, or sets for use as both green onions and dry bulbs.

Onions harvested in the green or immature stage are called "green bunching onions." These onions are sold in bunches tied with a rubber band. Any standard onion cultivar or hybrid can be used as green onions if grown under daylengths that do not induce the bulbing process and harvested at the proper stage of maturity.

FIGURE 21.1   Egyptian tree onion with cluster of bulbils near the top of the seedstalk.

The multiplier, or potato, onion (*Allium cepa* L. Aggregatum group) is composed of a compound bulb formed from the segregation of a large mother bulb. Each bulb in the compound bulb produces 6 to 12 plants. Their principle use is in the production of early green bunching onions. The Egyptian onion (*Allium cepa* L. Proliferum group) (Figure 21.1) produces clusters of small bulbs called *bulbils* at the top of the seedstalk. The bulbils can be planted to produce very early green onions.

Onions are native to Southern Asia and the Ural Mountains and have long been valued in China and India for their flavorings. They were a staple food of the workers at the pyramids in Egypt and were worshipped before the Christian era. Onions were also used in ancient Druid rites.

Onions, mentioned by Chaucer in about 1340, have been grown in England for centuries and were introduced into the West Indies by the Spanish. Onions soon spread to all parts of the Americas and were grown by the early colonists and soon thereafter by Native Americans. During the Civil War General Grant believed that consumption of onions by the soldiers prevented dysentery and other ills caused by hot climates. He would not move his army into battle without a generous supply of onions.

Onion for fresh market and processing was a $648.4 million industry in 1997 (Table 21.1). This was up from $581.5 million in 1996. Per capita consumption of onions has increased from 11.8 pounds per person in 1976 to 18.7 pounds per person in 1996 (Table 21.2). There are about 165,600 acres of onions in the United States (Table 21.3). The leading onion producing states are California, Oregon, Colorado, Texas, and Georgia.

Plant Characteristics

Height: 24 to 36 in.
Spread: 6 to 18 in.
Root Depth: 1½ to 3 ft

TABLE 21.1   *The production value of selected bulb crops for fresh market and processing in the United States*

	Value of production		
Crop	1995	1996	1997
	(1,000 dollars)		
Garlic (CA)	140,700	196,333	261,519
Onion	633,692	581,571	648,437

Source: USDA, NASS.

TABLE 21.2   *U.S. per capita consumption of commercially produced fresh and processing onions*

Use	1976	1986	1996
	(lb, farm weight)		
Fresh	11.8	13.7	17.9
Dehydrating	0.8	1.9	0.8
All	11.8	15.6	18.7

Source: USDA, Economic Research Service.

TABLE 21.3   *The top fresh market onion-producing states in the United States*

Rank	State	Planted acres (000)	Harvested acres (000)	Yield (cwt/acre)	Production (000)
1	California	37.40	36.20	441.00	15,964
2	Oregon	19.80	19.40	555.00	10,770
3	Washington	14.70	14.40	520.00	7,488
4	Idaho	8.40	8.20	690.00	5,658
5	Colorado	18.00	15.30	350.00	5,355
6	New York	12.50	12.20	300.00	3,660
7	Georgia	16.20	15.80	220.00	3,476
8	New Mexico	6.60	6.40	470.00	3,008
9	Texas	16.30	11.80	222.00	2,623
10	Michigan	6.20	6.10	320.00	1,952
.	United States	165.58	154.92	412.00	63,883

Source: USDA, NASS, 1997 data.

Climatic Requirements

Onions are a cool-season crop adapted to a wide range of temperatures and can withstand short exposures to temperatures well below freezing. Optimal onion production is obtained when cool temperatures (55° to 75°F) prevail over an extended period of time, permitting considerable foliage and root development before bulbing starts. After bulbing begins, high temperature and low relative humidity extending into the harvest and curing periods are desirable.

Onion bulbing is a daylength response that is controlled by genetic inheritance in each cultivar. Short-day cultivars bulb when daylengths are 10 to 11 h. Intermediate cultivars require 12- to 13-h daylengths to initiate bulbing and long-day cultivars require 14- to 16-h daylengths. Within daylength groupings, there are early and late maturing cultivars.

Maturing time for onions from seed to bulb stage varies, depending on temperature, daylength, and cultivar. Temperatures of 70° to 80°F are favorable for bulb development when daylength is also favorable. Bulbs will not form at temperatures of 50° to 60°F regardless of length of day.

Field Preparation

While onions can be grown on a wide range of soils, high organic soils are generally preferred for optimal production. Onions develop best in a loose, crumbly soil with a soil pH between 6.0 and 6.8. Prior to planting, the soil is prepared by deep plowing and disking. Fertilizer is typically applied in amounts that are generally twice as much as for most other vegetables. On low-potash soils, a complete fertilizer such as 5-10-10 at a rate of 1,500 to 2,100 lb per acre may be recommended. This is incorporated at time of planting. Onions respond to side-dressing with 20 to 25 lb of nitrogen 3 weeks after emergence and every 3 weeks thereafter.

Selected Cultivars

Onion cultivars are classified according to bulb formation. Photoperiod, along with temperature, regulates the bulbing process in onions. This effect of daylength results in some onion cultivars being unsuitable for growth in northern climates since they might bulb when the plants are considered too small for optimum production.

Some cultivars are short-day in response to daylength and form bulbs when there is 10 to 11 h or less of daylight. These cultivars are often called European or Bermuda onions and are typically very mild, soft fleshed, and not suitable for long storage.

Intermediate-day onion cultivars require a minimum of 12 to 13 h of daylight to bulb. They tend to be pungent or strong flavored and are relatively soft fleshed. They are sold for consumption soon after purchase and not stored.

Long-day onion cultivars form bulbs when there are 14 or more hours of daylight. They are called American or Spanish onions and are very pungent and hard. Long-day onion cultivars store well. They are planted in early spring and harvested during the long days of summer. These cultivars can be used to grow green onions if they are grown in areas where the daylength is not long enough to produce bulbs, such as in the southern United States during the late fall, winter, or early spring.

Onions are available in red, white, or yellow; mild or pungent; storing or non-storing; and deep-globed or flat-globed cultivars. There are many more yellow onion cultivars than red or white cultivars.

Planting

Onions can be planted using sets, transplants, and seeds. Onion sets are small bulbs (½ to 1⅛ in. in diameter) that are typically produced in the late winter and early

FIGURE 21.2   Planting of onions using sets.

spring for spring planting. If sets are used they are planted directly in the field ¾ in. deep and 3 in. apart in the row (Figure 21.2). In general, the larger the sets the sooner the onions are of marketable size. Large sets have a disadvantage because they tend to form seedstalks earlier than small sets.

If transplants are used they are often produced in protected beds or greenhouses and set in the field when the plants are about 6 in. tall. Four to 6 weeks are often required for the onion transplant to become large enough for transplanting to the field. Transplants placed in the field are spaced 2 to 4 in. in the row.

The practice of direct-seeding onions to a field is often done in late winter in the southern and western United States. Seeding is the least expensive method of planting onions, but requires a longer period of time in the field before onions are ready for harvest as compared with sets or transplants. Onion seeds germinate in cool soils and are planted early in the spring as soon as the soils can be prepared. Onions are seeded ¼ in. deep and emerging stands of seedlings are thinned when the plants are 2 to 3 in. high, leaving 3 to 4 in. between plants. Multiple rows or a broadcast pattern can be used as long as 3 to 4 in. are left between developing bulbs.

Green onions are often planted on raised beds 4 to 6 in. high to facilitate good drainage. Row spacing may be two rows per bed on 38-in. centers or four rows per bed on 60- to 76-in. centers. Onion rows are spaced 9 to 12 in. apart if two rows per bed are used and 9 to 18 in. apart with four rows per bed.

Irrigation

Onions have a shallow, poorly developed root system and require a relatively constant source of water. Soil moisture is especially important for the growth of new adventitious roots. It is suggested that the soil moisture must reach the base of the bulb periodically for the newly formed adventitious roots from the stem to grow into the soil. Irrigation is not used for bulbing onions when the bulbs begin to mature.

Culture and Care

Weeds must be controlled in fields because onions do not compete well with weeds. Shallow cultivations are used to control weeds. Two to 3 weeks before the harvesting of green onions, approximately 2 in. of soil is worked around the base of the stem. This is known as *blanching* and results in green onions that have a long, white, and tender stem.

Common Growing Problems and Solutions

Insects   Thrips and onion maggots are common pests of onions and must be controlled for successful onion production. Thrip feeding produces leaf scarring, which can reduce yields. Onion maggots are larvae that lay their eggs near the base of the plant. The small maggots burrow into the stem and bulb and kill the plant. Onion maggots tend to be more of a problem on moist soils with high levels of organic matter.

Diseases   Pink rot is caused by a soilborne fungus. This problem can be prevented by buying disease-free seed or transplants, growing in disease-free fields, and, when possible, planting disease-resistant varieties. Leaf diseases of onions include botrytis blast (numerous white specks), downy mildew (pale green, oval sunken spots), and purple blotch (purple lesions with yellow margins). White rot is identified as a white fungal growth on the base of the onion.

Neck rot occurs during or following harvest. This disease can be recognized by the grayish mold on the surface of the infected area. Bulbs that are well dried, such as in conditions of 90° to 120°F for 2 to 3 days, and stored at 32°F are less likely to succumb to the disease. Adequate ventilation is also needed to maintain a dry atmosphere during onion storage.

Physiological Disorders   Bolting, or seedstalk formation during the first year of growth of a biennial plant, can be a problem for fall-planted onions grown through the winter for spring harvest in the southern United States. The size of the overwintering plant and the exposure to cold temperatures appear to be the most critical factors determining whether the plant will bolt. An extended warm period following planting produces a large overwintering plant (over ¼-in. shank diameter), which results in a high percentage of bolting when exposed to extended temperatures below 50°F.

Splitting of bulbs is caused by erratic moisture and/or fertilization during growth. If onion bulb growth is slowed or checked from insufficient soil moisture during the growing period, the outer scales will begin to mature. Subsequently, when moisture or fertilization becomes available to the plant, the inner scales resume growth often causing the bulb to split.

Sunburning of onion bulbs occurs when the bulbs are exposed to high temperatures and bright sunlight. Damaged onion bulbs from excessive exposure to sunlight often have a bleached appearance with scales that may be soft and wrinkled. Onion leaves are often placed over bulbs after they have been harvested and windrowed for curing. This protects the harvested bulbs from direct sunlight.

Harvesting

Dry bulb onions are ready to harvest when the main stem begins to weaken and topple over (Figure 21.3). Harvest usually begins when 75% or more of the tops fall over. To hasten bulb drying, some growers use a subsurface knife to cut the roots a few inches below the bulb. For preliminary curing of harvested onion bulbs, the tops are cut 1 to 1½ in. from the bulb. This may be done by hand with shears or with a topping machine. Roots are trimmed off and the bulbs are placed in windrows in the field, or in slatted crates or field sacks. They may be left in the field for one to several days to allow the bulbs to dry.

Harvesting of green onions usually begins before bulbing occurs and when the basal diameters are ½ to 1 in. (Figure 21.4). The onions are undercut with a blade prior to harvest to make the harvesting easier. The discolored outside skin is removed, leaving the basal part of the plant white and clean. Green onions become stronger in flavor with age and increased size. The green onions are tied in small bunches (five to seven per bunch) with a soft string, tape, or rubber band and placed in crates (Figure 21.5) or boxes.

Postharvest Handling

Onion bulbs are in a state of rest and will not sprout for some time after harvest. The length of rest varies with cultivar and storage conditions. Bulbing onions store best under cool (close to 32°F), dry, well-ventilated conditions. Onions can also be braided with a cord and hung to dry.

Because the quality of green onions deteriorates very rapidly after harvest, the onions are kept cool for local sales or placed under refrigeration for distant markets. Green onions that are shipped are often packed with crushed ice during transit.

FIGURE 21.3   Bulbing onions almost ready for harvest.

Market Preparation and Marketing

Bulb onions are graded for quality and separated into various sizes. They are then often placed in 50-lb mesh sacks by size. Larger-sized onion bulbs may be sold to processors who make frozen french-fried onion rings. A few processors also dice and freeze onions for the institutional trade.

FIGURE 21.4   Harvesting a field of green onions.

FIGURE 21.5   Harvested green onions tied in small bunches and placed in wooden crates in the field.

The famed Vidalia onion is a Granex 33 cultivar grown in low-sulfur soils near Vidalia, Georgia. This combination of low sulfur and favorable climate is responsible for the sweet taste of these onions.

Uses

Onions are used primarily as flavoring agents. Bulb onions may be sliced and used on sandwiches, or dipped in batter and fried as onion rings or blooming onions. Green onions are eaten fresh, and can be chopped and added to salads.

Nutritional Value

Onions are low in calories and are a good source of vitamin C. One medium-sized onion has only 60 calories, with no grams of fat and 3 g of dietary fiber (Figure 21.6). This serving of onion also provides 20% of the RDA of vitamin C, 4% of the RDA for calcium, and 2% of the RDA for iron. Green onion tops have a higher vitamin A content than bulbing onions (Figure 21.7).

Nutrition Facts

Serving Size 1 medium onion (148g)

Amount Per Serving	
Calories 60	Calories from Fat 0
	% Daily Value*
Total Fat 0g	**0%**
Saturated Fat 0g	**0%**
Cholesterol 0mg	**0%**
Sodium 5mg	**0%**
Total Carbohydrate 14g	**5%**
Dietary Fiber 3g	**12%**
Sugars 9g	
Protein 2g	
Vitamin A 0%	• Vitamin C 20%
Calcium 4%	• Iron 2%

FIGURE 21.6 Nutritional value of bulb onions. Percent daily values are based on a 2,000 calorie diet. Used with permission from Dole Food Co., Inc.

Nutrition Facts

Serving Size 1/4 cup chopped (25g)

Amount Per Serving	
Calories 10	Calories from Fat 0
	% Daily Value*
Total Fat 0g	**0%**
Saturated Fat 0g	**0%**
Cholesterol 0mg	**0%**
Sodium 5mg	**0%**
Total Carbohydrate 2g	**1%**
Dietary Fiber 1g	**4%**
Sugars 1g	
Protein 0g	
Vitamin A 2%	• Vitamin C 8%
Calcium 0%	• Iron 0%

FIGURE 21.7 Nutritional value of green onions. Percent daily values are based on a 2,000 calorie diet. Used with permission from Dole Food Co., Inc.

Leek

Scientific Name

Allium ampeloprasum L. Porrum group

Common Names in Different Languages

leek (English); jiu cong (Chinese); poireau, porreau (French); Porree (German); porro, porretta (Italian); liiki (Japanese); luk porej (Russian); ajo porro, a. puerro (Spanish)

Classification, History, and Importance

Leeks, *Allium ampeloprasum*, are members of the Alliaceae family. Other members of the family include onion and garlic. The leek plant is a robust herbaceous biennial that has been cultivated for centuries but has not been found wild.

Leek plants resemble large onion plants with flat leaves. Unlike onion and garlic, leeks do not form bulbs or produce cloves. Leeks are made up of sheaths of basal leaves that can be 6 to 10 in. long and 2 in. in diameter. The taste of leeks is milder than that of either onion or garlic.

Plant Characteristics

Height: 18 to 24 in.
Spread: 6 to 15 in.
Root Depth: 18 to 24 in.

Climatic Requirements

Leeks will grow in any region that can produce onions and tend to be more frost and freeze tolerant than onions. The tendency of leeks to bulb is an undesirable characteristic that appears to be temperature controlled (with bulbing occurring between 60° and 65°F).

Field Preparation

Leeks require rich, loamy, well-drained soils with a pH ranging from 6.0 to 8.0. While most requirements for onions and garlic also apply to leeks, leeks tend to be more responsive to N fertilizer.

Selected Cultivars

There are several commercially important leek cultivars. Some of these include the following:

- Carina has heavy, thick, long white shanks and is good for overwintering.
- Titan is an extra long, early type that has vigorous growth. The leaves are dark green with white stems.

- Alaska has a dark blue-green foliage and is tolerant to subfreezing temperatures.
- Pancho is considered an early leek. It has thick white shanks, resists bulbing, and has freeze tolerance comparable to the later cultivars.
- American Flag has large, thick shanks. It blanches to a clear white with blue-green leaves. It is considered very hardy and cold resistant.

Planting

Leeks are generally direct-seeded into fields at rates of 10 to 15 seeds per foot of row. Emerging seedlings are thinned to 4 in. apart (Figure 21.8). Transplanting of leeks is also done and is often necessary for obtaining midsummer through fall harvests. Leeks are sown directly or transplanted into trenches 6 in. wide and 6 in. deep. As the leek plant grows, the trench is filled in. This results in the formation of long white stems, a desirable characteristic for marketing leeks. The deeper the leeks are trenched or hilled, the longer the tender white portion of the leaf stem becomes.

Irrigation

Uninterrupted growth is required for quality leeks and irrigation is often necessary in areas where moisture stress may occur.

Culture and Care

As leeks begin strong growth in the summer, the plants are cultivated and the soil is drawn toward the plant for blanching the edible portion of the plant. Blanching makes the leeks longer and whiter at harvest. Leeks are slow growers that require 120 days or more to reach 1 to 1½ in. in diameter.

FIGURE 21.8   Leek plants established in the field.

Common Growing Problems and Solutions

Weed, insect, and disease problems of leeks are essentially the same as those for onions.

Harvesting

Leeks are harvested when stalks reach a size of 1 in. or more in diameter. They are more tender and desirable before the stalk becomes too thick. Leeks can be harvested throughout late summer and fall in many regions of the United States. To harvest leeks the plants are undercut by hand, trimmed, and bunched. The roots and all but 2 in. of the green leaves are typically removed at harvest.

Postharvest Handling

Harvested leeks are cooled by hydrocooling, icing, or vacuum cooling to preserve freshness. If vacuum cooling is used, the leeks are often wrapped in ventilated polyethylene to prevent desiccation. Leeks held at near 32°F and about 90% relative humidity can be stored for 2 to 3 months.

Market Preparation and Marketing

Leeks are bunched for market, similar to green onions.

Uses

Leeks may be eaten raw or cooked. They are used primarily for flavoring soups and stews in place of onions. The sharp flavor of leeks often disappears upon boiling, leaving behind a very mild, pleasant-tasting product.

Nutritional Value

The leek does not offer a great deal of nutrient value besides bulk and a pleasant taste. The overall vitamin and mineral content corresponds roughly to that of onion.

Garlic

Scientific Name

Allium sativum L.

Common Names in Different Languages

garlic (English); da suan (Chinese); ail, a. ordinaire, a. blanc (French); Knoblauch (German); aglio, a. domestico (Italian); ninniku (Japanese); česnok (Russian); ajo, ajo comun (Spanish)

Classification, History, and Importance

Garlic, *Allium sativum,* is a bulbous perennial plant of the Alliaceae family related to onion, leek, and chive. The name garlic comes from the Welsh word *garlleg*, which is transformed into the English word *garlic*. The cultivated garlic is also called silver-skinned or Italian garlic. Elephant garlic, *Allium ampeloprasum* L. Ampeloprasum group, is marketed as a novelty item.

The leaves of garlic have solid, thin blades rather than the more rounded, tubular, hollow blades of onion leaves (Figure 21.9). As garlic matures, it produces a fairly smooth, round, dry bulb at the base. Each year garlic divides into cloves (usually 10 or more) soon after bulbing (Figure 21.10). Each clove is made of two modified mature leaves around an axis with a vegetative growing point. The outer leaf is a dry sheath, while the base of the inner leaf is thickened, making up the bulk of the clove. Garlic's most distinctive feature is its odor, produced by an organic sulfur compound known as allicin, which has potent antibacterial properties.

Garlic is native to middle Asia. There are written references to garlic from the writings of the Greeks, Egyptians, Romans, and Chinese. It is reported that Roman gladiators were instructed to eat garlic in order to make them capable of greater feats in the stadium. Ancient folk medicine says that a cold can be cured if a person rubs the soles of his feet with cut garlic cloves or if a tonic of honey and garlic is drank. In the world of occult, garlic is believed to be a protection against known and unknown evils. Garlic wreaths hung outside the door are said to ward off witches. Garlic was mentioned among garden esculents by American writers in 1806.

Garlic for fresh market was a $261.5 million industry in 1997 (see Table 21.1). This was up from $196 million in 1996. Per capita consumption of garlic has increased from 0.5 pounds per person in 1976 to 2.1 pounds per person in 1996 (Table 21.4). Most of the 37,000 acres of garlic grown in the United States are in California (Table 21.5).

Plant Characteristics

Height: 12 to 24 in.
Spread: 6 to 8 in.
Root Depth: up to 3 ft

FIGURE 21.9   Young garlic plants in the field. Courtesy of M. D. Orzolek.

FIGURE 21.10   Intact garlic bulbs (*top*) and bulbs separated into cloves (*bottom*). Courtesy of M. D. Orzolek.

TABLE 21.4   *U.S. per capita consumption of commercially produced fresh garlic*

Use	1976	1986 (lb, farm weight)	1996
Fresh	0.5	0.8	2.1

Source: USDA, Economic Research Service.

TABLE 21.5   *Garlic production in the United States*

Rank	State	Planted acres (000)	Harvested acres (000)	Yield (cwt/acre)	Production (000)
1	California	37.00	37.00	150.00	5,550
.	United States	37.00	37.00	150.00	5,550

Source: USDA, NASS, 1997 data.

Climatic Requirements

Garlic is a cool-season perennial that usually has its best growth under warmer and drier conditions than those favoring onion growth. Bulbing is initiated as temperatures and daylengths increase.

Field Preparation

Garlic plants have large, well-developed root systems that reach to depths of 3 ft or more. Well-drained clay loams are better for growing garlic than sandy soils. The preferred soil pH for garlic is 6.0 to 6.8. Surface drainage is improved by planting garlic on beds raised 6 to 8 in. above the natural level of the field.

Garlic is a heavy feeder and from 60 to 80 lb N/acre and 80 lb P/acre are used to fertilize the crop. Potassium is included if the soil is low in this element.

Selected Cultivars

There are two different types of garlic: those that send up a seedstalk (hardneck cultivars) and those that do not (softneck cultivars). Hardneck (ssp. *ophioscorodon*) types such as rocabole and continental usually do better in colder climates and are larger and easier to peel. Softneck (ssp. *sativum*) types such as silverskin and artichoke have been cultivated over a longer period of time and tend to be better adapted to a great range of climate conditions. Softneck types also hold up better in storage due to their tighter skins.

Rocabole type (hardneck)

- Spanish Roja produces 6 to 13 cloves per bulb. The cloves are brown to reddish purple and are easy to peel.
- Carpathian produces 6 to 10 cloves per bulb. The bulbs are large and uniform with bulb wrappers that have purple blotches. Carpathian has a hot and spicy flavor.
- German Red produces 10 to 15 cloves per bulb. It is very vigorous and has a deep green large bulb. The cloves are light brown with some purple at base.

Artichoke type (softneck)

- Inchelium Red produces 4 to 5 clove layers with 8 to 22 cloves. The bulbs can be over 3 in. in diameter and have a mild lingering flavor.
- California Early produces 4 clove layers with 10 to 22 cloves. The cloves can have a tan to off-white color with pinkish blush. It has a mild, slightly sweet flavor.

Siverskin type (softneck)

- Mild French produces 4 clove layers with 13 to 16 cloves. The color of the cloves is reddish pink blush on a yellow-white background. Mild French is better adapted to hot climates. It has a sharp taste when raw but has a smooth, nutty taste when cooked.
- Silverskin produces 15 to 20 cloves per bulb, usually in 5 layers. The clove color is off-white to tan with pink blush. It is a good producer of large bulbs that are mild and sweet tasting.

Elephant garlic or greathead garlic (*Allium ampeloprasum*) is closely related to leek but is not a true garlic. Elephant garlic grows optimally in mild to moderately cold regions and produces a segmented bulb that has a mild flavor similar to a garlic.

Planting

Garlic is usually planted from clean, dry bulbs. The bulbs are carefully separated into individual cloves. The bulbs will often separate naturally into two sizes of cloves: large and small. The large cloves are generally preferred for planting.

Typically in the southern and western United States, garlic cloves are planted in late summer and fall 3 to 4 in. apart in rows that are 6 to 8 in. apart. The cloves are placed 2 in. beneath the surface of the soil. Planting the cloves with the root end down is desirable. During the fall and winter the root system develops with a limited amount of top growth. By spring, the plant is generally well established and rapid top growth occurs as warmer temperatures prevail. Large vigorous tops are necessary to produce large bulbs.

Although garlic is commonly propagated from cloves, "top setting" cultivars may be propagated from bulblets or bulbils that form on the terminal end of a hollow seedstalk (scape) that develops from the main bulb before harvest. Bulbils form in a globe-shaped pod called a spathe. The outer whitish sheath of the spathe will eventually split, exposing a cluster of 10 to 40 brown, yellow, or purplish bulbils that can vary from the size of a grain of wheat to a kernel of corn.

Irrigation

Enough water during the spring and early summer to moisten the soil thoroughly to a depth of 2 ft every 8 to 10 days is required for quality production of garlic. Irrigation is usually stopped when garlic plants in the field begin to mature (the tops fall over naturally and begin to dry).

Culture and Care

Garlic is not a good competitor against weeds. Many roots of garlic are near the soil surface, so cultivations are done carefully.

Common Growing Problems and Solutions

Insects   While relatively few insects affect garlic in the field, thrips can be damaging to garlic if left uncontrolled. A silvery, streaked appearance on the leaves is an indication of injury from thrips.

Diseases   Garlic is susceptible to most onion diseases including botrytis, pink rot, powdery mildew, and purple blotch.

Harvesting

Garlic bulbs are ready to harvest after all the tops have fallen over and become fairly dry. The plants are carefully lifted and placed in the shade for a week or more. The bulbs are either pulled by hand or mechanically harvested. In areas of rainfall or dew, the bulbs are protected from moisture. After the bulbs are dry, the tops and roots are often removed. Alternately, the tops may be left on and braided with other garlic plants.

Postharvest Handling

Garlic must be handled carefully since bruised bulbs deteriorate rapidly in storage. It is often stored in mesh bags or slated crates. Garlic is best stored at 32°F at relative humidity below 60% and may be kept for several months under these storage conditions.

Market Preparation and Marketing

After being dried for a week or longer, the garlic plant is cured further if necessary, graded, and packed in 50- to 100-lb mesh bags. Those harvested for dehydration are bulk loaded for transportation out of the field to the processing plant.

Uses

Garlic is used fresh or dehydrated to produce garlic powder and other dehydrated products. Garlic is a basic flavoring for a wide variety of dishes including vegetable soups, meats, salads, tomato combinations, spaghetti, sausages, and pickles. Garlic butter has many uses in home and restaurant cooking and many restaurants offer garlic bread with meals. Garlic continues to be used medicinally, and belief in its mythical properties has survived since ancient times.

Nutritional Value

Garlic is of limited nutritional value since most consumers have relatively low daily intake of garlic. It is low in calories with a one-clove serving having only 5 calories and no grams of fat (Figure 21.11). This serving of garlic provides 2% of the RDA of vitamin C.

FIGURE 21.11   Nutritional value of garlic. Percent daily values are based on a 2,000 calorie diet. Used with permission from Dole Food Co., Inc.

Nutrition Facts

Serving Size 1 clove (4 grams)

Amount Per Serving

Calories 5   Calories from Fat 0

% Daily Value*

Total Fat 0g	0%
Saturated Fat 0g	0%
Cholesterol 0mg	0%
Sodium 0mg	0%
Total Carbohydrate 1g	0%
Dietary Fiber 0g	0%
Sugars 0g	
Protein 0g	

Vitamin A 0%	•	Vitamin C 2%
Calcium 0%	•	Iron 0%

Review Questions

1. What environmental conditions favor quality onion production?
2. How might the same onion cultivar be used for growing green onions and bulbing onions?
3. What factors contribute to splitting of onion bulbs?
4. How do you know when bulbing onions are ready for harvest?
5. What criteria are used to determine when leeks are ready for harvest?
6. What is the difference between hardneck and softneck garlic cultivars?
7. How is garlic typically propagated?

Selected References

Anderson, C. A. 1998a. *Garlic.* Univ. of Arkansas Coop. Ext. Serv. Publ.

Anderson, C. A. 1998b. *Leeks.* Univ. of Arkansas Coop. Ext. Serv. Publ.

Anderson, C. A. 1998c. *Onions.* Univ. of Arkansas Coop. Ext. Serv. Publ.

Ball, J. 1988. *Rodale's garden problem solver: Vegetables, fruits, and herbs.* Emmaus, PA: Rodale Press. 550 pp.

Cook, W. P., R. P. Griffin, and A. P. Keinath. 1995. *Commercial green onion production.* Clemson Univ. Coop. Ext. Serv. Hort. Leaflet 46.

Dickerson, G. W. 1996. *Garlic production in New Mexico.* New Mexico State Univ. College of Agriculture & Home Economics Guide H-234.

Ells, J. E., H. F. Schwartz, and W. S. Cranshaw. 1993. *Commercial bulb onions.* Colorado State Univ. Coop. Ext. Cir. 7.619.

Hayslip, N. C., D. D. Gull, V. L. Guzman, J. R. Shumaker, and R. M. Sonoda. *Bulb onion production in Florida.* 1987. Univ. of Florida Coop. Ext. Serv. Publ. 6-5M-87.

Kays, S. J., and J. C. Silva Dias. 1995. Common names of commercially cultivated vegetables of the world in 15 languages. *Economic Botany* 49:115–152.

Longbrake, T., J. Larsen, S. Cotner, and R. Roberts. 1974. *Keys to profitable onion production in Texas.* Texas A&M Univ. Ext. Publ. 4-2.

Mansour, N. S. 1993. *Garlic for the home garden.* Oregon State Univ. Ext. Serv. Publ. FS-138.

Marr, C. 1994. *Onions for the home garden.* Kansas State Univ. Coop. Ext. Serv. Publ. MF-761.

Nonnecke, I. L. 1989. *Vegetable production.* New York: Van Nostrand Reinhold. 656 pp.

Peirce, L. C. 1987. *Vegetables: Characteristics, production, and marketing.* New York: John Wiley & Sons.

Riofrio, M., and E. C. Wittmeyer. 1992. *Growing onions in the home garden.* Ohio State Univ. Ext. Serv. Fact Sheet HYG-1616-92.

Seelig, R. A. 1970. Dry onions. *Fruit & Vegetable Facts & Pointers.* United Fresh Fruit & Vegetable Association. Washington, DC.

Seelig, R. A. 1974a. Garlic. *Fruit & Vegetable Facts & Pointers.* United Fresh Fruit & Vegetable Association. Washington, DC.

Seelig, R. A. 1974b. Green onions. *Fruit & Vegetable Facts & Pointers.* United Fresh Fruit & Vegetable Association. Washington, DC.

Yamaguchi, M. 1983. *World vegetables.* New York: Van Nostrand Reinhold.

Selected Internet Sites

USDA Agricultural Marketing Service Standards for Grades

Fresh
Garlic   www.ams.usda.gov/standards/garlic.pdf
Onions (Bermuda-Granex-Grano type)   www.ams.usda.gov/standards/onsbgg.pdf
Onions (other types)   www.ams.usda.gov/standards/onionther.pdf
Onions (green)   www.ams.usda.gov/standards/onioncg.pdf
Onion sets   www.ams.usda.gov/standards/onionset.pdf

Processing
Onions   www.ams.usda.gov/standards/vponion.pdf

chapter
22

Legumes or Pulse Crops

I. Common Bean
II. Lima Bean
III. Garden Pea
IV. Southern Pea

Legumes or pulse crops are members of the Fabaceae or pea family. Legumes are primarily herbaceous plants in temperate climates but can exist as trees and shrubs in tropical climates. The fruit of members of the Fabaceae is a flattened dehiscent pod called a legume. Many members of the family can assimilate their own source of nitrogen as a result of a symbiotic relationship with nitrogen-fixing *Rhizobium* bacteria in nodules in their roots. Members of this group include the common bean, garden (or English) pea, Southern pea, and lima bean.

Common Bean

Scientific Name

Phaseolus vulgaris L.

Common Names in Different Languages

common, green, French or snap bean (English); cai dou (Chinese); haricot vert (French); Gartenbohne, Brechbohne, Salatbohne, Prinzeobohne (German); fagiolino, fagiolo (Italian); ingen mame (Japanese); fasol' obyknovennaja (Russian); judia común, j. verde, faséolo (Spanish)

Classification, History, and Importance

The common bean, *Phaseolus vulgaris,* is a member of the Fabaceae or legume family. It is an herbaceous annual with alternate trifoliate leaves. The crop is grown for its fleshy pods and immature seeds. The bean plant may be erect and bushy or twining (pole type). The edible-podded varieties are snap beans, green beans, and string beans.

The phrase "spilling the beans" comes from fortune-telling gypsies. They would spill a handful of beans on a flat surface and base their predictions on the results.

The common bean was not known prior to the discovery of America. Common beans were thought to have originated in the tropical southern part of Mexico, Guatemala, Honduras, and a part of Costa Rica. Beans were believed to have spread throughout North and South America prior to the arrival of European settlers and Native Americans were observed eating beans in 1524. Bean seeds were found in the outer wrappings of a mummified woman from a Peruvian cemetery in which many pre-Incan civilization artifacts were also found.

Early cultivars of common beans, then termed string beans, had tough fibrous strands in each suture of the pod. Crop improvement through breeding during the late 1880s eliminated this trait. Subsequent genetic improvements included development of fiberless pod walls, improved tenderness, and white seed (which ensures a clear liquid component of canned snap beans).

Common beans for fresh market were a $156.4 million industry in 1997 (Table 22.1). This was up from $155 million in 1996. The common bean for processing market is considerably smaller than for fresh market with a value of production of $35 million in 1997. Per capita consumption of common beans has decreased from 7.8 pounds of both fresh market and processing per person in 1976 to 7.1 pounds per person in 1996 (Table 22.2). This reduction in per capita consumption of common beans was primarily due to less consumption of canned beans since the amount of frozen common bean product increased during this time period. Of the 88,660 acres of common beans grown in the United States for the fresh market, 31,300 acres are in Florida (Table 22.3). Wisconsin is the largest bean-producing state for processing with about 63,500 acres of the total 204,480 acres planted for processing.

Plant Characteristics

Height: Bush, 10 to 15 in.; pole, 8 to 15 ft.
Spread: Bush, 4 to 8 in.; pole, 6 to 8 in.
Root Depth: Bush and pole, 36 to 48 in.

TABLE 22.1   *The production value of selected legume crops for fresh market and processing in the United States*

	Value of production		
Crop	1995	1996	1997
		(1,000 dollars)	
Fresh Market			
Beans, lima (GA)	5,280	4,216	4,950
Beans, snap	162,260	154,952	156,377
Processing[1]			
Beans, lima	31,589	33,105	35,006
Beans, snap	120,992	138,103	129,753
Peas, green	131,762	117,596	136,996

Source: USDA, NASS.

[1]Value at processing plant door.

TABLE 22.2 *U.S. per capita consumption of commercially produced fresh and processing common beans*

Use	1976	1986 (lb, farm weight)	1996
Fresh	1.4	1.3	1.4
Canning	4.9	3.9	3.8
Freezing	1.5	1.5	1.9
All	7.8	6.7	7.1

Source: USDA, Economic Research Service.

TABLE 22.3 *The top common bean-producing states for fresh market and processing in the United States*

Rank	State	Planted acres (000)	Harvested acres (000)	Yield (cwt/acre)	Production (000)
Fresh market					
1	Florida	31.30	28.70	43.00	1,234
2	California	6.30	6.30	100.00	630
3	Georgia	11.50	10.00	44.00	440
4	Tennessee	10.00	8.00	40.00	320
5	New York	5.10	4.90	62.00	304
6	North Carolina	7.10	6.50	45.00	293
7	Virginia	6.00	5.20	35.00	182
8	New Jersey	3.70	3.30	35.00	116
9	Ohio	1.70	1.60	54.00	86
10	Michigan	1.70	1.60	45.00	72
.	United States	88.66	80.06	47.00	3,790
Processing				———(tons)———	
1	Wisconsin	63.50	59.80	3.26	195
2	Oregon	23.70	23.30	6.36	148
3	Michigan	23.20	22.80	3.45	79
4	New York	23.50	22.80	3.40	78
5	Illinois	12.10	11.50	3.70	43
6	Pennsylvania	11.60	11.50	2.60	30
7	Indiana	4.00	4.00	2.24	9
.	United States	204.48	195.48	3.75	733

Source: USDA, NASS, 1997 data.

Climatic Requirements

The common bean is a warm-season crop with an optimum temperature range for growth of 60° to 70°F and an optimum soil temperature for seed germination of 86°F. Bean plants are injured or killed by freezing weather. Depending on the cultivar and temperature, 45 to 80 days are required from planting to pod harvest. Pod production and plant growth are reduced by hot temperatures and generally decline during the hottest part of the summer.

Field Preparation

Beans grow optimally in well-drained, relatively light soils with a pH of 5.5 to 6.7. While the bean root system is extensive, most of the roots are located close to the soil surface. Beans for early harvest are usually grown in sandy soils, which usually warm rapidly in the spring. However, sandy soils dry out quickly and do not hold the fertilizer needed by this relatively shallow-rooted crop.

Although common beans are members of the legume family, which can utilize nitrogen from the atmosphere through bacteria in the root system, they are inefficient "fixers" of nitrogen and usually require fertilizer at planting plus additional fertilizer as the plants mature. Although inoculation can provide bacteria to aid in the nitrogen-fixing capability of bean plants, little advantage has been observed in most areas where beans are grown.

Depending on soil-test recommendations, beans often require 250 to 300 lb of 12-12-12 (or similar analysis) or 150 to 200 lb of 18-46-0 per acre prior to planting. The rate can be reduced by banding fertilizer 4 in. deep and 3 to 4 in. from the row. Beans require a regular supply of fertilizer for uniform growth, so one or two side-dressings are often necessary through the season, especially in wet seasons or sandy soils. Typically, a side-dress application of 25 to 30 lb of nitrogen per acre in a band 3 to 4 in. from the row is applied 2 to 3 weeks after planting. A second application 2 to 3 weeks later is often necessary in sandy soils or when heavy rainfall occurs.

Selected Cultivars

Common beans are divided into two classes: bush (Figure 22.1) and vining (or "pole") (Figure 22.2) beans. Each of these classes can be further divided into green-podded and yellow-podded wax types. Beans of either color type may be classified by shape of the pod (flat, oval, or round) or color of the seed. Sprite and Strike are the most widely grown round-podded, green fresh market cultivars. Sprite is particularly well suited for mechanical harvesting and Strike has high yield capabilities. Green Gator and Provider are also suited for fresh market production. Bush Blue Lake 47 and Tendercrop are processing and home garden cultivars and Resistant Cherokee is a commonly recommended wax cultivar.

Planting

Beans are seeded when danger of frost is past, and at least 65°F soil temperature is needed for good germination and early plant growth. Additional plantings often continue at intervals until the summer.

Bean seeds are planted 1 to 1½ in. deep often with a multiple-row planter (Figure 22.3). Deeper plantings are used for fall plantings or in sandy soils. Bean size

FIGURE 22.1   Bush-type beans growing in the field.

FIGURE 22.2   An established planting of pole-type beans.

varies, so seeding rates are adjusted for the cultivar used. Seed rates may vary from 50 to 100 lb per acre, although 60 to 80 lb are most common. Seeds are usually spaced three to four seeds per foot of row, with rows usually 30 to 38 in. apart (36 in. being most common) (Figure 22.4).

A 3- to 4-year rotation that excludes other legume crops is generally practiced for common bean production to prevent the buildup of soilborne root rots. Corn, sorghum, and small grains are good rotation crops that are used with common bean.

FIGURE 22.3   A multiple-row seeder for planting beans.

FIGURE 22.4   An established planting of beans.

Irrigation

Beans require an even supply of water and usually 1 to 1½ in. of water every 7 days, from rain or irrigation, are considered essential for good yields. The period from blossom development through early pod set is the most critical time for adequate moisture to ensure quality yields, but deficiencies at any time during the season can adversely affect growth and yield.

Beans are subject to soil crusting after planting so some watering is usually needed early in the season to ensure rapid emergence and uniform stands of seedlings. In sit-

FIGURE 22.5   The Mexican bean beetle is a common pest of beans. Source: Clemson University Extension Service.

uations where soil crusting can be a problem, light applications of water (¼ to ½ in.) are sufficient to reduce soil crusting and encourage seedling emergence.

Culture and Care

Bean plant growth and yields are severely reduced by competition with weeds. Weeds compete with bean plants for water, light, nutrients, and space. An effective weed control program is essential for profitable yields of beans. Herbicides for beans provide excellent weed control alone or in combination with cultivation.

Common Growing Problems and Solutions

Insects   Insects that can affect beans include Mexican bean beetles (Figure 22.5), aphids, and spider mites.

Diseases   Diseases that may attack beans include bean root rots, powdery mildew, rust disease, anthracnose, and bacterial blight. Fusarium root rot is a common root rot of beans and affected plants will have leaves that turn yellow and then abscise. The progression of diseased leaves is from the base of the plant upward. Rust appears on both sides of leaves as reddish brown blisters filled with powdery spores (Figure 22.6). Powdery mildew appears on the upper leaf surface as white, powdery mold growth and can spread to the pods if not controlled. Bean roots are also susceptible to nematodes.

Physiological Disorders   Beans will not set pods under extremely wet conditions such as are often found in low-lying wet locations and heavy soils. Overfertilization can result in excessive foliage growth with little increase in yield and possible salt damage from the fertilizer. Some problems with blossom development may occur with hot, dry, windy conditions during the late summer.

FIGURE 22.6   Bean leaf infected with rust-exhibiting reddish-brown blisters. Source: Clemson University Extension Service.

Harvesting

It usually takes 15 to 18 days following full blossom for beans to reach harvest peak. Generally bush cultivar types require less time for flowering than the pole cultivar types. Most newer bean cultivars have been developed for a concentrated set of pods. This requires "staggered" plantings to ensure freshly harvested product over a longer period of time.

Beans are ready to harvest when the majority of pods are over ¼ in. in diameter, well formed, snap when bent, and are not fibrous (Figure 22.7). Harvesting is done by hand for fresh market or pick-your-own. Mechanical harvesters (Figure 22.8) are available for larger sale, commercial, fresh market production. A washing/sorting line is usually needed for the harvested beans since harvesters bring stems, leaves, and some trash along with the beans. Commercially grown beans are graded by a sieve size. The key desirable maturity characteristics are undersized seed development and low side-wall fiber and as the pods age both of these quality factors are reduced.

Postharvest Handling

Beans are perishable and must be cooled to about 41°F and moved quickly to market. Hydrocooling is preferred because water cools without the potential of dehydration of the pods, which could occur with vacuum cooling.

Beans are typically maintained in cool (40° to 45°F) storage conditions with high humidity until ready for sale. Beans in these storage conditions can be stored for 7 to 10 days. A high relative humidity in storage assists in preventing shriveling or darkening of the pods.

Market Preparation and Marketing

Typical market outlets for common beans include wholesale or retail grocers, farmer's markets, or roadside stands. The beans are packed in wirebound crates, waxed cartons, or bushel baskets for transportation to market. Some beans are

FIGURE 22.7   Harvested common beans.

FIGURE 22.8   Mechanical harvesting of beans.

repacked at retail destinations in trays overwrapped with perforated film, but most are sold from bulk containers.

Beans also make an excellent pick-your-own crop. Many newer cultivars develop a concentrated set of pods, allowing the customer to provide a once-over harvest.

Uses

Beans are consumed fresh, canned, or frozen. They may be served in salads, either hot or chilled. Beans are also marinated and served cold either alone or with other salads. Beans are also common ingredients of stews and soups.

Nutrition Facts

Serving Size 3/4 cup cut beans (83g)

Amount Per Serving	
Calories 25	Calories from Fat 0
	% Daily Value*
Total Fat 0g	**0%**
Saturated Fat 0g	**0%**
Cholesterol 0mg	**0%**
Sodium 0mg	**0%**
Total Carbohydrate 5g	**2%**
Dietary Fiber 3g	**12%**
Sugars 2g	
Protein 1g	
Vitamin A 4%	• Vitamin C 10%
Calcium 4%	• Iron 2%

FIGURE 22.9 Nutritional value of green beans. Percent daily values are based on a 2,000 calorie diet. Used with permission from Dole Food Co., Inc.

Nutrition Facts

Serving Size 3/4 cup cut beans (83g)

Amount Per Serving	
Calories 25	Calories from Fat 0
	% Daily Value*
Total Fat 0g	**0%**
Saturated Fat 0g	**0%**
Cholesterol 0mg	**0%**
Sodium 5mg	**0%**
Total Carbohydrate 4g	**1%**
Dietary Fiber 2g	**7%**
Sugars 2g	
Protein 1g	
Vitamin A 2%	• Vitamin C 20%
Calcium 2%	• Iron 2%

FIGURE 22.10 Nutritional value of yellow beans. Percent daily values are based on a 2,000 calorie diet. Used with permission from Dole Food Co., Inc.

Nutritional Value

Both green and yellow beans are low in calories and are good sources of vitamin C. A ¾-cup serving of chopped green beans has only 25 calories, with no grams of fat and 2 to 3 g of dietary fiber (Figure 22.9). This serving of green beans provides 10% of the RDA of vitamin C, 4% of the RDA of both vitamin A and calcium, and 2% of the RDA for iron. A ¾-cup serving of yellow beans provides 20% of the RDA of vitamin C, and 2% of the RDA for vitamin A, calcium, and iron (Figure 22.10).

Lima Bean

Scientific Name

Phaseolus lunatus L.

Common Names in Different Languages

lima or sieva bean (English); da li cai dou (Chinese); haricot de Lima, kissi, pois du Cap (French); Limabohne, Mondbohne (German); fagiolo di Lima (Italian); laima biin, raimame (Japanese); fasol'-lima (Russian); judia de Lima, poroto de Lima, p. de manteca (Spanish)

Classification, History, and Importance

The lima bean (*Phaseolus lunatus* L.) is a member of the Fabaceae family. Other members of the family include common beans, garden (or English) peas, and Southern peas. Lima beans are a tropical perennial crop in the wild, but have been developed as an annual crop for production in the United States. It is an important crop for canning and freezing. The lima bean growth habit is similar to common bean and commercial cultivars bear three to four seeds per pod. Seed coats can be white, creamy, buff, or light green.

Lima beans (named for the capital city of Peru) are also called sieva, civet, seewee, butter, and sugar bean. The lima bean is native to tropical Central and South America and was cultivated in South America before 5000 B.C. Guatemala is considered the general region of origin. Dispersal of the crop was a result of Native American trade and travel. Consequently, lima beans were well established in the southwestern and southeastern United States prior to the arrival of early European explorers.

Lima beans are an important vegetable crop in the United States and are grown for canning and freezing and for fresh consumption. Lima beans for processing was a $35 million crop in 1997 (see Table 22.1). This was up from $33 million in 1996. Lima beans for processing were planted on 52,000 acres in the United States in 1995. Delaware is the state with the greatest acreage with over 10,000 acres. Other major production areas for processing include California, Washington, Oregon, Illinois, and Minnesota. California is the only state to harvest dry lima beans with 23,000 acres of baby lima beans and 21,000 acres of large-seeded lima beans planted in 1995. The largest share of U.S. dry lima bean market is packaged for resale and most of the California dry lima bean production is exported to Japan. Lima beans for the fresh market was about a $5 million industry with production primarily in Georgia. The U.S. per capita consumption of processed lima beans has decreased slightly from 0.7 lb per person in 1976 to 0.5 lb per person in 1996 (Table 22.4).

Plant Characteristics

Height: Bush, 10 to 15 in.; pole, 8 to 15 ft.
Spread: Bush, 4 to 8 in.; pole, 6 to 8 in.
Root Depth: Bush and pole, 36 to 48 in.

Climatic Requirements

Lima beans require slightly warmer weather for growth than common beans. Mean monthly temperatures of 59° to 75°F and a frost-free period are necessary for crop growth. Lima beans germinate at 60° to 85°F with an optimum soil temperature of

TABLE 22.4 *U.S. per capita consumption of commercially produced green lima beans for processing*

Use	1976	1986 (lb, farm weight)	1996
Processing	0.7	0.5	0.5

Source: USDA, Economic Research Service.

80°F. Low soil moisture along with high temperatures and low relative humidity often result in reduced pod set and retention. Temperatures of 90°F or above also reduce pollination and pod set, while high humidity favors pollination and pod set. Depending on cultivar, first flowering generally occurs 35 days from planting with peak flowering at 60 days.

Field Preparation

Lima beans grow on a variety of soils with a well-drained, fertile sandy loam, loam, or silt loam often yielding the best production. A soil with some organic matter is desirable. Lima beans grow best in a pH range from 6.0 to 6.7 and lime, if needed, is applied 3 months before planting. Fertilizer is often applied 10 days to 2 weeks before planting. Fertilizer is applied according to soil-test recommendations and a typical application may be 800 to 1,000 lb per acre of an 8-8-8 fertilizer or its equivalent.

Selected Cultivars

The bush-type lima beans are generally less expensive to produce than the pole-type lima beans. The pole type bears longer and, because it is staked, is easier to harvest by hand. Lima bean cultivars for processing purposes are green cotyledon types. Before the discovery of genetically controlled cotyledon color, processors typically removed mature beans whose cotyledons turned white.

Bush

- Hendersons mature in 70 days and produce small light green seeds. They are productive plants that are medium-small, bush formed, and erect growing.
- Early Thorogreen mature in 60 days and produce small green seeds. They are early maturing and highly productive.
- Jackson Wonder mature in 65 days and produce speckled seeds.

Pole

- Carolina Sieva mature in 80 days and produce small greenish white seeds. They set well in hot weather.

Planting

Bush lima beans require 50 to 60 lb of seed per acre for planting. Pole lima beans require 25 to 30 lb of seed for sowing 1 acre. Bush lima beans are sowed four to five seeds per foot of row, about 1.5 to 2 in. deep. Rows are spaced 3 to 3½ ft apart. For pole limas staked with individual poles, three to four seeds are bunched 24 to 26 in. apart and 2 in. deep. When a wire or string trellis is used, seeds are drilled three to four seeds per row with rows spaced 4 to 5 ft apart.

Irrigation

Without irrigation, lima bean yields are inconsistent and quality may suffer. Soil moisture is particularly important during blossom and pod set. Lima beans grown in dry soils often experience blossom drop and smaller pods than typical. On clay soils, lima beans should receive about ½ to ¾ in. of rainfall or irrigation per week before blossom, and 1 in. per week during blossom and pod development. On sandy soils, lima beans require ½ in. twice a week or ¾ in. every 5 days during blossom and pod development.

Culture and Care

Good weed control is essential for satisfactory yields of lima beans and mechanical cultivation is a major component of a successful weed-control program for lima beans. Cultivations are typically shallow since a deep cultivation may cause a ridge that interferes with the ability of the mechanical harvester to effectively collect the beans. Weedy fields at harvest can also reduce raw product recovery by harvesters and generate more trash in the harvested product.

Common Growing Problems and Solutions

Insects   Common insect problems for lima beans include Mexican bean beetles, corn earworms, stinkbugs, thrips, and whiteflies.

Diseases   Following are some of the diseases that may attack lima beans:

- Bacterial wilt   Symptoms of bacterial wilt are small, water-soaked areas on leaves and pods.
- Mosaic virus   Symptoms of mosaic virus on lima beans are distorted leaves with green and yellow streaking.
- Powdery mildew   Powdery mildew is suggested by a grayish white mold on the leaf surface.
- Seedling disease and root rot   Plants that are stunted or die in seedling stage may indicate seedling diseases and/or root rots.
- Rust   Symptoms of rust include reddish-orange pustular spots on the lima bean leaves.
- Southern blight   Plants that suddenly wilt and die may be indicative of Southern stem blight. White mold at soil line can also be used as an indication of infection.
- Stem anthracnose   Reddish blotches on pods, stems, and leaves can be indicative of stem anthracnose.

Harvesting

Lima beans are typically harvested when the pods are well developed (Figures 22.11 and 22.12) and contain 65 to 75% moisture and 20 to 30% solids. As the beans mature, the seeds turn pale green then white. Lima beans harvested for fresh market are often hand-picked and almost all lima beans grown for processing are harvested mechanically with pod-stripper combines.

Postharvest Handling

Lima beans are highly perishable and sensitive to chilling injury. Precooling, preferably by hydrocooling, immediately after harvest to 41° to 43°F for unshelled beans or 37° to 39°F for shelled beans is typically practiced. Storage life, with 95% relative humidity and proper cooling conditions, is 5 days for unshelled or 7 days for shelled beans.

FIGURE 22.11   Lima bean pods on a plant at the approximate stage for harvesting.

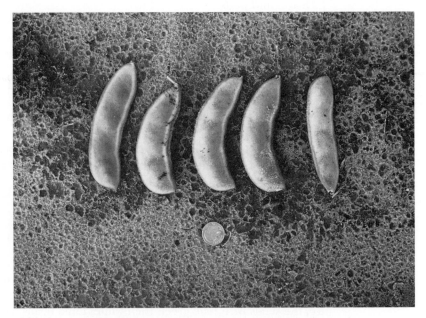

FIGURE 22.12   Harvested lima bean pods at proper maturity for harvesting.

Market Preparation and Marketing

Lima beans grown for processing are marketed as either a canned or frozen product and are often used as ingredients in soup products. Both canned and frozen lima beans are distributed for retail and institutional use, either as a single product or mixed with other vegetables.

Uses

Almost 50% of lima beans produced are used as frozen product, 30% are canned, and the rest are utilized fresh. Lima bean pods are not consumed. The seeds are boiled, steamed, or baked and served as a side dish, in succotash with corn, and in stews and salads.

Nutritional Value

Lima beans offer a good source of vital elements, especially vitamin A.

Garden Pea

Scientific Name

Pisum sativum L. ssp. *sativum*

Common Names in Different Languages

pea, garden pea (English); wan dou (Chinese); pois potager, petit pois (French); Erbse, Garten-Erbse (German); pisello coltivato (Italian); endou (Japanese); goroch ovoščnoj (Russian); guisante, g. comun, pésol, arveja (Spanish)

Classification, History, and Importance

Garden peas (*Pisum sativum*) are viny members of the Fabaceae family. Other family members include common beans, lima beans, and Southern peas. Garden peas are also known as English peas. Peas are viny, annual herbaceous plants that have tendrils at the tip of the compound leaves.

The garden pea appears to have originated in western parts of Asia and the eastern Mediterranean. Carbonized seeds discovered in Switzerland have been dated to about 7000 B.C. Peas were probably cultivated by the Greeks and Romans and became an important crop in the Middle Ages. The first detailed description of peas dates from the 16th century. Columbus brought the pea to North America where it quickly spread to all parts.

The pea was first grown only for its dry seeds and it was not until after the Norman conquest of England that "green peas" were mentioned. Peas were then used in the fresh green stage, cooked in whole pods and eaten from the pods. Peas were considered a delicacy in France in the 17th century. Because so many cultivars of peas were grown in England, they were often referred to as "English peas," a term that continues today. The name pea is derived from the Latin word *pisum*, which later became *pease*.

Commercial production of peas is primarily for processing, including canning and freezing. Garden peas were a $137 million processing industry in 1997 (see Table 22.1). This was up from $117.5 million in 1996. Per capita consumption of peas decreased from 4.8 lb per person in 1976 to 3.4 lb per person in 1996 (Table 22.5). Reduction in per capita pea consumption during this time period was recorded with canned but not frozen peas. Of the 292,600 acres of garden peas grown for processing in the United States, 90,000 are in Minnesota, 54,400 in Washington, 62,500 in Wisconsin, 28,100 in Oregon, and 18,900 in New York (Table 22.6).

TABLE 22.5   *U.S. per capita consumption of commercially produced garden peas for canning and freezing*

Use	1976	1986 (lb, farm weight)	1996
Canning	2.9	2.2	1.5
Freezing	1.9	1.9	1.9
All	4.8	4.1	3.4

Source: USDA, Economic Research Service.

TABLE 22.6   *The top garden pea producing states for processing in the United States*

Rank	State	Planted acres (000)	Harvested acres (000)	Yield (tons/acre)	Production (000)
1	Minnesota	90.00	75.20	1.48	111
2	Washington	54.40	53.70	1.95	105
3	Wisconsin	62.50	58.00	1.80	104
4	Oregon	28.10	27.80	1.54	43
5	New York	18.90	18.20	2.21	40
.	United States	292.60	268.90	1.77	476

Source: USDA, NASS, 1997 data.

Plant Characteristics

Height: Garden, 2 to 4 ft; snap, 4 to 6 ft
Spread: 6 to 10 in.

Climatic Requirements

Garden peas are frost-hardy vegetables that thrive in cool, moist weather. The optimum temperature for pea plant growth is between 55° and 65°F. Peas are planted when the soil temperatures are 45°F or higher and when the soil is dry enough to till. Pea plants may die if exposed to long periods above 78°F. Long days and low temperatures accelerate flowering.

Field Preparation

Peas prefer a well-drained sandy or silt loam soil with some organic matter. Soil pH for peas is critical with levels below 6.0 reducing yields. Lime, if needed, is typically applied at least 3 months before planting. Peas may require between 50 and 100 lb of P and K and 30 to 50 lb of N on a medium-fertile soil.

Selected Cultivars

Garden peas come in several types. The common shelled pea is divided into dry, canner, and freezer peas. Edible-podded peas are divided into snow peas and snap peas.

The garden cultivars have smooth or wrinkled seeds. The smooth-seed cultivars tend to have more starch than the wrinkled-seed cultivars. The wrinkled-seed cultivars are generally sweeter and are usually preferred for home use.

- Little Marvel—62 days to maturity. Medium-sized, dark green seed, good yield and quality, well adapted to home gardens
- Wando—60 days to maturity. Medium-sized, dark green seed, some heat tolerance
- Dwarf Gray Sugar—65 days to maturity. Light green, pods are edible
- Sugar Snap—70 days to maturity. Produces edible pod and shelled peas

Planting

Peas can be planted in early spring as soon as the soil can be tilled. Peas are planted 1 to 1½ in. deep and 1 in. apart in single or double rows. Row spacings of 18 to 24 in. between single rows and 10 to 18 in. between double rows are often used.

Irrigation

Peas are moisture sensitive and yields can be reduced if soil water is not sufficient during their growth. Irrigation up to bloom is important, but irrigation during the flowering period can cause pod drop or poor pod fill. However, irrigation after pods are set is usually beneficial and improves yield and quality of the crop.

Culture and Care

The germinating pea seeds and small seedlings are easily injured by direct contact with fertilizer or improper cultivation. Most dwarf and intermediate pea cultivars are self-supporting. The taller cultivars are more productive and easily picked when caged (Figure 22.13) or trained to poles or a fence for support. Mulching can be used to cool the soil, reduce moisture loss, and prevent soil rots.

Weed control is essential in the first 6 weeks after planting. Shallow cultivation is the preferred method of cultivation.

Common Growing Problems and Solutions

Insects Insect pests of peas include aphids, Mexican bean beetles, leafhoppers, and mites. Aphids are the major insect problem because of the damage they cause to young plants and for the disease they carry.

Diseases Diseases that may attack peas include fusarium wilt, powdery mildew, and root rot. The first signs of fusarium wilt and root rot diseases are the yellowing and wilting of the lower leaves and stunting of the plant. These diseases are not prevalent in well-drained soils.

Harvesting

Swollen pea pods (that appear round) are indicators that garden peas are ready to harvest. Peas are of best quality when they are fully expanded but not hard and starchy. Pea quality declines immediately after harvest (similar to sweet corn).

FIGURE 22.13   Wire cage used for growing tall cultivars of garden peas.

Edible-podded peas are generally harvested before the individual peas have grown to ¼ in. in diameter. The pods are usually ready for harvest 5 to 7 days after flowering. Snap peas are harvested when the pods are filled with plump seeds.

Postharvest Handling

Peas for fresh market are cooled to 32°F to reduce moisture loss and to preserve sugar. Often garden peas are packed with crushed ice to extend shelf life. Snap peas and snow peas are not typically iced.

Market Preparation and Marketing

Peas should be consumed as soon as possible after harvest or processed immediately to retain peak quality. Peas are generally shipped in 10-lb cardboard cartons.

Uses

Peas are grown for canning or freezing and are served as a major side dish or mixed in salads. They can be served alone or in sauces. Fresh peas may be marketed at roadside stands and are popular with home gardeners.

Nutritional Value

The nutritional value of the garden pea is related to its maturity. The contents of calcium, phosphorus, iron, sodium, and potassium increase substantially with seed maturity, as do those of protein and carbohydrate. Vitamins A and C decline with drying.

Southern Pea

Scientific Name

Vigna unguiculata (L.) Walp. ssp. *unguiculata* (L.) Walp.

Common Names in Different Languages

southern or black-eye pea, cowpea (English); pu tong jiang dou (Chinese); haricot à oeil noir, h. dolique, niébé, dolique chinois (French); Kuherbse, Kuhbohne, Augenbohne (German); fagiolino dall'occhio, veccia (Italian); sasage (Japanese); caupi, judia de vaca, frijol de vaca, f. de pinta negră, f. de ombligo negrŏ, caragilates (Spanish)

Classification, History, and Importance

The Southern pea (*Vigna unguiculata*) is a member of the Fabaceae or bean family. Other members of the family include common beans, lima beans, and garden (or English) peas. Southern peas originated in India in prehistoric times and were introduced into Africa and later into America. Southern peas can still be found growing wild in Africa. Historically, Southern peas were an important forage and green manure crop in the United States. Southern peas are known by 50 common names in India and are called "field peas," "crowder peas," "cowpeas," and "black-eyed peas" in the United States.

The Southern pea is a bean and not a pea. It is grown primarily for its green-shelled seeds. At the most popular edible stage of Southern pea, the pods and seeds have developed fully but the seeds have not dried.

Climatic Requirements

The Southern pea is a warm-season crop that grows best in hot, dry climates. It is very susceptible to injury from cold temperatures.

Field Preparation

Southern peas may be grown on a wide variety of soils from sandy loam to clay soils. In general, 400 to 600 lb of a complete fertilizer such as 4-12-12 or 5-10-10 is adequate. No added nitrogen is usually required. Too much fertilizer applied to Southern peas produces excessive vine growth and poor yields. Inoculants of nitrogen-fixing bacteria may increase yield especially in soils where Southern peas have not been grown. Neutral to slightly acidic soils (pH between 5.5 and 6.5) are preferred for growing Southern peas. Soils with pH above 7.5 are typically avoided. High calcareous soils may cause chlorosis (iron deficiency) of Southern pea, which reduces plant growth and yield.

Selected Cultivars

Southern peas are identified by color of hull, seed and embryo area, or by spacing (crowding) of seed within the pod. For example, there are purple hull peas (Figure 22.14), cream peas, pink, or black-eyed peas (Figure 22.15), and crowder peas

(Figure 22.16). Classifications also overlap the individual characteristics (e.g., pink eye, purple hull). The purple hull cultivars (Mississippi Purple, Purple Hull Pink Eye, and Purple Tip Crowder) are suited for auction or competitive markets. The larger cultivars such as Colossus, Hercules, and Big Boy are well suited for pick-your-own and local sale. Colossus is an excellent freezing cultivar, but it does not shell easily through a mechanical sheller.

Planting

In general, 30 to 40 lb of Southern pea seed are required per acre when seeding 36- to 42-in. spacing between rows. Seeds are spaced 3 to 4 in. within the row and planted 1 to 1½ in. deep. Seeding can begin when soil temperatures reach 60°F. Southern pea seeds will decay in cool, wet soils. Bush types are seeded 4 to 6 per foot or 30 to 50 lb per seed per acre for large seed. Vining types are seeded 1 to 2 per foot or 20 to 30 lb of seed per acre. Seeds are planted ¾ to 1¼ in. deep in rows spaced 20 to 42 in. apart depending on cultivation requirement.

Irrigation

Southern peas are among the more drought-hardy of the common vegetables. The critical irrigation period for Southern pea is during blooming. In many areas where they are grown, Southern peas do not require much supplemental irrigation water. In fact, reduced yields may result from too much water if vegetative growth is excessive. For highest yields, peas should receive 1 in. of water each week, either by rainfall or by irrigation.

FIGURE 22.14 A purple hull type of Southern pea. Source: Ogle, W. L., W. Witcher, and O. W. Barnett. 1987. *Descriptors for the Southern peas of South Carolina.* Clemson Univ. South Carolina Experiment Station Bull. 659.

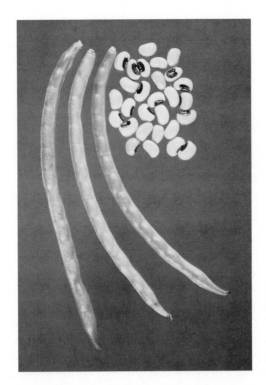

FIGURE 22.15 A black-eyed type of Southern pea. Source: Ogle, W. L., W. Witcher, and O. W. Barnett. 1987. *Descriptors for the Southern peas of South Carolina.* Clemson Univ. South Carolina Experiment Station Bull. 659.

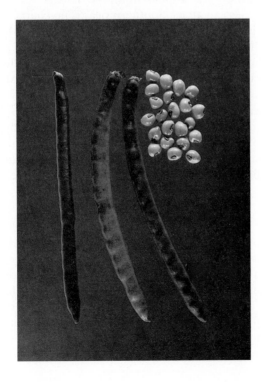

FIGURE 22.16 A crowder type of Southern pea. Source: Ogle, W. L., W. Witcher, and O. W. Barnett. 1987. *Descriptors for the Southern peas of South Carolina.* Clemson Univ. South Carolina Experiment Station Bull. 659.

Culture and Care

Weed control, especially during early growth in the field, is important for successful production of Southern peas. Besides application of herbicides to control weeds, shallow cultivations are commonly used.

Common Growing Problems and Solutions

Insects   Cowpea curculio is the most important insect on Southern peas. It causes stinging of the pods and is responsible for worms in the pod and dried seed. Thrips can appear early in the season and cause considerable crinkling of leaves of the seedling plants. Lesser cornstalk borers can be a problem for Southern peas, by boring into stems of young plants and reducing stands.

Diseases   Root rot and damping-off are caused by three different fungi. Symptoms vary and include rapid death of young, succulent plants, discoloration of taproots, longitudinal cracks of the stems, stunting, wilting, and poor yields.

Southern stem blight is caused by a fungus that attacks roots and stems of Southern peas. The first visible sign of southern stem blight is a progressive yellowing and wilting of the foliage, beginning with the lower leaves. The plant dies within a few days after the first symptoms appear. During warm, moist conditions, the coarse white mycelium of the fungus makes characteristic fan-shaped patterns of growth on the stem at the soil line. In this white-mat fungus, numerous smooth, round, light tan to dark brown mustard seedlike bodies called sclerotia are formed.

Fusarium wilt–infected plants have lower leaves that often turn yellow on one side of the plant. Infected plants are usually stunted and wilted. Brick red tissue can be observed in the stem when it is split lengthwise.

Several virus diseases can attack Southern peas. The characteristic symptom of mosaic virus disease is an intermixing of light and dark areas. Mottled areas are irregular in outline and may follow the main veins. Infected leaves are frequently stunted, and often there is a slight puckering and curling of the leaf edges. Infected plants are usually dwarfed and bunchy, and yields are reduced. Mosaic diseases also can result in malformed fruit or pea pods.

Roots infected with root-knot nematodes become galled and knotted. Aboveground plant symptoms of nematodes resemble nutrient deficiencies including stunting and often wilting. This occurs because infected root systems are incapable of absorbing adequate amounts of water and nutrients. Root-knot nematodes also increase the susceptibility of the plant to attack by fusarium wilt.

Harvesting

Southern peas are not harvested until some green has disappeared from the pod (Figure 22.17). Cultivars differ in their pod color, with some being purple and others yellow. They should be harvested before any browning is evident.

Harvest will begin 65 to 80 days after seeding and continue for 3 to 5 weeks, depending on weather and cultivar of Southern pea. Fresh-market peas are generally harvested by hand 16 to 17 days after bloom, depending on temperature. Mechanical harvesting of Southern peas is generally used for processing and growers may contract with a processor for harvesting.

FIGURE 22.17   Southern pea at the proper stage of maturity for marketing.

Postharvest Handling

Peas after harvest should be maintained in a shaded and well-ventilated area. Quality is maintained by using forced air to cool the peas to about 45°F. Peas cooled below 45°F may be injured.

Market Preparation and Marketing

A large amount of Southern peas are processed and canned or frozen. Fresh market peas are hand harvested, packed in crates, and shipped under refrigeration to local or distant markets. Dry peas are cleaned, graded, stored, and fumigated for future packaging in consumer-sized plastic bags.

Review Questions

1. When is the most critical time for moisture for common beans?
2. How can overfertilization reduce the yields of beans?
3. How do the preferred environmental conditions for lima beans differ from the preferred environmental conditions for common beans?
4. How does the production of bush-type lima beans differ from the production of pole-type lima beans?
5. What are the various cultivar types of garden peas?
6. What are the watering needs of garden peas during their growth and development?
7. What are some of the ways that Southern pea cultivars are classified?
8. Do Southern peas generally need supplemental irrigation?

Selected References

Anderson, C. R. 1998. *Peas.* Univ. of Arkansas Coop. Ext. Serv. Publ.

Ball, J. 1988. *Rodale's garden problem solver: Vegetables, fruits and herbs.* Emmaus, PA: Rodale Press. 550 pp.

Cook, W. P., R. P. Griffin, and C. E. Drye. 1988a. *Commercial Lima bean production.* Clemson Univ. Coop. Ext. Serv. Hort. Leaflet 13.

Cook, W. P., R. P. Griffin, and C. E. Drye. 1988b. *Commercial snap bean production.* Clemson Univ. Coop. Ext. Serv. Hort. Leaflet 1.

Cook, W. P., R. P. Griffin, and C. E. Drye. 1988c. *Commercial Southern pea production.* Clemson Univ. Coop. Ext. Serv. Hort. Leaflet 36.

Cook, W. P., R. P. Griffin, and A. P. Keinath. 1995. *Commercial pole bean production.* Clemson Univ. Coop. Ext. Serv. Hort. Leaflet 12.

Hagedorn, D. J., and D. A. Inglis. 1986. *Handbook of bean diseases.* Univ. of Wisconsin Coop. Ext. Serv. Publ. A3374.

Kays, S. J., and J. C. Silva Dias. 1995. Common names of commercially cultivated vegetables of the world in 15 languages. *Economic Botany* 49:115–152.

Kee, E., J. L. Glacey, and T. L. Wooten. 1997. The lima bean: A vegetable crop for processing. *HortTechnology* 7:119–128.

Marr, C. W., W. J. Lamont, and N. Tisserat. 1995. *Snap beans.* Kansas State Univ. Coop. Ext. Serv. Publ. MF-2076.

Menges, T., J. E. Larsen, S. Cotner, R. Roberts, B. D. McCraw, and T. Longbrake. *Keys to profitable Southern pea production.* Texas A&M Univ. Coop. Ext. Serv. Cir. L-1862.

Nagel, D. 1997a. *Commercial production of English peas in Mississippi.* Mississippi State Coop. Ext. Serv. Publ. 1503.

Nagel, D. 1997b. *Commercial production of lima beans in Mississippi.* Mississippi State Coop. Ext. Serv. Publ. 1520.

Nagel, D., W. L. Bateman, F. Killebrew, D. W. Houston, J. H. Jarret, B. Graves, and T. Jones. 1997. *Commercial production of Southern peas.* Mississippi State Coop. Ext. Serv. Publ. 1535.

Nonnecke, I. L. 1989. *Vegetable production.* New York: Van Nostrand Reinhold. 656 pp.

Ogle, W. L., W. Witcher, and O. W. Barnett. 1987. *Descriptors for Southern peas of South Carolina.* South Carolina Ag. Expt. Station Bull. 659.

Peirce, L. C. 1987. *Vegetables: Characteristics, production, and marketing.* New York: John Wiley & Sons.

Riofrio, M. 1992. *Growing peas and snap beans in the home garden.* Ohio State Univ. Ext. Serv. Fact Sheet HYG-1617-92.

Sanders, D. C. 1994. *Southern peas.* North Carolina Coop. Ext. Serv. Leaflet 20.

Seelig, R. A., and E. Roberts. 1960. Green or wax snap beans. *Fruit & Vegetable Facts & Pointers.* United Fresh Fruit & Vegetable Association, Washington, DC.

Yamaguchi, M. 1983. *World vegetables.* New York: Van Nostrand Reinhold.

Selected Internet Sites

USDA Agricultural Marketing Service Standards for Grades

Fresh
Beans (common)   www.ams.usda.gov/standards/beansnap.pdf
Beans (lima)   www.ams.usda.gov/standards/beanlima.pdf
Peas   www.ams.usda.gov/standards/peasfrh.pdf
Peas (Southern)   www.ams.usda.gov/standards/peasouth.pdf

Processing
Beans (common)   www.ams.usda.gov/standards/vpbeans.pdf
Beans (lima)   www.ams.usda.gov/standards/vpbeanl.pdf
Peas (canning or freezing)   www.ams.usda.gov/standards/vppeafz.pdf
Peas (Southern)   www.ams.usda.gov/standards/vppeas.pdf

23

Sweet Corn

Sweet Corn

Scientific Name

Zea mays L. ssp. *mays*

Common Names in Different Languages

sweet corn (English); yu mi (Chinese); maïs sucré, m. doux (French); Zuckermais (German); mais zuccherino, m. dolce, granoturco da zucchero (Italian); tomorokoshi (Japanese); sacharnaja kukuruza (Russian); maiz dulce, m. azucarado, m. tierno (Spanish)

Classification, History, and Importance

Sweet corn, *Zea mays,* is a member of the Poaceae (Gramineae) or grass family. Other members of the family include wheat, oats, barley, sorghum, and rice. Sweet corn is an annual, monocotyledonous cereal that can grow to 12 ft in height.

Sweet corn probably originated from a mutation of an ancient Peruvian corn called *Chuspillo* or *Chullpi.* Sugary forms of corn were not very popular in early culture because they were difficult to store. Corn (maize) is native to America and has been cultivated in Central America since 3500 B.C. It was an important food of the Incas of Peru, the Aztecs of Mexico, and the Mayas of Central America. The cliff dwellers of southwestern United States were also corn-growing and corn-eating people.

The first historical reference to sweet corn in the United States was in 1779 with the introduction of Papoon or Susquehanna, an eight-rowed red cob strain grown by the Iroquois. There were 63 known cultivars of sweet corn by the early 1900s including Golden Bantam (released in 1902), which became one of the more popular open-pollinated cultivars.

Most open-pollinated sweet corn cultivars have been replaced by improved hybrid cultivars that are easy to grow, produce good yields, taste sweeter, and store longer. Sweet corn cultivars are generally classified by seed color, maturity date, or nature of sweetness.

Sweet corn for fresh market was a $398.3 million industry in 1997 (Table 23.1). This was up from $384.4 million in 1996. The sweet corn for processing market is smaller than for fresh market with a production value of $247.8 million in 1997. Per capita consumption of sweet corn has increased from 25.8 lb of both fresh market and processing sweet corn per person in 1986 to 29.3 lb per person in 1996 (Table 23.2). Of the 241,000 acres of sweet corn grown for fresh market in the United States, 43,300 are in Florida, and 27,000 are in California (Table 23.3). Other large fresh market sweet corn-producing states include Georgia, New York, Ohio, Michigan, Pennsylvania, Colorado, and North Carolina. Of the 476,930 acres of sweet corn for processing grown in the United States, Minnesota grows 128,600 acres, Wisconsin 115,800 acres, and Washington 89,600 acres.

Plant Characteristics

Height: 6 to 12 ft
Spread: 1½ to 4 ft
Root Depth: Most roots are shallow, 18 to 36 in., but some are deep fibrous roots extending down as far as 7 ft.

Climatic Requirements

Sweet corn is a warm-season vegetable that is easily killed by frost. Seed germination is poor below 60°F and does not occur at soil temperatures below 50°F. The

TABLE 23.1   *The production value of sweet corn for fresh market and processing in the United States*

| | Value of production | | |
Crop	1995	1996	1997
		(1,000 dollars)	
Fresh market	389,288	384,445	398,279
Processing[1]	251,156	258,840	247,839

Source: USDA, NASS.
[1]Value at processing plant door.

TABLE 23.2   *U.S. per capita consumption of commercially produced fresh and processing sweet corn*

Use	1976	1986	1996
		(lb farm weight)	
Fresh	8.0	6.1	8.3
Canning	13.1	12.1	10.5
Freezing	5.9	7.6	10.5
All	27.0	25.8	29.3

Source: USDA, Economic Research Service.

TABLE 23.3   *The top fresh market and processing sweet corn-producing states in the United States*

Rank	State	Planted acres (000)	Harvested acres (000)	Yield	Production (000)
Fresh market					(cwt/acre)
1	Florida	43.30	41.30	140.00	5,782
2	California	27.00	27.00	145.00	3,915
3	New York	23.00	21.20	73.00	1,548
4	Georgia	20.00	19.00	130.00	2,470
5	Pennsylvania	19.50	16.00	55.00	880
6	Ohio	15.10	14.20	74.00	1,051
7	Michigan	12.50	11.50	85.00	978
8	Illinois	7.70	7.50	94.00	705
9	North Carolina	6.60	6.50	110.00	715
10	Colorado	6.00	5.80	165.00	957
	United States	241.00	222.80	101.00	22,587
Processing					(tons/acre)
1	Minnesota	128.60	127.20	6.40	814
2	Wisconsin	115.80	109.40	6.56	718
3	Washington	89.60	87.70	8.88	779
4	Oregon	41.50	41.00	8.61	353
5	New York	40.40	39.30	6.40	252
6	Illinois	15.30	15.00	5.90	89
7	Idaho	15.20	14.30	9.50	136
8	Pennsylvania	2.80	2.70	7.11	19
	United States	476.93	464.22	7.16	3,324

Source: USDA, NASS, 1997 data.

optimum soil temperature for sweet corn is 70° to 85°F and optimum plant growth occurs between 70° and 86°F. In general, the warmer the air temperatures the faster the corn will grow to maturity.

Field Preparation

Sweet corn can be grown on a wide variety of soil types, but grows best on well-drained soils that have good water-holding capacity. Sandy soils are often used in the early spring for early season production. Soil pH for growing sweet corn should be between 5.8 and 6.5.

Sweet corn with long green leaves and dark green husks is required by the commercial market (Figure 23.1). An adequate supply of nitrogen must be maintained in the soil to produce corn of this quality. While fertilizer is applied according to soil-testing results, a general recommendation in North Carolina is to apply a total of 150 to 180 lb of nitrogen, 50 to 60 lb of phosphate (P_2O_5), and 70 to 90 lb of potash (K_2O_5) per acre. Fertilizer is often applied in one or two bands approximately 3 in.

FIGURE 23.1   Good quality sweet corn ears as indicated by the long green outer leaves and the dark green husks. Courtesy of W. J. Lamont, Jr.

FIGURE 23.2   Yellow sweet corn type. Courtesy of M. D. Orzolek.

to the side and 2 to 3 in. below the seed. Between 50 to 60 lb of actual N is applied preplant or at planting, while a side-dressing of 90 to 100 lb N is applied when the plants are 18 to 24 in. tall.

Selected Cultivars

Sweet corn cultivars are classified by seed color and maturity date. Sweet corn comes in three colors: yellow (Figure 23.2), white (Figure 23.3), and bicolor (yellow and

FIGURE 23.3   White sweet corn type.

FIGURE 23.4   Bicolor (yellow and white) sweet corn type. Courtesy of M. D. Orzolek.

white) (Figure 23.4). Cross-pollination of yellow kernel cultivars with white kernel cultivars results in bicolor corn. There is no relationship between color and sweetness.

The genetics of kernel sweetness is increasingly more important in selecting sweet corn cultivars. The difference in the sweetness in normal sweet corn is due to specific genes. Regular sweet corn has the recessive version of the *sugary-1* gene *(su1)*. Cultivars of the *su* type have traditional sweet corn flavor and texture. They require careful harvesting and handling procedures to preserve sweetness and flavor. Cultivars of the *su* type include Merit (yellow), Silver Queen (white), and Sweet Sue (bicolor).

Modified sugary sweet corn (e.g., Snowbell) has the *sugar-enhancer* gene (*se*), which increases the sugar content of the kernel by modifying the *sugary-1* gene. Cultivars of the *se* type feature high sugar, creamy texture, natural corn flavor, and thin pericarp (making it easy to chew), and are best for local markets. Cultivars of the *se* type include Kandy Korn (yellow), Platinum Lady (white), and Double Delight (bicolor).

Supersweet corn has a higher sugar content because of its *shrunken-2* gene (*sh2*) than normal sweet corn or modified sugary sweet corn. Supersweet types compared to normal or modified sugary types are higher in sugar, tougher skinned, lower in starch, and retain their quality longer after harvest. The seed of supersweet types have lower seed germination rates as compared to normal or modified sugary types, particularly in cold soils. They also have smaller, more brittle seeds that crack easily if mishandled. Cultivars of *sh2* types include Summer Sweet 7600 (yellow), Summer Sweet 8601 (white), and Summer Sweet 8502 (bicolor).

Genes affecting sweetness in sweet corn are recessive. If cultivars having these "sweet" genes are pollinated by cultivars having the dominant forms of these genes, sweetness will be lost. The risk of cross-pollination is minimized by separating stands of different types of sweet corn by at least 250 ft or by planting so that flowering of different types of cultivars does not occur at the same time.

Early maturing cultivars of sweet corn often produce smaller ears and have poorer eating quality than late maturing cultivars. The number of days to maturity are listed for cultivars by seed companies but are only used as guides. The actual number of days required to reach harvest quality will be greater in cool weather than in warm weather.

Planting

Sweet corn is typically planted 1 or 2 weeks before the frost-free date in the spring in many regions of the United States. Sweet corn is planted when soil temperatures reach at least 60°F and the possibility of hard frosts (24°F or less) has passed. Sweet corn is often planted in successive plantings to extend the harvest period later into the season. Early maturing cultivars are typically used for early spring production since they are normally more vigorous under cool temperatures.

Generally, sweet corn is mechanically planted in rows 36 to 38 in. apart. Early cultivars may be spaced 8 to 10 in. apart in the row and later cultivars 10 to 12 in. apart in the row (Figure 23.5). Seeds are planted about 1 in. deep on loam soils, ¾ in. deep on clay soils, and 1¼ in. deep on sandy soils. Sweet corn is typically planted in blocks (multiple rows) that are at least three to four rows wide to ensure good pollination. Pollination is accomplished by wind or gravity.

Irrigation

Adequate soil moisture is needed during early stages of germination and development and to activate herbicides used for growing sweet corn. Adequate soil moisture is also critical during silking and ear development. Good yields of high-quality sweet corn are rarely attained without irrigation, especially during these critical periods.

FIGURE 23.5   An established planting of sweet corn. Courtesy of M. D. Orzolek.

Culture and Care

Weeds are controlled in sweet corn plantings using shallow cultivations until the plants are 18 to 25 in. high. Cultivations are shallow to prevent damage to corn roots. Suckers or side shoots of sweet corn may develop off the main stem of plants during the season. The number of suckers a sweet corn plant produces depends on the cultivar. Suckers are not usually removed because their removal does not increase yields and may even reduce yields.

Common Growing Problems and Solutions

Insects   Major insect pests on sweet corn include soil insects (cutworms, wireworms), foliar insects (armyworms, flea beetles, corn earworms, aphids), and stem borers (corn borers). Corn earworms are generally the most destructive of the insect pests. Damage is usually more severe in warmer areas. Eggs of the corn earworm are laid on the young silks where they hatch and the larvae feed on the silks and tips of the ears (Figure 23.6). Flea beetles are small black jumping insects that eat small holes in the leaves of young plants. They may carry and transmit Stewart's wilt. Corn borers are white worms with dark heads. First-generation borers feed on whorl leaves during the early summer and later generations invade stalks, ears, and ear shanks.

Diseases   Stewart's wilt is caused by a bacterium transmitted by the flea beetle. Symptoms include plant wilting and leaves with long pale green to yellow or brown irregular streaks. Stewart's wilt is usually more severe on smaller, early maturing cultivars.

Corn smut is a fungus disease characterized by whitish galls usually erupting from ear tips around silking time. Corn smut can be a problem brought on by drought, high nitrogen levels, or plant damage from insects, cultivation, or hail.

Physiological Disorders   Incomplete pollination results in kernel skips on the ear (Figure 23.7). Poor pollination can result when air temperatures are above 96°F, when plants are exposed to hot winds, or when plants are under moisture stress.

FIGURE 23.6   Corn earworm damage to sweet corn. Courtesy of W. J. Lamont, Jr.

FIGURE 23.7   Kernel skips on a sweet corn ear as a result of poor pollination.

Harvesting

Sweet corn has a very short period of optimum harvest maturity, and quality deteriorates rapidly prior to and following this optimum maturity. Sweet corn ears are generally ready for harvest approximately 3 weeks after silk emergence (depending on cultivar and air temperatures).

The condition of the silk can be used as an indicator of harvest maturity and sweet corn is not usually harvested until the silks are dry and brown. Ears harvested immaturely will be of small diameter, have poor cob fill, and kernels that are watery and lack sweetness. The kernels are plump, sweet, milky, tender, and nearly maximum size at optimum harvest maturity.

After optimum harvest maturity has been reached, the eating quality of sweet corn begins to decrease rapidly while husk appearance changes very little. Over-mature corn is starchy (rather than sweet), tough, and the kernels are often "dented." Ears are often harvested with a few immature kernels at the tip because these ears tend to be sweeter and more tender than ears in which the tip kernels are full size.

Sweet corn may be harvested either by hand or by mechanical harvester (Figure 23.8). Selection of harvest method depends on the grower, availability of labor, size of operation, and intended market. If harvested by hand, the ears are usually pulled from the stalk. If the shank at the base of the harvested ear is too long it is broken off. For commercial marketing, all ears should have about the same length shank and the same amount of husk (Figure 23.9).

Postharvest Handling

Freshly harvested sweet corn is highly perishable and eating quality deteriorates rapidly after harvest. The loss of sweetness in stored harvested sweet corn is due to conversion of sugars to starch and this conversion is most rapid at high temperatures. At 86°F, 60% of the sugar may be converted to starch in 24 hours; whereas, at 32°F, sugar content would decrease only 6%. For optimum keeping quality, sweet corn should be cooled to 32°F within 1 hour after harvest and held at 32°F until consumed (Figure 23.10).

Temperature management in postharvest handling of sweet corn begins in the field. Sweet corn is often harvested in the early morning when air and ear temperatures are coolest. Ears harvested in the morning require less cooling to reduce the ear temperatures to 32°F and tend to have a longer shelf life than ears harvested during the warmer times of the day. Sweet corn is placed in the shade after harvest to prevent heating by the sun.

FIGURE 23.8   Mechanical harvesting of sweet corn. Courtesy of M. D. Orzolek.

FIGURE 23.9   Uniformly harvested sweet corn ears placed in boxes. Courtesy of W. J. Lamont, Jr.

FIGURE 23.10   Forced-air cooling of harvested sweet corn in boxes. Courtesy of W. J. Lamont, Jr.

Market Preparation and Marketing

Sweet corn ears are typically marketed in their husks. Long shanks and flag leaves of the harvested sweet corn ears are trimmed to reduce moisture loss from the ears and subsequent denting of the kernels. Icing or spraying with water helps reduce ear temperatures and moisture loss from the kernels.

Some sweet corn ears may have their husk leaves removed and several ears may be wrapped in cellophane for marketing. This allows the consumer to see the characteristics of ear fill, plumpness of kernels, and any damage from corn earworms without having to peel back the husk as is so often done when sweet corn ears are marketed in their husks.

High-quality sweet corn is sold on farms, at roadside stands, city fruit and vegetable stands, or farmer's markets. If sweet corn is marketed locally, frequent small harvests are usually practiced so that holding time is minimal. Fresh market sweet corn traditionally is sold from open bulk containers or by the dozen in paper or cellophane bags.

If sweet corn is to be shipped to distant markets, it is usually packed in wire-bound crates, precooled to remove field heat, top iced, and held under refrigeration at 32°F throughout the distribution system. Even under optimum conditions, normal sweet corn cultivars will not maintain marketable quality for more than 5 to 8 days. Harvested ears of the supersweet cultivars will maintain good quality for 10 to 12 days postharvest. Large-acreage sweet corn growers often utilize mechanical harvesting, hydrocooling, and truckload shipments to terminal markets, supermarket distribution centers, or processing plants.

Uses

Corn on the cob (boiled, steamed, or roasted) is one of the most popular uses of sweet corn. It is also canned or vacuum packed as whole kernel, creamed and canned as cream style, and frozen as whole kernel or on the cob. Sweet corn is part of numerous dishes ranging from specialty items to pickled products.

Nutrition Facts

Serving Size kernels from 1 medium ear (90g)

Amount Per Serving

Calories 80   Calories from Fat 10

	% Daily Value*
Total Fat 1g	**2%**
Saturated Fat 0g	**0%**
Cholesterol 0mg	**0%**
Sodium 0mg	**0%**
Total Carbohydrate 18g	**6%**
Dietary Fiber 3g	**12%**
Sugars 5g	
Protein 3g	

Vitamin A 2%	•	Vitamin C 10%
Calcium 0%	•	Iron 2%

FIGURE 23.11   Nutritional content of sweet corn. Used with permission from Dole Food Co., Inc.

Nutritional Value

Sweet corn is low in calories and is a source of vitamin C and vitamin A. One medium ear has only 80 calories with 1 g of fat and 3 g of dietary fiber (Figure 23.11). This serving of sweet corn also provides 10% of the RDA of vitamin C and 2% of the RDA for both vitamin A and iron.

Review Questions

1. How are sweet corn cultivars generally classified?
2. Explain the importance of the *su1, se,* and *sh2* genes in sweet corn production and marketing.
3. When should sweet corn be planted into a field in the spring?
4. What is generally the most destructive pest of sweet corn?
5. How do you determine when sweet corn is ready for harvest?
6. What are some storage considerations for sweet corn?

Selected References

Dickerson, G. W. 1996. *Home and market garden sweet corn production.* New Mexico State Univ., College of Agriculture and Home Economics Guide H-223.

Els, J. E. 1996. *Sweet corn for the garden.* Colorado State Univ. Coop. Ext. Serv. Cir. no. 7.607.

Gaus, A. E., J. B. Lower, and H. F. DiCarlo. 1993. *Fresh-market sweet corn.* Univ. of Missouri Coop. Ext. Serv. Ag. Publ. G06390.

Hall, R. H. 1968. *Sweet corn. Fruit & Vegetable Facts & Pointers.* United Fresh Fruit & Vegetable Association. Washington, DC.

Kays, S. J., and J. C. Silva Dias. 1995. Common names of commercially cultivated vegetables of the world in 15 languages. *Economic Botany* 49:115–152.

Mansour, N. S., and C. Raab. 1996. *Grow your own sweet corn.* Oregon State Univ. Ex. Serv. Publ. EC 1260.

Motes, J. E., W. Roberts, and B. O. Cartwright. 1994. Sweet corn production. Oklahoma State Univ., Oklahoma Coop. Ext. Serv. Publ. F-6021.

Orzolek, M. D., G. L. Greaser, and J. K. Harper. 1997. *Sweet corn production.* Penn State Univ. Coop. Ext. Serv. Publ.

Shultheis, J. R. 1994. *Sweet corn production.* North Carolina State Univ. Coop. Ext. Serv. Leaflet 13.

Selected Internet Sites

USDA Agricultural Marketing Service Standards for Grades

Fresh
Sweet corn   www.ams.usda.gov/standards/cornswt.pdf

Processing
Sweet corn   www.ams.usda.gov/standards/vpcornsw.pdf

chapter

24

Solanum Crops

I. Tomato
II. Pepper
III. Eggplant
IV. Irish Potato

Solanum crops are members of the Solanaceae or nightshade family. Many species contain alkaloids such as solanine, nicotine, and atropine. While crops in this family can grow as perennials in a protected or noncommercial situation, they are typically grown as annuals for commercial production. Solanum crops include tomato, pepper, eggplant, and Irish potato.

Tomato

Scientific Name

Lycopersicon lycopersicum (L.) Karsten

Common Names in Different Languages

tomato (English); shu fan qie (Chinese); arbe a tomates, tomate arbustive (French); Tomatobaum, Zbaumtomate, Baumtomatenstrauch (German); pomodoro (Italian); tomato (Japanese); pomidor (Russian); tomate (Spanish)

Classification, History, and Importance

Tomato, *Lycopersicon lycopersicon (Lycopersicon esculentum),* is an annual shrubby member of the Solanaceae or nightshade family. Other members of the family include potato, pepper, eggplant, tobacco, petunia, and jimson weed. In a protected environment, tomato is a short-lived herbaceous perennial.

Tomato is grown for its commercially important fruit. Colors of domesticated tomato fruit may be red, orange, or yellow depending on the genetic makeup of the cultivar.

TABLE 24.1 *The production value of selected solanum crops for fresh market and processing in the United States*

Crop	Value of Production		
	1995	1996	1997
	(1,000 dollars)		
Fresh Market			
Eggplant	16,225	18,146	17,558
Peppers, bell	452,786	474,801	502,595
Tomatoes	891,343	966,679	1,246,843
Processing			
Tomatoes	713,544	711,121	605,350

Source: USDA, NASS.

TABLE 24.2 *U.S. per capita consumption of commercially produced fresh and processing tomatoes*

Use	1976	1986 (lb, farm weight)	1996
Fresh	12.6	15.8	16.6
Canning	65.7	63.6	74.2
All	78.3	79.4	90.8

Source: USDA, Economic Research Service.

The tomato originated in either Peru or Bolivia. It was introduced to Central America and Mexico and was used by the Aztec and Toltec people, whose custom was to intersow tomatoes and corn. The name tomato is derived from the South American word *xitomate* or *zitotomate* and the Mexican word *tomati*.

Tomato was introduced to the United States in 1710 and catsup was produced in New Orleans in 1779. It was not until about the 1830s that the tomato became generally cultivated for culinary use in the United States. Early use of tomato was hampered by the notion that it was poisonous. The tomato industry made rapid strides during the latter half of the 19th century. Tomatoes are grown for fresh market in the field or greenhouse and for processing as whole pack, juice, or puree.

Tomato for fresh market was a $1.2 billion industry in 1997 (Table 24.1). This was up from $966.7 million in 1996. The tomato for processing market is about one-half the size of the fresh market with a production value of $605 million in 1997. Per capita consumption of tomatoes has increased from 78.3 lb of both fresh market and processing tomatoes per person in 1976 to 90.8 lb per person in 1996 (Table 24.2). This increase in per capita consumption was in both fresh tomatoes and canned tomatoes. Of the 129,100 acres of fresh market tomatoes grown in the United States, 40,800 acres are in California and 38,300 acres are in Florida (Table 24.3). California leads the nation in amount of acreage of processing tomatoes grown with 270,000 of the 293,700 acres of processing tomatoes grown in the

TABLE 24.3 *The top fresh market and processing tomato producing states in the United States*

Rank	State	Planted acres (000)	Harvested acres (000)	Yield (cwt/acre)	Production (000)
Fresh market					
1	California	40.80	40.80	285.00	11,628
2	Florida	38.30	38.10	415.00	15,812
3	Georgia	5.50	5.40	420.00	2,268
4	New Jersey	4.20	4.20	180.00	756
5	South Carolina	3.80	3.70	190.00	703
6	Ohio	3.50	3.40	190.00	646
7	Virginia	3.40	3.30	395.00	1,304
8	Tennessee	3.40	3.20	240.00	768
9	Texas	3.40	3.20	200.00	640
10	North Carolina	2.20	2.00	300.00	600
	United States	129.18	125.37	302.00	37,809
Processing				(tons/acre)	
1	California	270.00	260.00	35.96	9,350
2	Ohio	9.10	8.90	28.39	253
3	Indiana	6.40	6.40	24.60	157
4	Michigan	3.80	3.80	32.50	124
5	Pennsylvania	1.20	1.10	24.55	27
	United States	293.70	283.37	35.19	9,973

Source: USDA, NASS, 1997 data.

United States. Other states that grow processing tomatoes include Ohio, Indiana, Michigan, and Pennsylvania.

Plant Characteristics

Height: Determinate, 3 to 4 ft; indeterminate, 7 to 15 ft
Spread: 24 to 36 in.
Root Depth: Most of the roots are found in the top 8 in., but some fibrous spreading roots extend 4 to 6 ft deep.

Climatic Requirements

Tomato is a warm-season crop, killed by freezing temperatures and injured by light frosts. Tomato blossoms will not set fruit effectively if day temperatures rise above 94°F or if night temperatures remain above 70°F or drop below 60°F. The optimum range of day temperatures for growth is 70° to 75°F. The minimum soil temperature for germination is 50°F, with an optimum of 86°F and a maximum of 95°F. From planting of seed or transplant to harvest requires 60 to 90 days, depending on the cultivar and growing conditions (i.e., temperature and daylength).

Field Preparation

Tomatoes can be grown on almost any moderately well-drained soil type from a deep sand to a clay loam. Highest production is usually achieved on a well-drained loamy soil. Regardless of soil type, a good supply of organic matter typically increases yield and reduces production problems.

The soil is well prepared before planting tomatoes. Any existing cover crops are plowed under at least a month in advance of fumigation and planting. Bottom plowing or deep disking the field is often done to break up or destroy any soil clods or plant residue. Subsoiling is also used if needed to break up any hardpan layers in the field.

Soil pH for growing tomatoes should be between 6.0 to 7.0 and fertilizer needs are best determined by a soil test. In general, tomatoes require fairly high levels of several fertilizer elements and a tomato crop requires a continuous fertilizer supply to be productive over a period of several months. Sandy loam soils often require a split application of fertilizer to reduce nutrient losses due to leaching. From one-third to two-thirds of the fertilizer is typically applied preplant and the remaining fertilizer is applied after the first cultivation. On loams and silt loams, all fertilizer can be applied before transplanting. A calcium nitrate fertilizer is often used for tomatoes to help prevent blossom-end rot.

Plastic Mulches   Plastic mulch is commonly used in many areas of the United States for fresh market tomato production. The plastic is typically 1.25 mil thick and commonly embossed. The plastic is laid on the soil as tightly as possible, since loose plastic may damage young plants or be displaced in high winds. The sides of the mulch are covered with soil to prevent removal by winds. A well-prepared, loose, friable soil generally ensures smooth plastic mulch application and good soil coverage. Fumigation with methyl bromide-chloropicrin is often done during mulch application to sterilize and free the soil of weeds, diseases, and nematodes.

Organic Mulches   Organic mulch, such as straw, leaves, grass clippings, or compost, can be used for growing tomatoes and is applied after plants are set. These mulches are applied 4 to 6 in. thick to provide weed control, maintain uniform moisture levels, reduce certain disease problems, and improve fruit quality.

Selected Cultivars

Tomatoes are either determinate or indeterminate (Figure 24.1). Determinate tomatoes are referred to as self-topping or low-growing types and may reach a height of 3 to 4 ft. When the terminal bud of a determinate tomato forms a flower, the plant will not grow any taller. Numerous fruit are set on determinate types over a very few weeks and ripen over a short commercial harvest interval, usually 4 to 5 weeks.

Indeterminate cultivars continue to grow taller throughout the growing season unless they are killed by insects or disease. If planted in the spring the plants will set and produce fruit throughout the summer and fall. Indeterminate cultivars require 5- or 6-ft stakes to provide good support. Fruit of indeterminate tomatoes is usually softer and has more gel and thinner walls than determinate types.

Commercial fresh market tomatoes are typically determinate cultivar types, while greenhouse tomatoes are indeterminate types. Home garden tomatoes may be either determinate or indeterminate types.

Tomato cultivars are available that provide a range of colors, sizes, shapes, and maturities. Ripe fruit colors may be yellow, orange, pink, or red. Shapes may

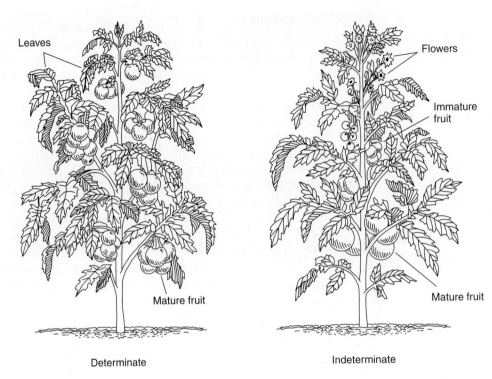

FIGURE 24.1   Diagram of determinate and indeterminate tomato plants. Source: Iowa State University.

vary from globe or round to slightly flattened, or pearlike. Cherry tomato types are very prolific, producing many small fruit. While most tomatoes can be used for fresh eating and for cooking, the paste- or Roma-type tomatoes are generally considered the best suited for cooking into sauces and ketchup due to their lower water content.

Processing tomato cultivars are usually determinate and ready for harvest about 75 days after field setting. With the concentrated fruit set of new cultivars, most fruit is harvested in two to three harvests. Ethephon, an ethylene-releasing compound, can be sprayed on the plants near harvest time (when a proportion of the fruit is red or pink) to enhance ripening. Tomatoes will ripen fully 10 to 14 days after spraying.

Planting

Row width of tomatoes is typically 6 ft wide for fresh market tomato production if plastic mulch is used. Rows may be as close as 4½ to 5 ft for bare-ground culture. In-row plant spacing may be between 18 to 24 in. Transplants for field setting should be 5 to 6 weeks old, 8 to 10 in. tall with thick, stocky stems, dark green leaves, and have a well-developed root system. Tomatoes are highly susceptible to injury from frost and are slow growing when they are set in cold soils.

For processing tomato production, raised beds are often used to reduce fruit rotting during maturation. Plant spacing can range from 8,000 to 12,000 plants per acre. Early production can be enhanced by using transplants, while direct seeding can be successfully used for later production.

FIGURE 24.2   Wire cages used to grow tomatoes.

Irrigation

Tomatoes require a steady supply of water for top yields. The tomato root system is relatively deep and if unobstructed will penetrate to a depth of 4 ft or more. Tomatoes require the equivalent of 1 to 1½ in. of rain per week depending on soil type, plant age, and temperature. If plastic mulch is applied a drip irrigation system is often used.

Culture and Care

Fresh market tomato producers may use cultural practices such as mulching, pruning, staking, and caging (Figure 24.2) to enhance production. Processing tomatoes are usually grown on bare ground with a minimum or no additional production practices besides pest control and/or application of a ripening agent (ethylene) prior to harvesting.

Pruning   Pruning or the removal of axillary leaves or stems (suckers) reduces the amount of foliage and the number of tomatoes the plant produces. It increases the size of fruits produced and facilitates disease control as well as tying and harvesting of staked tomatoes. The short stake system requires one major pruning and a second light ground sucker pruning. In this system, the large sucker produced directly below the first fruit cluster is located and all suckers below this are removed. During the second pruning, new ground suckers and suckers missed during the first pruning are removed. Pruning is done as soon as possible after the suckers are visible. Allowing suckers to become too large before pruning makes them harder to remove, increases the possibility of damage to the plant, and delays maturity. All pruning tends to delay maturity. Ground tomatoes are not usually pruned, but trellis tomatoes are pruned continuously.

Staking   Short stakes, 1 by 1½ in. in diameter and 4 to 4½ ft long, are driven 10 to 12 in. into the soil midway between every other plant (Figure 24.3). The plants are supported

FIGURE 24.3   Growing tomatoes in the field using stakes.

FIGURE 24.4   Staked tomatoes tied with string for support. Strings are added as the plant grows.

by strings that run down each side of the row and loop around each post. The stringer proceeds down one side of the row, looping the string around each stake. When the end of the row is reached, the stringer returns along the opposite side of the row, again looping the string around each stake. Pruning is typically done before the first string is attached and the first string should be in place before the plant falls over. As the plants grow and need additional support subsequent strings are attached 6 to 10 in. above the preceding string (Figure 24.4). A total of three or four strings is often required for determinate tomatoes that are pruned, staked, and tied.

FIGURE 24.5   Leaf miner injury to tomato leaves.

Caging   Tomato plants can be supported on all sides by enclosing them in a wire cage. Caged tomatoes tend to be more productive because suckers are not removed. Caged tomatoes are also less prone to sunscalding injury because of greater foliage cover. Caging of tomatoes is often practiced by home gardeners.

Common Growing Problems and Solutions

Insects   Cutworms often cut off tomato plants close to the soil surface, particularly early in the season. Hornworms are large 4-in. worms that have a horn on one of their ends and eat foliage and fruit. Fruitworms eat holes in fruits and buds and stalk borers tunnel in the stem causing the plant to wilt and die. Leaf miners tunnel inside the leaves, leaving white curly trails behind (Figure 24.5). Stinkbugs suck juices from the plant and often cause whitish spots to develop just below the skin of the fruit. Spider mites are tiny, barely visible, and cause yellowish speckling on the foliage and very fine webbing.

Diseases   Several wilts can affect tomato production. These include fusarium wilt and bacterial wilt. The fungus for fusarium wilt is soilborne and can remain in the soil for several years. The first symptom of fusarium wilt on tomatoes is a yellowing of the lower leaves. Later development of the wilt results in a brown discoloration of vascular tissue and eventual wilt and death of the plant. Plants infected with bacterial wilt will wilt rapidly but retain their green color. Dark brown discoloration of the central pith is usually observed when the stem is cut.

Early blight, late blight, and bacterial spot are foliar diseases that can affect tomatoes. They often begin as spots on either the leaves or the fruits. Tobacco mosaic and cucumber mosaic are two common viruses that can affect tomato yields.

Physiological Disorders   Blossom-end rot is a leatherlike decay of the blossom end of the fruit (Figure 24.6). There are several ways of reducing the amount of blossom-end rot on tomatoes that are based on maintaining proper calcium levels in the

FIGURE 24.6   Blossom-end rot of tomatoes.

tomato fruit. The soil for growing tomatoes should be limed the proper amount to raise the soil pH to a suitable range for optimum calcium uptake. Lime is calcium carbonate, which supplies calcium to the soil making it easier for the plant to absorb calcium. Adequate soil moisture must also be maintained to reduce blossom-end rot. Calcium must dissolve in soil moisture to be assimilated into the plants. Tomatoes will also be less susceptible to blossom-end rot if they are not pruned too heavily and/or if they are not fertilized too heavily with ammonium nitrate.

Blossom drop is caused by temperature extremes. Fruit set occurs only when night temperatures are between 55° and 75°F. When fruits are not set, blossoms fall off.

Fruit cracks usually occur during hot (above 90°F), rainy periods, especially when preceded by a long dry period. Fruits exposed to the sun are the most susceptible. Radial cracking is the most common, but concentric cracks also occur on some cultivars.

Sunscald appears first as a yellow or white patch on the side of the fruit facing the sun. The spot may blister and dry, forming a paperlike surface. Poor foliage cover allows exposure to sun on pruned, staked, sprawling, or unhealthy plants.

Blotchy ripening or the uneven development of color on the fruit may be due to temperatures below 60°F, root stress from compacted or soggy soil, or low levels of potassium in the soil. Catfacing is the puckering and scarring at the blossom end of the fruit. Cool and cloudy weather at blooming time may cause the blossom to stick to the young, developing fruit, resulting in the malformation.

Harvesting

Tomato fruits undergo a color change (Table 24.4) as they ripen and mature. These changes in color are recognized by the commercial industry and growers as indicators of maturity required for a specific market (i.e., mature green fruit vs. vine-ripened red fruit). Fresh market tomatoes are commonly harvested at the mature green stage and ripened later either during shipping or storage. At the mature

TABLE 24.4   *USDA ripeness classification for fresh market tomatoes*

Ripeness stage	Description of tomato surface
1 (Green)	Surface is completely green
2 (Breaker)	Definite break in tan, pink, or red color up to 10% of surface
3 (Turning)	10 to 30% tan, pink, or red color
4 (Pink)	30 to 60% pink or red color
5 (Light red)	60 to 90% pink or red color
6 (Red)	More than 90% red color

Source: Sargent, S. A. 1997. *Tomato production guide for Florida: Harvest and grading.* Univ. of Florida Coop. Ext. Serv. Publ. SP-214.

FIGURE 24.7   Fully ripened tomatoes.

green stage of maturity, the pulp surrounding the seed has softened and become jellylike. Harvesting tomatoes at the immature green stage reduces yield and results in poor-quality fruit that may not ripen properly after harvest. Harvest of mature green tomatoes is usually spaced a week apart. Other markets may require pink or vine-ripened tomatoes that are harvested only after color has developed on the fruit (Figure 24.7). Pink or vine-ripened fruit needs to be harvested much more frequently than mature green fruit.

Processing tomatoes can be harvested by hand or by mechanical harvesters (Figure 24.8). To facilitate uniform fruit maturity of processing tomatoes, the plants may be sprayed with an ethylene-releasing compound to enhance the fruit red color and acidity.

Postharvest Handling

Harvested tomato fruits are sensitive to injury from chilling temperatures. Storage temperatures will also affect the ripening that occurs after harvest. Mature green

FIGURE 24.8   Mechanical harvesting of processing tomatoes.

FIGURE 24.9   Ripening generator that releases ethylene to ripen mature green tomatoes in storage.

tomatoes will begin to ripen for packing in 7 to 14 days if held at 57° to 61°F, but will not ripen normally at temperatures above 81°F. For storage, light red tomatoes with 60 to 90% red color can be held up to 7 days at 50°F. Ethylene fumigation during storage of harvested mature green tomatoes is often used to accelerate and promote a uniform ripening (Figure 24.9).

FIGURE 24.10   Nutritional value of tomatoes. Percent daily values are based on a 2,000 calorie diet. Used with permission from Dole Food Co., Inc.

Market Preparation and Marketing

Mature green tomato fruits are bulk packed in ventilated fiberboard or wooden containers for shipping. They may later be regraded and packed in tubes or trays covered with cellophane acetate. Fruits that are pink can be packed in boxes in two or three layers.

Contract prices for processing tomatoes vary with processors, grades, and years. Contract prices are seldom as high as fresh market prices but the cost of producing processing tomatoes is not as great as the cost of fresh market tomatoes.

Uses

Besides being eaten fresh, tomatoes can be boiled, stewed, fried, juiced, or pickled and used in soups, salads, and sauces. Tomatoes are an important component in salsa along with onions, garlic, peppers, cilantro, cumin, and lime juice. Cherry tomatoes are often served with a dip as an appetizer. Vine-ripened tomatoes are good for stuffed tomato entrees or for wedges in salads. Mature Roma or cherry tomatoes are best for slicing, garnishing, or chopping.

Nutritional Value

Tomatoes are nutritious and low in calories. One medium-sized tomato has only 35 calories with 0.5 g of fat and 1.0 g of dietary fiber (Figure 24.10). This serving of tomato provides 40% of the RDA of vitamin C, 20% of the RDA of vitamin A, and 2% of the RDA for iron and calcium.

Pepper

Scientific Name

Capsicum annuum L. Grossum group (bell pepper or pimiento pepper)
Capsicum annuum L. Longum group (cayenne, chili or hot pepper)
Capsicum frutescens L. (tabasco pepper)

Common Names in Different Languages

bell, green, sweet or pimiento pepper (English); tian jiao (Chinese); poivre d'Es-
pagne, poivron (French); Gemüse-Paprika, Paprika, Gewürzpaprika, Spanischer
Pfeffer (German); peperone, p. dolce, peperoncino (Italian); piiman, shishi-tô-
garashi (Japanese); chile, guidnilla, pimento picante (Spanish)
cayenne, chili or hot pepper, chili (English); tian jiao (Chinese); poivre de Cayenne,
piment de Cayenne, p. d'oiseau (French); Cayenne, Chilli, Vogel or Spanischer
Pfeffer, Gewürzpaprika (German); pepe di Cayenna, peperone rabbioso, peperoncino piccante (Italian); nagami-togarashi (Japanese); chile, guidnilla, pimento
picante (Spanish) tabasco pepper (English); xiao mi jiao (Chinese); piment,
p. d'oiseau, pilli, tabasco pepper (French); Tabasco Pfeffer, Gewurzpaprika (German); tabasco, peperone ornamentale (Italian); tabasco, chile, guindilla (Spanish)

Classification, History, and Importance

Peppers are warm-season crops that belong to the Solanaceae family. The pepper that
is grown as a vegetable is related to eggplant, tomato, and potato but not to *Piper ni-
grum* (Piperaceae family), the species that furnishes the black and white peppers that
are ground and used as condiments. Peppers are typically grown as annuals in the
United States, but many will grow as perennials in areas where there is no killing frost.

While it is not certain whether peppers are native to the Americas or if they were
brought into the Americas during pre-Columbian times, the greatest diversity of cultivated forms occur in Central Mexico with a secondary center in Guatemala. *Capsicum*
was described in tropical America in 1514 and was referred to in Peruvian writings in
1609. Peppers were mentioned in England in 1548 and by the 1920s the growing of
peppers had become an important industry throughout the southern United States.

Pepper fruits are actually berries and pepper types are classified by fruit characteristics (i.e., pungency, color, fruit shape) and by their use. Peppers come in a greater variety of sizes, shapes, colors, and tastes than most other vegetables. The most popular
peppers are the mild bell, banana types, and the pungent Hungarian wax types.

Bell peppers were a $502.6 million industry in 1997 (see Table 24.1). This was
up from $474.8 million in 1996. Per capita consumption of green peppers and chili
peppers has increased about 50% since 1986 (Table 24.5). Of the 67,600 acres of bell
peppers grown in the United States, 25,000 acres are in California and 19,200 acres
are in Florida (Table 24.6). Other states that grow bell peppers include New Jersey,
Texas, North Carolina, Michigan, Louisiana, Ohio, and Virginia. New Mexico is the
leader in the United States in pungent pepper production.

Plant Characteristics

Height: 15 to 36 in.
Spread: 2 ft

TABLE 24.5 *U.S. per capita consumption of commercially produced fresh and processing bell and chili peppers*

Crop	Use	1976	1986 (lb, farm weight)	1996
Bell peppers	All	2.7	4.0	7.1
Chili peppers	Processing	—	4.6	6.6

Source: USDA, Economic Research Service.

TABLE 24.6 *The top bell pepper (combined fresh and processing) producing states in the United States*

Rank	State	Planted acres (000)	Harvested acres (000)	Yield (cwt/acre)	Production (000)
1	California	25.00	25.00	300.00	7,500
2	Florida	19.20	18.50	295.00	5,458
3	New Jersey	5.00	4.80	260.00	1,248
4	Texas	5.40	5.10	175.00	893
5	North Carolina	7.30	7.00	120.00	840
6	Michigan	1.80	1.70	220.00	374
7	Louisiana	0.90	0.85	200.00	170
8	Ohio	1.00	1.00	145.00	145
9	Virginia	2.00	1.70	85.00	145
	United States	67.60	65.65	255.00	16,773

Source: USDA, NASS, 1997 data.

Root Depth: The fibrous spreading roots of peppers are generally confined to the top 8 in., but sometimes extend 4 ft deep.

Climatic Requirements

Peppers are warm-season crops that require somewhat similar conditions as tomatoes but generally have a little higher temperature requirement. Peppers are very sensitive to light frost and grow poorly in cool temperatures (40° to 60°F). Pepper fruit set is also sensitive to extreme temperatures and will not set fruit above 90°F during the day or below 60°F at night. Best yields occur when temperatures during fruit set are between 65° and 80°F. The first bell pepper fruit for harvest are typically ready 8 to 10 weeks after transplanting.

Field Preparation

Peppers grow optimally on a well-drained sandy loam soil. Sandy soils may be used for early production of peppers. While optimum soil pH for peppers is between 6.0 and 6.5, peppers are fairly tolerant to soil pH as low as 5.5. Peppers generally require 50 lb/acre of preplant N with recommended P and K. A topdressing or side-dressing of

TABLE 24.7   *Fruit characteristics of some commercially important groups of peppers*

Species and group	Fruit characteristics
Capsicum annuum	
Bell group	Large three- to four-lobed peppers with blocky sweet fruits. The fruit color is green when immature, turning red, yellow, gold, or purple when mature.
Pimiento group	Large sweet conical or heart-shaped thick-walled fruits green when immature and turn red when mature.
New Mexican group	Long, green fruits that can be sweet or pungent depending on the cultivar. The fruits turn red when mature.
Cayenne group	Long, crescent, or irregular shaped, very pungent fruits that turn red characteristically wrinkled when mature.
Jalapeno group	Very pungent fruits, thick walled, and conical shaped. Dark green skin that turns red when mature.
Wax group	Sweet or pungent fruits depending on the cultivar that are small or large and yellow when immature, turning red, orange-red, or orange when mature. The fruits are usually conical to round shaped.
Capsicum frutescens	
Tabasco group	Thin yellow or yellow-green fruits that become red when mature. The fruits are highly pungent.

Based on: Bosland, P. W., A. L. Bailey, and J. Iglesias-Olivas, 1996 *Capsicum* pepper varieties and classification. New Mexicao State Univ. Coop. Ext. Service Circular 530.

N is often needed when the first pepper fruit on the plants are set. Additional N may be needed later in the season if loss of N by leaching has occurred.

Overfertilizing peppers can have negative effects on fruit earliness, yield, or quality. For example, excessive N leads to vigorous plants that grow vegetatively at the expense of early fruit set and total yields. High N also increases the severity of blossom-end rot and reduces fruit firmness by reducing wall thickness and causing blossoms and small fruits to abscise. High K can increase shriveling of harvested peppers and reduce shelf life.

Selected Cultivars

Although there are numerous common or commercial names for peppers, most cultivars can be classified into two main types: those with mild- or sweet-fleshed fruit, and those with hot- or pungent-fleshed fruit (Table 24.7).

Pepper fruit pungency is determined by the amount and location of the chemical compound capsaicin. Nonpungency (sweetness) is determined by a single gene, but the varying levels of pungency (heat) are conditioned by many genes. Hybrid cultivars, regardless of type, generally set fruit earlier and are more consistent producers than nonhybrid cultivars and the seed is more expensive.

The following is a list of some commercially important types of peppers.

- Bell types—Bells are sweet peppers that are grown primarily for fresh market. Their shape is blocky with three to four lobes and thick flesh. Most bell-type

FIGURE 24.11   Pepper transplants grown in Styrofoam flats ready for transplanting.

cultivars are green when immature and red when ripe. Commonly grown cultivars require 75 to 80 days from transplanting to harvest and include Keystone Resistant Giant, Yolo Wonder, California Wonder, Lady Bell, Hybelle, and Pip.
- Cherry peppers—Cherry peppers are globe shaped and grow on long upright stems, usually above the leaves of the plant. They range from orange to deep red when harvested and may be sweet or hot and large or small. Cultivars include Sweet Cherry, Bird's Eye, Red Cherry Small, and Red Cherry Large.
- Chili type—Chili peppers are pungent and thin fleshed. Chili cultivars vary in size and shape from cherry size to slender fruits up to 8 in. long. They require 100 days from transplanting to mature green fruit and 140 days to red ripened fruit. Cultivars include El Paso and Anaheim M.
- Pimiento peppers—Pimiento peppers are sweet with very thick walls. The fruit are conical, 2 to 3 in. wide and 3 to 4 in. long, and slightly pointed. Pimientos are red when ripe. Popular cultivars include Perfection Pimiento and Pimiento L.
- Tabasco peppers—Tabasco peppers are 1 to 3 in. long, slim, tapered, and very hot. Tabasco peppers are grown commercially for making tabasco sauce.

Planting

Peppers are usually transplanted but can also be direct-seeded. Transplanting offers several advantages over field seeding: weed control is much easier, fruit set occurs before high summer temperatures develop, a field stand is much easier to obtain using transplants, and less seed may be required.

Pepper transplants are usually greenhouse grown and require 6 to 8 weeks from seeding to transplant size (Figure 24.11). Ideal transplants have five to six true leaves and are 6 to 8 in. tall. They have a stiff but not woody stem strong enough to withstand transplanting, an intact root system (but are not root-bound), and should not have suffered shock from storage or handling.

Transplants can be set by hand or by machine after soil temperatures in the spring are above 55°F and danger of frost is past. Transplanting is usually delayed until the danger of late spring frost has passed. Transplanting peppers through black plastic mulch will generally increase growth and promote earlier production. Peppers are usually transplanted 14 to 24 in. apart on 36-in. rows.

Irrigation

Irrigation may be needed to maintain uniform soil moisture for optimum pepper production. Plants recover slowly from drought and long dry periods may cause pepper plants to shed flowers and small fruits.

Peppers that have been transplanted to the field root only to a depth of about 2 ft and water stress during flowering and fruit development can cause poorly developed, small, misshapen fruit, or blossom-end rot. Overwatering can promote diseases such as phytophthora and other root-rotting organisms.

Culture and Care

Peppers respond favorably to plastic mulch and row covers. Plastic mulches are typically black. Row covers, if used, are removed when air temperatures inside the cover exceed 90°F for 2 or 3 consecutive days at midday. Shallow mechanical cultivation and hand hoeing are often needed to control weeds.

Common Growing Problems and Solutions

Insects   Cutworms can be a problem for peppers early in the season. Pepper seedlings can also be attacked by flea beetles when the cotyledons emerge and aphids can attack peppers during the summer.

Diseases   Bell peppers are susceptible to several diseases. Seeds and seedlings may become infected with damping-off fungus. Phytophthora root rot causes rotting of roots and underground portions of the stem. Leaves may become infected by anthracnose fungi, by *cercospora* leaf spot fungus, or bacterial spot disease.

Physiological Disorders   Blossom-end rot is characterized by small areas at or near the tip of pepper fruit that become light brown and sunken. Affected areas on the fruit develop a leathery texture as the fruit matures. Blossom-end rot usually results from an irregular or insufficient supply of moisture and/or calcium with the first pepper fruit most often being affected.

Sunscald is caused by exposure of the fruit to direct sunlight, especially after being shaded by foliage. Sunscald is characterized by a light-colored area that becomes slightly sunken with a papery appearance.

Harvesting

Bell peppers grown for commercial marketing are harvested when the fruit are full sized and green. Fruits are snapped off by hand and carried from the field in buckets or sacks. The size for the pepper fruit should be 2½ in. in diameter and 2½ in. in length. The first, or "crown," set of fruit of bell peppers located at the first main fork of the stem usually produces a large percentage of large fruit. If the crown set is lost, early yield is reduced. Yields will be greater if peppers are picked as they mature rather than if mature fruit are allowed to fully ripen on the plant.

FIGURE 24.12   Jalapeno pepper type.

As pepper fruits mature their sugar content increases and their color changes from green to yellow, red, or chocolate. While green bell peppers are harvested before any colors (besides green) develop on the fruit, there is an increasing market for bell peppers that are left to mature on the plant until the fruit matures to a color such as chocolate or red. Depending on the type of pepper and its intended use, hot peppers may be harvested either at the green or red ripened stage.

Postharvest Handling

Mature green pepper fruits are stored best at temperatures between 45° and 50°F. Peppers, regardless of type, are not to be held at temperatures below 45°F. Under best conditions peppers can be stored for about 2 weeks.

Market Preparation and Marketing

Peppers marketed fresh are often carefully wiped with a soft cloth to remove soil and dust. Bell peppers for market are usually waxed, graded by size, and packed in 1- or 1½-bu baskets, crates, or boxes.

Uses

Sweet bell peppers are eaten raw or cooked. Raw peppers are cut and placed in salads. Additionally, bell peppers are canned or pickled in brine for use in salads or other foods. Diced green or red peppers are sometimes mixed with sweet corn or other vegetables. Pimiento peppers are canned and used in preparing such foods as pimiento cheese and the red stuffing for olives. The mild spice paprika is the finely ground fruit walls of paprika peppers.

Pungent peppers such as the jalapeno type (Figure 24.12) and the cayenne type (Figure 24.13) vary from mildly pungent to very hot. They are most commonly used in making chili, salsa, or similar dishes, and are canned or dried.

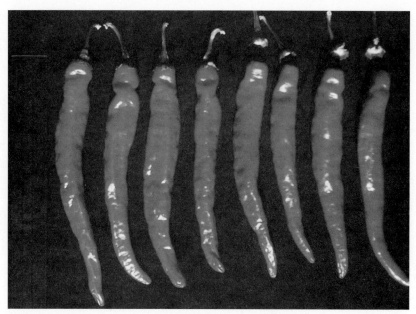

FIGURE 24.13   Cayenne pepper type fully ripened and colored.

FIGURE 24.14   Ornamental pepper grown in the field.

Although peppers are mainly grown for human consumption, they can also be used as ornamentals in containers or among flowers in the garden. Ornamental peppers are increasing in popularity, especially as potted plants during the holiday season, and are available in many forms with small fruits of various shapes and colors (Figure 24.14). They are usually purchased when already in fruit.

Nutrition Facts
Serving Size 1 medium pepper (148g)

Amount Per Serving
Calories 30   Calories from Fat 0

% Daily Value*

Total Fat 0g **0%**
  Saturated Fat 0g **0%**
Cholesterol 0mg **0%**
Sodium 0mg **0%**
Total Carbohydrate 7g **2%**
  Dietary Fiber 2g **8%**
  Sugars 4g
Protein 1g

Vitamin A 8%   •   Vitamin C 190%
Calcium 2%   •   Iron 2%

FIGURE 24.15   Nutritional value of green peppers. Percent daily values are based on a 2,000 calorie diet. Used with permission from Dole Food Co., Inc.

Nutrition Facts
Serving Size one pepper (45 grams)

Amount Per Serving
Calories 20   Calories from Fat 0

% Daily Value*

Total Fat 0g **0%**
  Saturated Fat 0g **0%**
Cholesterol 0mg **0%**
Sodium 10mg **0%**
Total Carbohydrate 3g **1%**
  Dietary Fiber 0g **0%**
  Sugars 2g
Protein 1g

Vitamin A 80%   •   Vitamin C 170%
Calcium 0%   •   Iron 0%

FIGURE 24.16   Nutritional value of chili peppers. Percent daily values are based on a 2,000 calorie diet. Used with permission from Dole Food Co., Inc.

Nutritional Value

Both green and chili peppers are low in calories and are excellent sources of vitamin C. A medium bell pepper has only 30 calories with no grams of fat and 2 g of dietary fiber (Figure 24.15). This serving of bell pepper provides 190% of the RDA of vitamin C, 8% of the RDA of vitamin A, and 2% of the RDA for calcium and iron. One chili pepper has only 20 calories with no grams of fat (Figure 24.16). This amount of chili pepper provides 170% of the RDA of vitamin C and 80% of the RDA of vitamin A.

Eggplant

Scientific Name

Solanum melongena L.

Common Names in Different Languages

eggplant, aubergine (English); qie zi (Chinese); aubergine (French); Eirfrucht, Aubergine (German); melanzana, petonciano (Italian); nasu (Japanese); baklažan (Russian); berenjena (Spanish)

TABLE 24.8   *U.S. per capita consumption of commercially produced fresh eggplant*

Use	1976	1986 (lb, farm weight)	1996
Fresh	0.5	0.5	0.4

Source: USDA, Economic Research Service.

TABLE 24.9   *The top fresh market eggplant producing states in the United States*

Rank	State	Planted acres (000)	Harvested acres (000)	Yield (cwt/acre)	Production (000)
1	Florida	1.70	1.70	320.00	544
2	New Jersey	0.90	0.90	215.00	194
	United States	2.60	2.60	284.00	738

Source: USDA, NASS, 1997 data.

Classification, History, and Importance

Eggplant, *Solanum melongena* L., belongs to the Solanaceae or nightshade family. Other members of the family include pepper, tomato, potato, tobacco, horse nettle, petunia, and Jerusalem cherry. Eggplant was so named because early cultivars had egg-shaped fruits. The fruit is the edible part of the plant and can vary in shape from oval to round and long to oblong. The color of the mature fruit is typically purple to purple-black, but can also be red, yellowish white, white, or green.

Eggplant is native to the tropics and the oldest records on eggplant are found in a Chinese book written in the 5th century. Although cultivated in India, China, and adjacent areas, eggplant became known to the Western world about 1,500 years ago. Eggplants first grown in Europe were small fruited types that resembled eggs in shape and were probably used for ornamental purposes. Yellow and purple cultivars were introduced into Germany in 1550 while the Spaniards are suspected of having introduced eggplants into America. Eggplants were grown in Brazil before 1650. Purple and white cultivars grown for ornamental purposes were described in 1806 in the United States.

Eggplant was a $17.6 million industry in 1997 (see Table 24.1). This was down from $18.1 million in 1996. Per capita consumption of eggplant has remained relatively stable through the past three decades at about 0.5 lb per person (Table 24.8). Commercial eggplant acreage is primarily located in Florida (1,700 acres) and New Jersey (900 acres) (Table 24.9).

Plant Characteristics

Height: 24 in.
Spread: 3 to 4 ft
Root Depth: 4 to 7 ft

Climatic Requirements

Eggplant is a warm-season crop that is very susceptible to frost and requires a relatively long warm growing season (approximately 100 to 140 days from seeding or 75 to 90 days from transplanting). Day temperatures of 80°F are considered optimum and eggplant growth is inhibited at temperatures below 60°F. The optimum soil temperature for seed germination is 86°F with a good range for seed germination being 68° to 95°F. Eggplant is considered more susceptible to injury by low temperatures than are tomatoes or peppers.

Field Preparation

Well-drained sandy loam soils are ideal for eggplant production. Adequate soil preparation helps promote the growth and development of an extensive root system. Poorly drained soils usually result in reduced functional root area, poor plant growth, and low yields. Soils are typically tilled at least 8 in. deep. Eggplant is intolerant of poorly drained soils; therefore, it is often common (especially on heavier soils or in low-lying areas) to transplant eggplant on raised beds.

Satisfactory soil pH for eggplant growing is between 6.0 to 6.5 and high fertilizer amounts, especially phosphorus, are necessary for good eggplant growth. A typical fertilizer recommendation for eggplant may be about 130 lb/acre each of N, P, and K.

Selected Cultivars

Eggplant cultivars differ in earliness as well as in size, shape, and color of mature fruit. Some commonly grown cultivars and their characteristics are the following:

- Black Beauty—large, smooth, purplish fruit
- Dusky—medium-sized plant adapted to container growing
- Ichabod or Ichiban—long, slender oriental-type fruit, adapted to container growing
- Burpee Hybrid—oval, medium-sized dark glossy purple fruit
- White Italian—medium-sized white fruit, slightly milder flavor than purple types
- Golden Yellow—produces yellow lemon-sized fruit
- Morden Midget—small, bushy plants that bear smooth, medium-sized deep purple fruit
- Applegreen—apple-green-colored fruit
- White Beauty—white medium-sized fruit

Planting

Eggplant transplanted to the field often provides the best opportunity for optimum production under most conditions in the United States. Transplants are grown in a greenhouse for about 8 weeks before field setting. Strong, healthy eggplant transplants are typically 6 to 8 in. tall and are placed in the field after danger of frost has passed at a spacing of 2 to 2½ ft apart in rows that are 3 to 4 ft apart.

Irrigation

Eggplant requires a relatively stable amount of available soil moisture for a steady growth pattern that is essential for good eggplant production. Trickle irrigation is often used if black plastic mulch is applied.

FIGURE 24.17   Eggplants grown with plastic mulch. Courtesy of M. D. Orzolek.

Culture and Care

Eggplants fruit earlier and are more productive when grown with plastic mulch (Figure 24.17). For this system all fertilizer is placed in the bed prior to laying the plastic, or drip irrigation is used and fertigation is practiced. Eggplants require insect pollination for good yields and beehives are placed in or near the field if wild bee activity is suspected of being inadequate for pollination needs.

Common Growing Problems and Solutions

Insects   Flea beetle and Colorado potato beetle are two of the major insects of eggplants.

Diseases   Root-knot nematodes cause major damage to eggplant in the southeastern United States. Southern stem blight is also a common disease.

Physiological Disorders   Fasciation of fruit, caused by stressful growing conditions, is characterized by flattened eggplant fruit.

Harvesting

Eggplant fruits are harvested when they are full sized and before the seeds mature (Figure 24.18). Maturity for harvest can be determined by pressing the thumb against the side of the fruit. The fruits are considered immature if the indentation from the thumb pressure on the fruit surface springs back to its original shape. The flesh softens as the fruit matures so at proper harvest stage the thumb pressure leaves an indentation in the fruit.

A sharp knife or small pruning shears are typically used to harvest eggplant. Eggplant is harvested at least once a week, and preferably twice a week, before the seeds within the fruit begin to harden. Fruits that are marketed commercially are generally 4 to 6 in. in diameter.

FIGURE 24.18   Harvested eggplant fruit at the proper stage of maturity desired for marketing. Courtesy of M. D. Orzolek.

A high-quality fruit is one that is heavy in relation to size with a glossy uniform color, free from surface cuts or bruises, and shows no decay. Overmature fruit lose their gloss.

Postharvest Handling

Field heat of harvested eggplant fruits must be removed quickly so that the fruits do not lose moisture and quality. Forced-air and/or room cooling is often used to remove the field heat from the eggplant fruit. Eggplant fruit is typically stored at 45° to 50°F with a relative humidity of at least 90%. Storage of eggplant fruit below 50°F could result in chilling injury. Fruits that have been chilled exhibit pitting and decay several days after removal from storage. Eggplant is not typically stored for longer than 10 to 14 days.

Market Preparation and Marketing

Eggplants are packed in several types of containers. The crop is not suitable for prolonged storage and will be injured in temperatures below 50°F.

Uses

Eggplant fruit is usually baked, sautéed, cut into strips or cubes and fried, or stuffed. In some cases the fruit is boiled and hollowed out and stuffed with various fish and meat dishes. Eggplant is a good substitute for meat in dishes such as lasagna.

Nutritional Value

Eggplants do not offer substantial amounts of vitamins or minerals. In a serving size of one-fifth of an eggplant fruit there are only 25 calories if eaten raw (Figure 24.19).

FIGURE 24.19　Nutritional value of eggplants. Percent daily values are based on a 2,000 calorie diet. Used with permission from Dole Food Co., Inc.

This serving of eggplant also provides 2 g of dietary fiber and 2% of the RDA for both vitamin C and iron.

Irish potato

Scientific Name

Solanum tuberosum L.

Common Names in Different Languages

potato (English); ma ling shu (Chinese); pomme de terre, patata (French); Kartoffel (German); pomi di terra, patata (Italian); jagaimo (Japanese); kartofel (Russian); patata, papa (Spanish)

Classification, History, and Importance

Potatoes are native to the Americas and were cultivated from Chile to New Granada before the discovery of the New World. The Spaniards introduced the potato to Europe soon after 1580, and the popularity of potatoes spread all over Europe and the British Isles by the end of the 17th century. First introduced into New England by Irish immigrants in 1719, the white potato is now referred to as the "Irish potato" because of its association with the potato famine in Ireland in the 19th century.

Potatoes are not roots but specialized underground storage stems called "tubers" (Figure 24.20). Potatoes are ranked with wheat and rice as one of the most im-

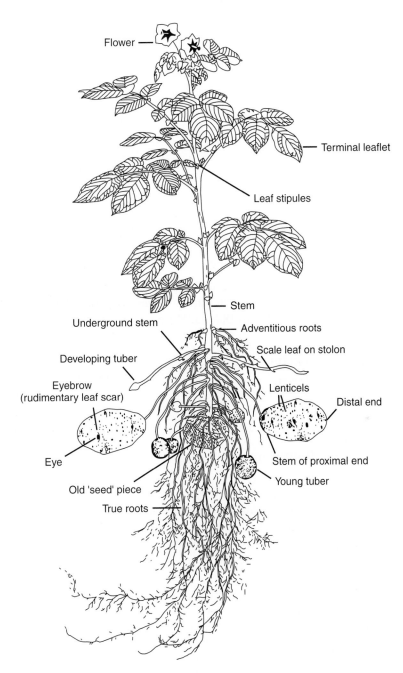

FIGURE 24.20   Diagram of Irish potato plants. The potato tubers are underground storage stems. Used with permission from the Potato Association of America.

portant staples in the human diet. Per capita consumption of Irish potatoes has increased from 125.2 pounds of both fresh market and processing potatoes per person in 1976 to 142.8 pounds per person in 1996 (Table 24.10). Per capita consumption of fresh market potatoes has remained relatively stable during this time period, while the amounts of potatoes that have been processed have increased. The largest

TABLE 24.10   *U.S. per capita consumption of commercially produced fresh and processing potatoes*

Use	1976	1986 (lb, farm weight)	1996
Fresh	49.4	48.8	48.8
Processing	75.8	77.2	94.0
Chips			
Shoestrings	15.8	18.2	16.9
Dehydrated	16.3	10.9	15.2
Canning	1.9	1.8	2.1
All	125.2	126.0	142.8

Source: USDA, Economic Research Service.

potato producing states are Idaho with 140 million cwt/acre and Washington with 95 million cwt/acre in 1996. Other potato producing states include Oregon, Wisconsin, Colorado, North Dakota, Minnesota, and Maine.

Plant Characteristics

Height: 24 to 30 in.
Spread: 24 in.
Root Depth: 2 ft

Climatic Requirements

Potato is a cool-season vegetable that exhibits maximum tuber formation at soil temperatures between 60° and 70°F. Potatoes will withstand light frost in the spring and generally require a growing season of 90 to 120 frost-free days. Tubers fail to form when soil temperatures reach 80°F. Potatoes can be grown in areas with shorter seasons, especially in the northern climates because the shortness of the season is often compensated for by long days.

Field Preparation

Good water penetration and aeration are necessary for proper potato plant growth and tuber formation. Excessive tillage and land preparation can cause compaction and reduce potato growth and yield. The soil is typically plowed below any compacted layer within the normal root zone and then disk harrowed before planting. Spike-tooth harrowing just prior to planting may be used to break clods and level the soil.

While potatoes can grow well on a wide variety of soil types, well-drained, fertile sandy loam soils are considered best. Potatoes can withstand soil pH as low as 5.0. The tubers are less susceptible to scab, a major soilborne disease of potatoes, when the soil pH is between 5.0 and 5.5.

A typical fertilization program for potato is to apply 75 lb/acre of N along with recommended P_2O_5 and K_2O by either broadcast preplant-incorporated or one-half broadcast and one-half with the planter in bands placed 3 to 4 in. to each side and 1 to 2 in. below the seed piece. Additional N is top-dressed when the tubers begin to form. Two or three top-dress applications of 30 to 40 lb/acre of N may be needed. Potassium sulfate is often preferred to potassium chloride as the potassium source of the fertilizer for potatoes since skin color and specific gravity of the tubers may be adversely affected by potassium chloride.

Selected Cultivars

Potato cultivars are chosen according to yield potential, cooking characteristics, time to maturity, skin and flesh color, and storage life. Cultivars have been classified as white, red, or russet based on skin color and texture. Increased interest in new gourmet potato cultivars has added purple, blue, orange, yellow, and other colors to the more traditional available colors of potato tubers.

Tubers of standard white and red cultivars generally are round to oblong and relatively thin skinned compared to russets. Russet tubers tend to be oblong and relatively dark colored and thick skinned at maturity. Because of their thick skins, russet cultivars are less susceptible to injury during harvest than reds or whites. The thick russet skins generally are resistant to common scab. Gourmet cultivars are available in many color combinations and tuber shapes.

The time span from planting to maturity is important when selecting a cultivar for a region and market. Norland, for example, may mature in 80 to 90 days as compared with the 120 days or more required for maturity of Russet Burbank. Early maturing cultivars are preferred for "new" potatoes, but late-maturing cultivars store better because they resist sprouting and shriveling in storage.

Cultivars are also chosen according to their method of preparation for consumption. Russet Burbank, the leading cultivar grown in the United States, is considered excellent for frying and baking, but is often considered inferior to Kennebec or Red Pontiac for boiling. Many home gardeners prefer to grow an all-purpose cultivar such as Kennebec.

The following are some commonly grown cultivars and their characteristics.

Early-Maturing Cultivars
- Norland, Dark Red Norland—Red skinned with white-fleshed cultivar that produces low to medium yields of average-sized tubers. Good for boiling and frying, fair for baking. Dark Red Norland produces darker, more colorful skins than Norland.
- Norland Russet—Yield slightly more than Norland, but also matures later. Fair to good for baking, French frying, and boiling.
- Russet Norkotah—Early russet type that forms oblong tubers with excellent medium russet skins. Fair to good for baking, frying, and boiling.
- White Rose—A white cultivar type with tubers that are generally long, thin skinned, and deep eyed. Will not bake well but is satisfactory for most other uses.

Midseason Cultivars
- Kennebec—A white cultivar type that is an excellent all-purpose potato. It is considered excellent for French frying.
- Red LaSoda—Tubers have bright red skins but may be slightly rough in shape with deep eyes. Satisfactory to good for most home uses.
- Red Pontiac—Midseason to late type with large, deep eyes. Good cooking quality for most uses and is a good red storage potato. Generally considered too late for early "new" potatoes.

Late-Maturing Cultivars
- Russet Burbank—Is the leading cultivar grown in the United States and is also known as Netted Gem, the Idaho Baker, and Russet. Medium to high yields of large, long, and often knobby and misshapen tubers. Excellent for baking and French frying, fair for boiling, but tends to lose shape and fall apart because of high starch content. Stores extremely well.
- Century Russet—Smooth, large and oblong tubers that are good for baking, boiling, and microwaving. Has less moisture than Russet Burbank when baked because of lower starch content.
- Ranger Russet—Tubers similar to those of Russet Burbank but sometimes longer and less susceptible to knobs. Good for all table uses, but does not store as well as Russet Burbank.

Gourmet and Specialty Cultivars
Many gourmet cultivars are grown for special features including skin and flesh color or unusual flavor. Often some of these gourmet cultivars produce low yields and may not store well as compared with standard cultivars.

- Purple-skinned cultivars—All Blue, also called Purple Marker, and Blue Cristie are commonly available. All Blue produces smooth, oblong, late-maturing tubers with dark purple skin and flesh. Blue Cristie tubers are purple skinned with some red-to-pink stripes, prominent eyes, and white flesh.
- Yellow skins and flesh—These include Yukon, Yellow Finn, and Bintje. Yukon Gold tends to produce smoother tubers and matures earlier than the other two. Yellow Finn stores well.
- Red skin, yellow flesh—Desiree is probably the best known. Others include Saginaw Gold and Iditared.

Planting

Potato growers use only certified seed tubers as the planting material (Figure 24.21). If the seed tuber is small then the whole tuber is planted, whereas if the tuber is large then seed pieces are used for planting. In general, seed pieces are 1½ to 2 oz in size. Using smaller pieces often results in lower yields. It is recommended that seed pieces be cut prior to planting and allowed to suberize (heal over) before planting. Planting fresh cut seed is often practiced since growers usually do not have the time and the place to store large quantities of cut seed before planting.

There are several types of planters available that place the potato seed pieces in the soil along with fertilizer and systemic insecticides in one operation. Seed depth is usually about 4 in. below the top of the planted bed (Figure 24.22). The soil is ridged over the row by mounding or hilling soil around the plants during early cultivation so that about 6 in. of soil covers the seed piece (Figure 24.23). Distance between rows of potatoes is commonly 36 in.

Irrigation

Soil moisture is one of the most important factors in determining potato yield and quality. When irrigation is needed to supplement rainfall it is usually applied in frequent small amounts. Secondary growth and growth cracks occur when irrigation or rainfall occurs after a period of moisture stress. The soils used for growing potatoes are generally maintained uniformly moist until tubers have reached full size. The ef-

FIGURE 24.21   Only certified seed is used by potato growers.

FIGURE 24.22   Potato seed pieces placed in a trench about 4 in. deep prior to covering with soil.

fective root depth of potatoes is about 2 ft, and the soil is typically not allowed to dry below 65% of field capacity. Moisture levels significantly above field capacity also seriously affect yield and quality.

Culture and Care

Loose, friable soils improve tuber set and the development of smooth, well-shaped and even-colored potatoes. Potatoes develop larger and more extensive root systems

FIGURE 24.23   Hilling or bedding of a potato field.

in response to proper cultivation. Cultivation is often necessary to control weeds and the soil is typically hilled around the plants. Deep cultivation is avoided because many roots may be destroyed. Cultivation is no longer used after the plant reaches full bloom.

Weeds are controlled in potato fields because they cause many problems besides being hosts for insects and diseases. An effective weed-control program takes into account the weed problem, cultivation, and herbicides. Fields containing perennial weeds are generally avoided for growing potatoes.

Common Growing Problems and Solutions

Insects   Potatoes are never planted in fields that have been in sod or grass the previous year. By avoiding this situation, it decreases the chance of having wireworm and grubworm problems.

Once emerged, potatoes are susceptible to cutworms, flea beetles, and leafhoppers. Colorado potato beetle is a major insect problem as adults overwinter in the soil and emerge about the same time the potatoes are emerging. Colorado potato beetles feed on young foliage and deposit eggs on the lower one-third of the plant on the underside of the leaves (Figure 24.24). Defoliation by Colorado potato beetles reduces potato yields the greatest when the tubers are sizing. While the larvae can cause extensive defoliation, they can be controlled with systemic soil insecticides, foliar sprays, or appropriate crop rotations. Potato fields planted after nonhosts (peanuts, wheat, sorghum, etc.) have fewer beetle problems than fields planted to potatoes the previous year(s).

Diseases   Common fungal problems of the potato foliage are early blight (alternaria), fusarium wilt, and verticillium wilt. Blackleg, a bacterial disease, is characterized by blackening of the stems and yellowing and curling of leaves. Various fusarium species and blackleg bacterium can cause tuber rots. *Rhizoctonia solani*

FIGURE 24.24   Colorado potato beetle is a major pest of Irish potatoes. Source: Clemson University Extension Service.

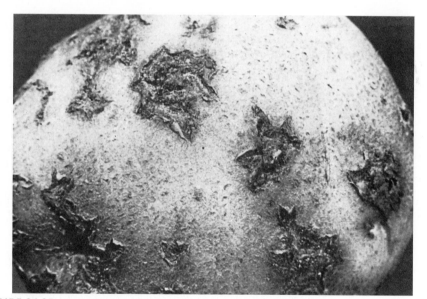

FIGURE 24.25   Irish potato tubers infected with the scab organism. Source: Clemson University Extension Service.

forms black sclerotia on the surface of the tubers. Another fungal disease of the tubers' surface is common scab (streptomyces) (Figure 24.25).

Root-knot nematodes form irregular bumps on the tubers. The potato plant is also susceptible to a variety of virus diseases such as potato leaf roll and purple-top wilt.

Physiological Disorders   Nonparasitic diseases include sunscald, sunburn, and tipburn.

FIGURE 24.26   Potato harvester.

Harvesting

Potato tubers must be fully mature before digging for table stock. Vines are often killed by vine beaters or chemicals to promote good skin set. Potatoes for storage are typically not dug until 2 weeks after vines have died and have been removed. Early potatoes are sometimes dug before optimum maturity to take advantage of certain limited market demands and high prices. Potatoes for processing may be tested for reducing sugars to determine if the tubers are in the acceptable range for chipping into the light-colored potato chips that consumers prefer.

Potato harvest is almost fully mechanized (Figure 24.26). The harvester digs and loads the potatoes on trucks for transport to a shed where tubers are washed, graded, and sized for bulk marketing or packed in bags or boxes. Tubers bruise easily during harvest especially at temperatures above 85°F and below 50°F. Soil condition, tuber condition, and harvest operation are important factors that influence bruising. Other common market defects besides bruising are rots, cracks, skinning, enlarged lenticels, heat sprouts, greening, and numerous diseases. Tubers that have decay or bad cuts are never stored. Potatoes are usually dug only when the soil is relatively dry in order to prevent excess mud and soil on the tubers.

Postharvest Handling

Trucks and equipment at the packinghouses for potatoes are usually padded to minimize skinning and bruising. Potatoes are precooled to 40°F for fresh market use to remove field heat quickly. A curing period of 4 to 5 days at 60° to 70°F in high humidity may be used, although not typically by commercial growers, to suberize or heal cuts of the tubers. The best temperatures for holding potatoes are between 40° to 42°F.

Potatoes are stored in a cool, dark, well-ventilated area where there is no danger of freezing. Tubers exposed to light during storage form chlorophyll and turn green. Shrinking and sprouting increases at storage temperatures above 50°F. To reduce shriveling, potatoes are stored in a humid but not wet area. Also, sprouts are removed from the tubers to assist in reducing water loss and shriveling during storage.

Nutrition Facts

Serving Size 1 medium (148g)

Amount Per Serving

Calories 100   Calories from Fat 0

% Daily Value*

Total Fat 0g	0%
Saturated Fat 0g	0%
Cholesterol 0mg	0%
Sodium 0mg	0%
Total Carbohydrate 26g	9%
Dietary Fiber 3g	12%
Sugars 3g	
Protein 4g	

Vitamin A 0% • Vitamin C 45%

Calcium 2% • Iron 8%

FIGURE 24.27   Nutritional value of Irish potatoes. Percent daily values are based on a 2,000 calorie diet. Used with permission from Dole Food Co., Inc.

Market Preparation and Marketing

Summer-harvested potatoes are not stored or held any longer than necessary before marketing. Chipping potatoes are normally sold at contract prices. Season-long potatoes grown for table use may be stored for several months and marketed in bags of various sizes. These storage bags may be paper, vented plastic, or net bags for consumers, or in 50- to 100-lb mesh bags or cardboard boxes for institutional sales.

Uses

Potatoes are served baked, boiled, mashed, or roasted, and as French fries, shoe-strings, chips, and home fries. Potatoes are served whole, canned, or scalloped and used in soups, stews, chowders, and dumplings. They are also used in hot and cold potato salads. Processing potatoes are dehydrated, flaked, and granulated and used in instant mashed potatoes or reconstituted into chips.

Potatoes are being used more as a processed product and less as a fresh commodity. The popularity of potato chips and French fries is increasing and more potatoes are being used for these purposes.

Nutritional Value

Potatoes are less fattening than other foods in the diet and are an excellent source of vitamin C and a good source of potassium, phosphorus, and iron. The manner in which potatoes are prepared for eating often increases their caloric value. For example, French fries have about five times the calories of mashed potatoes. A medium-sized potato has 100 calories with no grams of fat and 3 g of dietary fiber (Figure 24.27). This serving of potato also provides 45% of the RDA of vitamin C, 8% of the RDA for iron, and 2% of the RDA for calcium.

Review Questions

1. What is the role of air temperature on tomato fruit set?
2. What are the differences in growth habit and fruit production between determinate and indeterminate tomato cultivars?
3. What effect does pruning have on tomato fruit yield?
4. What environmental factors contribute to the occurrence of blossom-end rot in tomatoes?
5. What is a consequence of using excessive amounts of nitrogen fertilizer for growing bell peppers?
6. How is pungency determined in peppers?
7. What factors determine when to harvest peppers?
8. What are the advantages of growing eggplants on raised beds?
9. What are some of the characteristics of a high-quality eggplant fruit at harvest?
10. What is the planting material used for potato production?
11. Describe "new" potatoes.

Selected References

Anderson, C. R. 1998. *Potatoes.* Univ. of Arkansas Coop. Ext. Serv. Publ.

Ball, J. 1988. *Rodale's garden problem solver: Vegetables, fruits and herbs.* Emmaus, PA: Rodale Press. 550 pp.

Benner, B. 1964. Eggplant. *Fruit & Vegetable Facts & Pointers.* United Fresh Fruit & Vegetable Association. Washington, DC.

Bosland, P. W., A. L. Bailey, and J. Iglesias-Olivas. 1996. *Capsicum pepper varieties and classification.* New Mexico State Univ. Coop. Ext. Serv. Cir. 530.

Cook, W. P., D. O. Ezell, R. P. Griffin, C. E. Drye, and P. J. Rathwell. 1995. *Commercial tomato production in South Carolina.* Clemson Univ. Coop. Ext. Serv. Cir. 625.

Granberry, D. M. 1990a. *Commercial eggplant production.* Univ. of Georgia Coop. Ext. Serv. Cir. 812.

Granberry, D. M. 1990b. *Commercial pepper production.* Univ. of Georgia Coop. Ext. Serv. Bull. 1027.

Kays, S. J. and J. C. Silva Dias. 1995. Common names of commercially cultivated vegetables of the world in 15 languages. *Economic Botany* 49: 115–152.

Lerner, B. R. 1993. *Tomatoes.* Purdue Univ. Coop. Ext. Serv. Publ. HO-26.

Lindgren, D. T. 1982. *Eggplant.* Univ. of Nebraska Coop. Ext. Serv. Publ. G82-603-A.

Lindgren, D. T., and L. Hodges. 1990a. *Peppers.* Univ. of Nebraska Coop. Ext. Serv. Publ. G81-540-A.

Lindgren, D. T., and L. Hodges. 1990b. *Tomatoes in the home garden.* Univ. of Nebraska Coop. Ext. Serv. Publ. G80-496-A.

Mansour, N. S. 1993. *Grow your own peppers.* Oregon State Univ. Coop. Ext. Serv. Publ. EC 1227.

Marr, C. 1992. *Irish potatoes.* Kansas State Univ. Coop. Ext. Serv. Publ. MF-488.

Marr, C., W. J. Lamont, N. Tisserat, B. Baurenfeind, and K. Gast. 1995. *Tomatoes.* Kansas State Univ. Coop. Ext. Serv. Publ. MF-1124.

Mosley, A., O. Gutbrod, S. James, K. Locke, J. McMorran, L. Jensen, and P. Hamm. 1995. *Grow your own potatoes.* Oregon State Univ. Coop. Ext. Serv. Publ. EC 1004.

Motes, J. E., and J. T. Criswell. 1994. *Potato production.* Oklahoma State Univ. Coop. Ext. Serv. Publ. F-6028.

Motes, J. E., J. T. Criswell, and J. P. Damicone. 1992. *Pepper production.* Oklahoma State Univ. Coop. Ext. Serv. Extension Fact F-6030.

Nagel, D. 1997. *Commercial production of tomatoes in Mississippi.* Mississippi State Univ. Coop. Ext. Serv. Publ. Information sheet 1514.

Nonnecke, I. L. 1989. *Vegetable production.* New York: Van Nostrand Reinhold. 656 pp.

Peirce, L. C. 1987. *Vegetables: Characteristics, production, and marketing.* New York: John Wiley & Sons.

Pleasant, B. 1996. Technicolor taters. *Carolina Gardener.* (February):34.

Roberts, W., J. Motes, and R. J. Schatzer. 1994. *Commercial production of fresh market tomatoes.* Oklahoma State Univ. Coop. Ext. Serv. Publ. F-6019.

Sanders, D. C. 1994a. *Home garden eggplant.* North Carolina State Univ. Coop. Ext. Serv. Leaflet 8015.

Sanders, D. C. 1994b. *Pruning and supporting tomatoes.* North Carolina State Univ. Coop. Ext. Serv. Leaflet 28-G.

Sanders, D. C. 1996a. *Eggplant.* North Carolina State Univ. Coop. Ext. Serv. Leaflet 15.

Sanders, D. C. 1996b. *Fresh market tomato production—Piedmont and Coastal Plain of North Carolina.* North Carolina State Univ. Coop. Ext. Serv. Leaflet 28-A.

Sanders, D. C. 1996c. *Tomatoes for processing in eastern North Carolina.* North Carolina State Univ. Coop. Ext. Serv. Leaflet 28-F.

Sanders, D. C., and N. G. Creamer. 1996. *Commercial potato production in eastern North Carolina.* North Carolina State Univ. Coop. Ext. Serv. Leaflet 22.

Seelig, R. A. 1968. Peppers. *Fruit & Vegetable Facts & Pointers.* United Fresh Fruit & Vegetable Association. Washington, DC.

Sieczka, J. B., and R. E. Thornton. 1992. *Commercial potato production in North America.* Orono, Maine: Potato Association of America Handbook.

Yamaguchi, M. 1983. *World vegetables.* New York: Van Nostrand Reinhold.

Yanta, J. P., and C. Tong 1993. Commercial postharvest handling of potatoes (*Solanum tuberosum*). Minnesota Coop. Ext. Serv. Publ. FS-6239-GO.

Selected Internet Sites

USDA Agricultural Marketing Service Standards for Grades

Fresh
Eggplant   www.ams.usda.gov/standards/eggplant.pdf
Peppers (sweet)   www.ams.usda.gov/standards/peperswt.pdf
Potatoes   www.ams.usda.gov/standards/potatoes.pdf
Potatoes (seed)   www.ams.usda.gov/standards/potatosd.pdf
Tomatoes   www.ams.usda.gov/standards/tomatfrh.pdf
Tomatoes (greenhouse)   www.ams.usda.gov/standards/tomatogr.pdf

Processing
Peppers (sweet)   www.ams.usda.gov/standards/peperswt.pdf
Potatoes   www.ams.usda.gov/standards/vppot.pdf
Potatoes (chipping)   www.ams.usda.gov/standards/potatoch.pdf
Tomatoes   www.ams.usda.gov/standards/vptom.pdf
Tomatoes (canning)   www.ams.usda.gov/standards/vptomita.pdf

chapter
25

Cucurbits

I. Cucumber
II. Watermelon
III. Muskmelon
IV. Squash
V. Pumpkin

The cucurbits are members of the Cucurbitaceae or gourd family. Members in this family typically thrive in hot weather and will not withstand frost. Cucurbits have tendrils, leaves that are often rough to the touch, and fleshy fruits with many seeds. Other members in the family include watermelon, muskmelon and other melons, cucumber, pumpkin, and the squashes.

Cucumber

Scientific Name

Cucumis sativus L.

Common Names in Different Languages

cucumber (English); hu gua (Chinese); concombre (French); Salat-Gurke, Schlangengurke, Schälgurke (German); cetriolo, cetriolino (Italian); kyuuri (Japanese); ogurec, ogurtzy dlinno plodny, ogurtzy korotka (Russian); pepino, cohombro (Spanish) pickling cucumber (English); cornichon, c. au vinaigre (French); Essiggurke, Gewürzgurke, Einlegegurke (German); cetriolino sotto acete (Italian); pepinillo en vinagre (Spanish)

Classification, History, and Importance

Cucumber (*Cucumis sativus* L.) is a member of the Cucurbitaceae. Other family members include pumpkins, squashes, gourds, melons, watermelons, and chayotes. Cucumber is an herbaceous trailing an-

TABLE 25.1   *The production value of selected cucurbit crops for fresh market and processing in the United States*

Crop	Value of production		
	1995	1996 (1,000 dollars)	1997
Fresh market			
Cucumber	166,333	186,325	185,194
Cantaloupe	350,698	400,795	417,859
Honeydew melon	89,193	80,405	109,394
Watermelon	357,062	275,684	309,230
Processing			
Cucumber (for pickles)	135,803	139,330	146,043

Source: USDA, NASS.

TABLE 25.2   *U.S. per capita consumption of cucumbers*

Use	1976	1986 (lb, farm weight)	1996
Fresh	3.1	4.6	6.0
Pickling	6.1	5.3	4.6
All	9.2	9.9	10.6

Source: USDA, Economic Research Service.

nual capable of spreading in all directions. It is a monoecious plant that is cultivated for its immature fruits.

Cucumber is believed to be indigenous to India and was grown in Western Asia as far back as 3,000 years ago. Cultivation of cucumbers soon followed in Greece, Italy, and then China. Cucumbers were introduced into France in the 9th century, England in the 1300s, and America by 1539.

Cucumber for fresh market was a $185.2 million industry in 1997 (Table 25.1). This was down slightly from $186.3 million in 1996. The cucumber for processing market is slightly smaller than for fresh market with a production value of $146 million in 1997. Per capita consumption of fresh cucumbers has increased from 3.1 lb per person in 1976 to 6.0 lb per person in 1996 (Table 25.2). The amount of cucumbers consumed as pickles declined during that time period from 6.1 lb per person in 1976 to 4.6 lb per person in 1996. Of the 58,300 acres of cucumbers grown in the United States for fresh consumption, 13,500 acres are in Georgia and 9,500 acres are in Florida (Table 25.3). Other states that grow fresh-market cucumbers include California, Michigan, North Carolina, Virginia, New York, New Jersey, Texas, and South Carolina. There are about 107,780 acres of cucumbers grown for processing in the United States. Michigan, North Carolina, and Texas lead in the production of cucumbers for processing.

TABLE 25.3   *The top fresh market and processing cucumber producing states in the United States*

Rank	State	Planted acres (000)	Harvested acres (000)	Yield (cwt/acre)	Production (000)
Fresh market					
1	Georgia	13.50	13.00	170.00	2,210
2	Florida	9.50	9.20	310.00	2,852
3	North Carolina	6.20	6.00	120.00	720
4	Virginia	6.00	5.60	125.00	700
5	California	5.70	5.70	315.00	1,796
6	Michigan	5.70	5.60	200.00	1,120
7	New York	3.40	3.30	200.00	660
8	New Jersey	2.30	2.20	135.00	297
9	South Carolina	2.00	1.70	110.00	187
10	Texas	1.80	1.70	170.00	289
	United States	58.25	56.05	195.00	10,957
Processing				**(tons/acre)**	
1	Michigan	26.00	25.00	5.20	130
2	Florida	6.50	6.10	13.00	79
3	North Carolina	18.30	18.00	4.00	72
4	Texas	11.50	10.10	6.50	66
5	Ohio	3.00	2.90	12.26	36
6	California	4.40	4.20	7.19	30
7	Wisconsin	3.70	3.50	7.90	28
8	South Carolina	4.00	3.80	3.00	11
9	Indiana	2.00	2.00	4.97	10
10	Colorado	0.78	0.72	8.45	6
	United States	107.78	103.57	5.98	619

Source: USDA, NASS, 1997 data.

Plant Characteristics

Height: 6 to 8 in.
Spread: On ground, 12 to 20 sq ft; on trellis, 12 to 15 in.
Root Depth: Most roots are in the top 12 in. of the soil with a taproot that can grow from 2 to 3 ft deep.

Climatic Requirements

Cucumber is a warm-season crop that requires soil temperatures of at least 60°F for satisfactory germination. Cucumbers grow best in the 65° to 85°F range and suffer chilling injury below 50°F. The crop growth rate increases steadily as the temperatures rise to 90°F. Cucumber crops planted in early spring when soils are cool may take 8 to 9 weeks or more to reach harvest stage, while later plantings during warmer soil temperatures may require only 6 weeks.

FIGURE 25.1   A green slicing-type of cucumber.

Field Preparation

Cucumbers can be grown on most fertile, well-drained soils. Loam-textured soils well supplied with organic matter are often best suited for high yields. Cucumbers can be grown in light, sandy soils provided they are adequately fertilized and irrigated. Soil pH for growing cucumbers should be 6.0 to 7.0.

Cucumbers respond favorably to high optimum levels of fertilizer and about 60 lb N/acre, 100 lb P/acre, and 60 lb K/acre are generally recommended. In some areas it is recommended that part of the fertilizer be broadcast before disking and the remainder banded at planting time. In other areas it is recommended that the entire amount be broadcast and disked in before planting. Crops that are grown for hand-harvesting over a prolonged period require a side-dressing with nitrogen. In some areas phosphates and potash are necessary for cucumber crops grown for once-over mechanical harvesting. Heavy, leaching rains on light, sandy soils often make it necessary to apply additional nitrogen during the season.

Selected Cultivars

Cultivars of cucumber have been developed for specific uses and can be categorized into types.

Fresh Market   Slicing-type cucumbers are used almost exclusively for consumption as fresh product. These fruits are more attractive in appearance and distinctly darker green and longer (Figure 25.1) than pickling cucumbers that are grown for processing. Slicer types are hand-harvested to minimize physical damage.

Greenhouse-type cucumbers for protected cultivation are of two fruit types: the Dutch type and the Japanese type. Dutch types are long, dark green, fruited, parthenocarpic, and are grown exclusively in Europe and Canada and to a lesser extent in the United States (Figure 25.2). Japanese types differ from Dutch types in

FIGURE 25.2   A greenhouse cucumber type.

that most hybrids are monoecious with some degree of apparent parthenocarpy. This type is also grown in the open field on trellises.

Processing   Pickling cucumbers are processed from both fresh and brined cucumbers. Pickling types are cylindrical in shape with blocky ends and a medium-green color (Figure 25.3). Color intensity of the exterior of the fruit, ranging from light green to dark green, is dependent on cultivar and the environment. The preferred color varies according to processor specifications and the area where the crop is grown. Generally, lighter-colored cultivars are used in cool environments and darker green cultivars are used in warmer environments. Uniform color among fruits is highly desirable regardless of the intensity. Internal fruit structure is extremely important in processing cucumber types. Superior cultivars have relatively strong carpel sutures, small seed cavity, and slow seed development.

Planting

Seedbed preparation is one of the most important steps in growing cucumbers. The soil is prepared to form a firm, level seedbed of uniform texture.

Precision spacing of cucumber seeds in the row to produce a desired stand of plants saves seed and eliminates the need for thinning after the crop is established. Precision in-depth seed placement in the soil affects the uniformity of maturity at harvest time. This is particularly important in fields that are to be harvested by machine. Cucumbers are normally planted at a depth of ½ to 1 in., depending on soil type, time of season, and weather conditions.

Coated seed aids significantly in obtaining uniform spacing of direct-seeded cucumbers and works well in plate planters as well as in precision planters. Cucumber transplants have typically performed poorly in the field due to rapid suberization of the broken root hairs caused by the transplanting process.

FIGURE 25.3   Processing type of cucumber made into pickles. Used with permission from Seminis Vegetable Seed Co., Inc.

FIGURE 25.4   An established planting of cucumbers for fresh market.

General recommendations for fresh market cucumbers are in-row spacings of 9 to 12 in. with between-row spacings of 36 to 72 in. (Figure 25.4). Pickling cucumbers have smaller vines and are grown at higher populations. Maximum yields for hand-harvested cucumbers may be obtained with in-row spacings of 6 to 12 in. with 36 to 72 in. between rows. High yields of pickling cucumbers grown for once-over harvest are obtained with in-row spacings of 4 to 6 in. with 12 to 28 in. between rows.

Irrigation

Although cucumbers are a deep-rooted crop, adequate soil moisture is needed for good yields. Cucumbers require large amounts of water because fruits typically are

FIGURE 25.5   Beehive on the edge of a field provides bees necessary for pollination of the crop.

about 95% water. The water requirements for cucumbers in some areas have been estimated to be as high as 15 acre-inches during the crop-growing season. The most critical need for water for cucumbers is during the fruiting period when drought conditions seriously reduce the yield of marketable fruit and cause "nubbing." Cucumber plants should never be allowed to wilt from water stress if maximum yields of high-quality, marketable fruit are desired. The last irrigation for cucumber crops grown for mechanical harvest is usually timed to allow the soil to dry out sufficiently to support heavy equipment without subjecting the plants to water stress. Irrigation is usually applied early in the day whenever possible to permit the soil and leaf surfaces to dry before nightfall.

Both furrow and overhead sprinkler irrigation is commonly used for cucumbers. Furrow irrigation is limited to areas where land is level enough to be graded. Furrow irrigation is often preferred to overhead since overhead irrigation is more apt to spread diseases to the foliage and the fruit by splattering of diseased soil.

Culture and Care

Bees are extremely important for maximum production of field-grown cucumbers (Figure 25.5). At least one colony of bees is often necessary for each 50,000 plants. Wet, cool weather reduces bee activity and pollination resulting in lower yields and/or misshapen cucumber fruit.

Plastic mulches are often used for fresh market cucumbers. Mulches conserve moisture, prevent soil compaction and rotting of the fruit, and help suppress weeds. Season-long weed control is difficult in cucumber plantings (Figure 25.6).

Common Growing Problems and Solutions

Insects   Common insect problems of cucumbers are cucumber beetles, aphids, cabbage loopers, cutworms, pickleworms, spider mites, and squash bugs. Two types of

FIGURE 25.6 Weeds can be a problem in cucumber fields.

FIGURE 25.7 Spotted cucumber beetle.
Source: Clemson University Extension
Service.

cucumber beetles are harmful to cucumbers. One is yellow and black spotted (Figure 25.7) and the other is striped (Figure 25.8). Both beetles feed on the plants early in the season and transmit bacterial wilt that causes the plants to wilt and die. Aphids also feed on plants, sucking juices and causing a twisted, distorted growth.

Diseases Cucumbers are affected by a number of leaf and fruit diseases. Downy mildew appears initially as pale yellow angular spots on upper leaf surfaces. Under wet conditions, a soft, white, or purplish mold forms directly under these spots on the lower leaf surface. Anthracnose starts as light brown round spots, often with split, ragged centers

FIGURE 25.8   Striped cucumber beetle. Source: Clemson University Extension Service.

on the leaves. Sunken lesions can also form on the lower leaf surface. Gummy stem blight appears as large, irregular tan spots on leaves, usually at the margins and as watery spots, up to several inches long, on the vines. Belly rot appears as sunken, reddish brown crusty spots on the undersides or ends of the fruit touching the soil. Powdery mildew appears as white powdery spots on both upper and lower leaf surfaces.

Physiological Disorders   Bitter-tasting cucumbers can result from moisture stress, temperature, inappropriate soils, or heredity. Bitter taste usually occurs during the hotter part of the summer or later in the growing season. Cucumbers are only grown on well-drained soils with a pH between 6.0 and 6.5 to avoid bitter-tasting fruit. A mulch that conserves moisture and keeps roots cool during hot, dry weather tends to reduce the occurrence of bitter-tasting fruit.

Harvesting

Fresh Market Types   Fresh market cucumber fruits are harvested before they are fully elongated and while the seeds are succulent. Fresh market cucumbers are ready for harvest about 50 to 70 days after planting or about 7 to 10 days after pollination. Cucumbers for fresh market are harvested on the basis of size with the fruits typically being 6 to 10 in. long. The fruits are of highest quality when they are dark green, firm, and crisp. Cucumber fruits become bitter with size and age and are not allowed to reach the yellowish fruit color stage or they will become unmarketable. Fresh market cucumbers are typically harvested several times per week during the peak of the fruiting season.

Processing Types   Several stages of immature fruits are harvested for processing cucumbers. Traditionally, approximately 90% of the acreage of processing-type cucumbers are hand-harvested with the remainder harvested by machine in a once-over operation. Hybrids with a high and concentrated initial fruit set are best suited for mechanical once-over harvest.

Nutrition Facts

Serving Size 1/3 medium cucumber (99g)

Amount Per Serving

Calories 15   Calories from Fat 0

% Daily Value*

Total Fat 0g	**0%**
Saturated Fat 0g	**0%**
Cholesterol 0mg	**0%**
Sodium 0mg	**0%**
Total Carbohydrate 3g	**1%**
Dietary Fiber 1g	**4%**
Sugars 2g	
Protein 1g	

Vitamin A 4%	•	Vitamin C 10%
Calcium 2%	•	Iron 2%

FIGURE 25.9   Nutritional value of cucumbers. Percent daily values are based on a 2,000 calorie diet. Used with permission from Dole Food Co., Inc.

Postharvest Handling

Cucumbers are very sensitive to chilling injury and yellowing. At 50°F or lower chilling injury occurs, while at 60°F or above yellowing begins. The best temperature range for storage is 54° to 56°F. Hydrocooling is often used to remove the field heat and lower fruit temperatures. Cucumbers are held at 95% relative humidity to prevent water loss.

Market Preparation and Marketing

Cucumber fruits shipped long distances are often waxed to reduce moisture loss. The additional wax can reduce transpiration loss by 50%, primarily by covering the stem scar or petiole base. English cucumbers are film wrapped to retain moisture.

Uses

Cucumbers are widely used as fresh and processed products. Fresh cucumbers are sliced, diced, or cubed in salads. They are fresh pickled, brine pickled with dill, and pickled with sweet, sour, or herb flavoring in varied ways depending on ethnic and regional taste preferences. They are also made into relish and used in hot and cold soups.

Nutritional Value

Cucumber is low in calories and is a source of vitamin C and vitamin A. One medium cucumber has only 15 calories with no grams of fat and 1 g of dietary fiber (Figure 25.9). This serving of cucumber also provides 10% of the RDA of vitamin C, 4% of vitamin A, and 2% of the RDA for both calcium and iron. Cucumbers both fresh and pickled are prized for their flavor and texture.

FIGURE 25.10   Watermelon fruit maturing on the plant in the field.

Watermelon

Scientific Name

Citrullus lanatus (Thunb.) Natsum & Nakai

Common Names in Different Languages

watermelon (English); xi gua (Chinese); pastèque, melon d'eau (French); Wassermelone (German); cocomero, anguria, melone d'acqua (Italian); suika (Japanese); arbuz stolovyi (Russian); sandia (Spanish)

Classification, History, and Importance

Watermelon, *Citrullus lanatus,* is a member of the Cucurbitaceae family. Other members of the family include cucumber, gourds, muskmelon, pumpkin, and squash. Watermelon is a monoecious spreading annual vine with large pinnately lobed leaves. It is grown for its large juicy fruits (Figure 25.10). Watermelon fruits may be oblong, ellipsoidal, or spherical with a thick but fragile rind. The fruit flesh at maturity may be white, greenish white, yellow, pink, or red.

Watermelons are indigenous to tropical Africa. Some evidence also indicates possible American origins since Native Americans were discovered growing watermelons in the Mississippi Valley as early as French exploration to the United States. Cultivation of watermelons dates back 4,000 years to the ancient Egyptians as proven by artistic records, and watermelons were also widely distributed throughout the remainder of the world by African and European colonists. Watermelons were grown in the Americas either before or soon after the arrival of the first European explorers.

TABLE 25.4   *U.S. per capita consumption of watermelon*

Crop	1976	1986 (lb, farm weight)	1996
Watermelon	12.6	12.8	17.4

Source: USDA, Economic Research Service.

TABLE 25.5   *The top watermelon producing states in the United States*

Rank	State	Planted acres (000)	Harvested acres (000)	Yield (cwt/acre)	Production (000)
1	California	18.20	18.20	450.00	8,190
2	Florida	33.00	30.00	250.00	7,500
3	Georgia	34.00	31.00	210.00	6,510
4	Texas	42.00	37.50	170.00	6,375
5	Arizona	7.70	7.50	310.00	2,325
6	Indiana	7.30	6.80	290.00	1,972
7	North Carolina	9.80	9.50	150.00	1,425
8	South Carolina	9.60	8.50	150.00	1,275
9	Missouri	4.80	4.70	240.00	1,128
10	Oklahoma	11.00	9.00	100.00	900
Total	United States	204.30	184.60	221.00	40,734

Source: USDA, NASS, 1997 data.

Watermelons were a $309.2 million industry in 1997 (see Table 25.1). This was up from $275.7 million in 1996. Per capita consumption of watermelon has increased from 12.6 lb per person in 1976 to 17.4 lb per person in 1996 (Table 25.4). There were 204,300 acres of watermelons grown in the United States in 1997 (Table 25.5). California, Georgia, Florida, and Texas are the largest watermelon producing states.

Plant Characteristics

Height: Vining cultivars, 12 in.; bush cultivars, 12 to 18 in.
Spread: Vining cultivars, 10 to 20 sq ft; bush cultivars, 36 to 48 in.
Root Depth: Most roots are shallow but some extend to 4 ft and beyond.

Climatic Requirements

Watermelon is a tender warm-season vegetable that requires up to 4 months of frost-free weather. Optimum temperatures for watermelon seed germination are 77° to 95°F. Watermelon plant growth occurs optimally at daytime temperatures of 70° to 80°F with nighttime temperatures of 65° to 70°F.

Field Preparation

Watermelons can be grown on many different soil types but generally grow best on sandy loam soils. Satisfactory watermelon yields can be produced on clay soils if the plants are mulched with plastic to conserve soil moisture. Sandy soils generally warm faster in the spring, are easier to plant and cultivate, and allow deep root penetration.

Soils are deep plowed or tilled, disked, and harrowed to make a fine seedbed for watermelon production. Windbreaks, usually composed of rye strips spaced every 20 to 40 ft, may be used in very sandy areas where blowing sand can damage vines. Windbreaks reduce abrasion injury to the leaves and stems from windblown sand particles and reduce whipping of the vines by the wind. Enhanced plant growth and earlier fruit production (by 1 to 2 weeks) are often observed with the use of windbreaks since air temperatures around the plants are greater because the plants are sheltered from the cold winds.

Fertilizer P and K levels for watermelons are applied as suggested by the soil-testing recommendations. A typical recommendation may include a moderate application of nitrogen, 50 to 70 lb, which produces good yields without excessive vine growth. An additional quantity of nitrogen is often used when vines are 8 to 12 in. long.

Selected Cultivars

Both standard (open-pollinated) and hybrid cultivars of watermelons are available. Many newer cultivars are hybrids. Some common cultivars and their characteristics are the following:

- Crimson Sweet produces a round, striped fruit that averages 25 pounds and is finely textured and sweet. This cultivar is resistant to some strains of anthracnose and fusarium wilt.
- Blackstone or Improved Black Diamond are smaller and more uniform than Crimson Sweet. They are dark green colored with coarse red flesh.
- Charleston Gray—Several strains such as Charleston Gray 133, Calhoun Gray, Sweet Charlie, and Sweet Princess are available that have improved disease resistance or more uniform size and shape than the original Charleston Gray cultivar. The fruit of this type is long, light green in color, and resists injury from the sun. Some of these cultivars may be slightly resistant to fusarium wilt and tolerant to some strains of anthracnose.
- Jubilee is a popular cultivar grown in the southern United States with slight resistance to fusarium wilt and tolerance to anthracnose. It produces a long, striped fruit that weighs up to 40 lb and has dark green stripes and good flavor.
- Oasis has deep red flesh and is earlier than Crimson Sweet. It is striped and slightly oval in shape.
- Bluebell has an early, round, small dark green fruit (approximately 15 lb).
- Royal Sweet has a long, striped fruit with good flavor and bright red flesh.
- Royal Jubilee has a long, blocky, striped fruit and is a late-maturing cultivar.
- Icebox or small melons—A limited market exists for small melons especially in urban areas where their smaller size and ease of carrying or storing is useful to some consumers.
  Sugar Baby or Sugar Doll are early and smaller sized (8 to 10 pounds). The fruit is dark green and round.
  Mickey Lee is an early icebox type with light green fruit weighing 8 to 15 pounds.

FIGURE 25.11   Seedless watermelon fruit. Courtesy of B. B. Rhodes.

Minilee has small, round, light green fruit weighing 8 to 15 pounds.
- Seedless melons—A premium market is often available for seedless watermelons (Figure 25.11), although the seed is expensive and emergence and vigor are less than the other watermelon types. King of Hearts, Tiffany, or Crimson Trio are some of the more common seedless cultivars.

Planting

Watermelons are planted after the soil is warm (60° to 65°F) and when danger of frost is past. Watermelons generally require large amounts of space for optimum growth, so seeds are planted 1 in. deep in hills within the row spaced 6 ft apart. Often there are 7 to 10 ft between rows. After seedlings are established they are thinned to the best three plants per hill. Closer plant spacings are used in some areas and have provided earlier and more concentrated production than with larger plant spacings.

Some growers gain an earlier advantage by starting transplants in hotbeds, cold frames, or greenhouses. Plastic cells (1½ in. or larger) or peat pots are often used and seeds are planted for transplant production 10 to 14 days before field setting. Melon transplants should have two to three leaves when they are planted in the field.

Irrigation

Watermelons are deep rooted and fairly drought tolerant, but adequate water early in the season generally improves vine growth and fruit yield. A critical time in determining yield potential is when the first blossoms appear; thus adequate moisture is important at this time. When water is applied, the soil is thoroughly wet to a 2- to 3-ft depth. Irrigation is never used after the fruit are near full size since watering during the ripening season may delay harvest, cause fruit to split, and reduce sweetness of the fruit flesh. Watermelons are also well adapted to drip or trickle irrigation.

Culture and Care

Mulching with black plastic film generally promotes earliness in watermelon production by warming the soil and conserving moisture. Black plastic absorbs heat and warms the soil. A 7- to 10-day earlier harvest of watermelon is possible in most years with the use of black plastic mulch. The mulch also prevents weed growth, reduces evaporation of soil moisture, and lessens nutrient losses from leaching.

Mulches are typically applied with a mulch applicator that rolls out the mulch over the soil surface. The edges of the mulch are covered with soil. Plants or seeds are planted through the mulch with a small trowel or "bulb planter." Mechanical "punch" planters are also available to set plants through the plastic. Row covers such as clear plastic with wire hoops or open frabrics are also used.

Usually native or local bees are sufficient to provide adequate pollination for watermelon. Bees are not active during cloudy days or rainy weather. When local bees are not sufficient a beehive may be needed. One active hive should pollinate 8 to 12 acres of watermelons. Since bees pollinate early in the morning and melon flowers typically close by midafternoon and remain closed all evening, insecticides are applied only when flowers are closed so that no spray material will cause injury to the bees.

Common Growing Problems and Solutions

Insects   Striped or spotted cucumber beetles (yellow and black in color), squash vine borers, pickleworms, squash bugs, and melon aphids are serious pests of watermelons. Cucumber beetles and aphids feed on the undersides of leaves.

Diseases   Bacterial wilt (spread by cucumber beetles), fusarium wilt, leaf spot, powdery mildew and downy mildew, and alternaria blight are common diseases of watermelons.

Anthracnose is a major foliar disease of watermelons and is most severe during relatively warm periods with frequent rainfall and a high relative humidity. Symptoms first appear on leaves as light yellow or tan lesions that eventually turn black. The roughly circular lesions expand rapidly during favorable weather conditions. Elongated lesions also may form on the petioles and stems, causing death of plant tissue beyond the lesion. Severely blighted plants appear scorched.

Anthracnose also affects the fruits and young fruits are either killed or become malformed. Larger fruits develop sunken, roughly circular lesions that are water-soaked. During humid weather, masses of pink- to salmon-colored spores are extruded from fruiting structures in the center of the lesions. This spore matrix is visible during hot weather. The incidence of anthracnose can be reduced through the use of clean seed and crop rotation. Rotation periods are typically not shorter than 3 or 4 years since the fungus can survive in the crop debris.

Alternaria leaf blight is incited by *Alternaria cucumerina* and is more severe on weakened plants. Symptoms first appear on leaves nearest the center of the hill. Small yellow spots on the leaves enlarge rapidly, turning brown, and often coalesce to kill large portions of the leaf surface. Dark concentric rings form within the lesions, resulting in a target board appearance of leaf spot. Defoliation causes a reduction in plant vigor and exposes the fruit to sunscald injury. Field rotation should be followed to reduce alternaria. The fungus tends to attack weakened plants; therefore, it is important to maintain optimal growing conditions (including balanced fertility, uniform moisture, and good soil texture).

Fusarium wilt is one of the most important diseases of watermelon. The host range of this pathogen (*Fusarium oxysporum* f. sp. *niveum*) is thought to be restricted to watermelon. The disease may affect plants in all stages of growth. Damping-off is common on young seedlings. Young plants may develop a soft cortical rot that results in chlorotic leaves and stunting. The most common symptom in older plants is wilting. The leaves become flaccid, wither, and turn brown. Gradually the wilting progresses until the entire plant is killed. Vascular discoloration often is associated with wilted vines. This is best seen by slicing vertically through a vine near the soil surface. A yellowish brown discoloration may be seen in tissue directly beneath the outer layer of the vine. A white fungal growth also may be formed at the base of the dead vines. Because the fungus can survive in soil for as long as 10 years, it is often impractical to control the disease through rotation. The best way to control fusarium wilt is through resistant cultivars.

Weeds can be a serious problem because melons are widely spaced, planted late, and develop a poor crop canopy. Cultivation between and within rows is done with cultivation equipment if hills are cross-spaced. Herbicides offer benefits, but are sometimes variable in providing weed control under different seasonal conditions.

Physiological Disorders   Blossom-end rot is a very common disorder in watermelons. The affected fruits are misshapen with a brown, leathery, rotted lesion on the blossom end. This disorder is most prevalent during and following extended dry periods and is caused by insufficient calcium in the fruit. This problem can be reduced by liming the soil with dolomitic lime before planting and applying well-timed irrigations to alleviate prolonged drought periods. Calcium nitrate can also reduce the incidence of blossom-end rot.

Poor flavor and lack of sweetness in the fruit may be due to poor soil fertility (low potassium, magnesium, or boron) and cool temperatures. Additional causes may be wet weather, poorly adapted cultivar, loss of leaves from disease, or harvesting melons before they are ripe. Poor pollination and fruit set of watermelons are caused by wet, cool weather, insufficient bee pollinators, and excessive vegetative growth as a result of closely spaced plants. Split fruits in the field are often caused by heavy rains.

Harvesting

A combination of the following indicators can be used to determine if watermelons are ripe for harvest: (1) the light green curly tendrils on the stem near the point of attachment to the fruit turn brown and dry when the fruit is ripe, (2) the surface color of the fruit turns dull, (3) the skin becomes resistant to penetration by the thumbnail and is rough to the touch, (4) the bottom of the fruit (where it lies on the soil) turns from light green to a yellowish color, and (5) a dull thud (as compared to a metallic sound when the fruit is not ripe) is heard when the fruit is tapped.

A few watermelons of marketable size are often sampled from a field to ensure that the indicator of maturity used was suggestive of full ripeness for that cultivar. Good quality watermelons have a soluble solids reading of at least 10.5% at the center or core of the fruit. The stem is cut so that 1½ in. of the stem is left on the fruit during harvest. Fruits are never pulled from the vines and two to five harvests through a field might be necessary.

FIGURE 25.12 Nutritional value of watermelons. Percent daily values are based on a 2,000 calorie diet. Used with permission from Dole Food Co., Inc.

Postharvest Handling

Watermelons are usually stacked on trailers, trucks, or in pallet boxes in the field. Bruising or rough handling is avoided to minimize damage to the fruit. Freshly harvested watermelons are never stored in direct exposure to the sun (to avoid excessive warming of the fruit) or on their ends (the rind is the thinnest on the ends).

Watermelons do not store for long periods and are best kept for 2 weeks or less. Old watermelon fruit develops a dark orange flesh and has reduced sugar content and flavor with a mealy texture. Refrigeration at 45° to 50°F is optimum for storage of watermelons.

Market Preparation and Marketing

Watermelons are sold to a variety of sources including roadside stands, farmer's markets, or to shippers. The demand for melons is greatest on hot summer days and slows after cooler weather approaches.

Uses

Watermelon is used fresh as a dessert fruit and alone or together with other fruits in fruit salads. The watermelon rind can be pickled as a preserve.

Nutritional Value

Watermelon is low in calories and is a good source of vitamin C and vitamin A. A 2-cup serving size of watermelon has only 80 calories with no grams of fat and 2 g of dietary fiber (Figure 25.12). This serving of watermelon also provides 25% of the

RDA for vitamin C, 20% of the RDA for vitamin A, 4% of the RDA for iron, and 2% of the RDA for calcium.

Muskmelon

Scientific Name

Cucumis melo L. Reticulatus group (muskmelon, Persian melon, or netted melon)
Cucumis melo L. Cantaloupensis group (cantaloupe)
Cucumis melo L. Inodorous group (honeydew, winter, or casaba melon)

Common Names in Different Languages

muskmelon, Persian melon, netted melon (English); wang wen tian gua (Chinese); melon brodé, m. réticulé (French); zuckermelone, Netz-melone (German); melone retado (Italian); netto melon, makuwauri (Japanese); dynja (Russian); melón bordado (Spanish)

cantaloupe (English); ying pi tian gua (Chinese); cantaloup, melon cantaloup (French); Kantaloupe-melone (German); melone cantalupo (Italian); kantarohpu (Japanese); mélon cantaloupe (Spanish)

winter, honeydew, or casaba melons (English); song tian gua (Chinese); melon, m. d'hiver (French); melone (German); melone (Italian); wintaa melon (Japanese); melon (Spanish)

Classification, History, and Importance

Muskmelons (*Cucumis melo* var. *reticulatus*) belong to the Cucurbitaceae or cucurbit family. Other members of the family include cucumber, gourds, watermelon, pumpkin, and squash. The muskmelon is also called cantaloupe in the United States. The name cantaloupe comes from the Italian *cantaluppi*, the name of a former papal residence near Rome where a melon was introduced from America. *Muskmelon* is derived from a Persian word that means "a kind of perfume." *Melon* is derived from the Latin *melopepo* meaning "apple shape melon."

The muskmelon is native to the Near East and probably originated in Persia (Iran). There are numerous botanical varieties of muskmelons including netted melons, cantaloupe melons, winter (casaba) melons, snake or serpent melons, and mango or lemon melons. Technically cantaloupes are only those muskmelons with a rough, warty surface and a hard rind; however, the name cantaloupe is also applied to the netted varieties of muskmelons.

The first record of muskmelons being grown in the United States was made by Columbus during his second voyage in 1494. Cartier reported the presence of muskmelons among Native Americans near the present city of Montreal in 1535.

Muskmelon was a $417.9 million industry in 1997 (see Table 25.1). This was up from $400.8 million in 1996. Per capita consumption of muskmelon has increased from 5.3 lb per person in 1976 to 10.6 lb per person in 1996 (Table 25.6). Of the 118,500 acres of muskmelons grown in the United States, 65,500 acres are in California (Table 25.7). Other large muskmelon-producing states include Arizona (20,500 acres), Texas (12,500 acres), and Georgia (10,000 acres).

Honeydew melons were a $109 million industry in 1997 (see Table 25.1). Per capita consumption of honeydew melons is substantially less than for muskmelons,

TABLE 25.6 *U.S. per capita consumption of melons*

Crop	1976	1986 (lb, farm weight)	1996
Cantaloupe	5.3	9.4	10.6
Honeydew	1.0	2.4	2.3

Source: USDA, Economic Research Service.

TABLE 25.7 *The top cantaloupe producing states in the United States*

Rank	State	Planted acres (000)	Harvested acres (000)	Yield (cst/acre)	Production (000)
1	California	65.50	65.50	220.00	14,410
2	Arizona	20.50	20.50	255.00	5,228
3	Texas	12.50	9.70	140.00	1,358
4	Georgia	10.00	9.00	140.00	1,260
5	Indiana	3.80	3.70	130.00	481
6	Colorado	2.00	1.50	220.00	330
7	Maryland	1.80	1.70	110.00	187
8	Pennsylvania	1.10	1.00	120.00	120
9	Michigan	0.80	0.75	150.00	113
10	Ohio	0.45	0.42	165.00	69
Total	United States	118.45	113.77	207.00	23,556

Source: USDA, NASS, 1997 data.

but has increased from 1.0 lb per person in 1976 to 2.3 lb per person in 1996 (Table 25.6). California leads the nation in acres of honeydew melons grown with 23,200 acres, followed by Arizona with 4,200 acres and Texas with 3,300 acres (Table 25.8).

Plant Characteristics

Height and Spread: If grown vertically on a trellis muskmelons require little room. If grown on the soil surface a plant can cover 30 to 40 sq ft, depending on the cultivar. Bush cultivars will stand 24 in. tall and spread 36 to 48 in.
Root Depth: Roots are usually shallow, but some go down 4 ft.

Climatic Requirements

The muskmelon originated in the hot valleys of Iran (formerly Persia) and prefers a hot, dry climate. The plants require warm weather, ample soil moisture, and a dry atmosphere during the early stages of growth.

Muskmelons are adapted to mean monthly temperatures of 65° to 75°F and require 85 to 120 days from planting to harvest depending on the cultivar and growing conditions. The best-quality melons are produced with hot, dry conditions during the ripening period.

TABLE 25.8   *The top honeydew melon producing states in the United States*

Rank	State	Planted acres (000)	Harvested acres (000)	Yield (cwt/acre)	Production (000)
1	California	23.20	23.20	200.00	4,640
2	Arizona	4.20	4.20	175.00	735
3	Texas	3.30	2.00	210.00	420
Total	United States	30.70	29.40	197.00	5,795

Source: USDA, NASS, 1997 data.

FIGURE 25.13   Eastern type of muskmelon.

Field Preparation

Muskmelons grow optimally on well-drained sandy loam or silt loam soils. Warmth, good drainage, an abundance of available plant food, and plenty of humus are the chief soil requirements for successful growing of muskmelons. Soil pH should be between 6.0 and 6.5. As a general recommendation, 500 lb of 10-10-20 fertilizer per acre with a side-dress of an additional 50 lb of actual nitrogen is often applied.

Selected Cultivars

Muskmelons can be broadly based into eastern and western types. Eastern-type melons are often called muskmelons in the commercial trade and are round, netted, orange-fleshed, sutured, or ribbed melons that are green when immature, but turn yellow when ripe (Figure 25.13). Eastern-type melons have course netting and prominent ribs and are typically utilized for local markets. Western-type melons are often called cantaloupes or shipping types in the trade and are small, round, netted, nonsutured melons with orange flesh. (Figure 25.14). Western

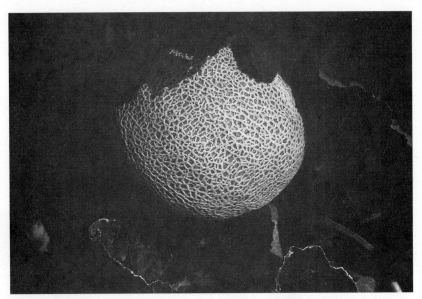

FIGURE 25.14   Western type of muskmelon.

FIGURE 25.15   Honeydew type of melon.

types are green when immature, but turn yellow when ripe and have little or no ribbing. They are suitable for long-distance markets.

There are several specialty melons including honeydew and casaba melons. Honeydew melons are oval-round, creamy white, smooth-skinned, late-ripening melons with a flesh color of green or orange (Figure 25.15). Casaba melons are oval-long, non-netted, corrugated, white-fleshed, late-ripening melons (Figure 25.16).

FIGURE 25.16   Casaba type of melon.

FIGURE 25.17   An established planting of muskmelons.

Planting

Seeds or transplants are used for planting muskmelons. Transplants are generally preferred if an early harvest is desired. Melon plants are not seeded or transplanted in the spring until all danger of frost has passed because they are very sensitive to freezing temperatures. For a direct-seeded crop, the muskmelon seeds are not planted deeper than 1 in. to ensure quick emergence. Rows are typically spaced 5 to 6 ft apart and plants are spaced 2 to 3 ft apart in the row (Figure 25.17). Alternately, hills of two plants or seeds are spaced 3 ft apart on 5-ft centers.

At recommended plant spacings, a muskmelon plant will set one to two fruits. If more than two fruits are set and develop to maturity on a single plant, these fruits are likely to be smaller with lower soluble solids than fruits from plants in which one or two fruits are set and develop to maturity.

Irrigation

On sandy soils, muskmelons generally require ¾ in. of water twice a week. On heavier soils, 1½ in. of water can be applied once a week. Overhead irrigation is most commonly used; however, drip irrigation with plastic is becoming more common. Drip irrigation provides the plants with a more uniform application of water, placing it near the root zone and using less water. Drip irrigation also minimizes the amount of diseases on foliage and fruit as compared with overhead irrigation, and it does not interfere with honeybees and subsequent pollination. If overhead irrigation is used it is typically applied early in the afternoon so the foliage will dry before nightfall. The key time to provide adequate moisture for muskmelons is from the time the fruit begins to form until it is half grown. Typically, if too much water is made available to the plant near harvest it will cause the fruit to crack.

Culture and Care

Melons require bees for satisfactory flower pollination. Growers are careful to apply insecticides when bees are least active, to avoid killing them during the insecticide applications.

The use of black plastic is often practiced when growing muskmelons. The black plastic conserves moisture, controls weeds in the row, and promotes early fruiting.

Common Growing Problems and Solutions

Insects   Cucumber beetles are vectors or carriers of bacterial wilt that affects muskmelons. Bacterial wilt often causes rapid wilting of leaves on a single vine, but the entire plant may wilt and die. Protecting muskmelon plants with insecticides (or row covers until flowering) is necessary because there are no control measures for bacterial wilt once the plants are infected. Other pests of muskmelons are aphids, flea beetles, and melon worms.

Diseases   Muskmelons in the field are affected by several diseases. These include powdery mildew (Figure 25.18), downy mildew (Figure 25.19), alternaria leaf spot, anthracnose, and fusarium wilt. Crop rotation, resistant cultivars, and fungicides are important control measures.

Physiological Disorders   Poorly flavored melons can result from too much water near harvest or not enough boron and magnesium available in the soil.

Harvesting

Fruits of both eastern and western types of muskmelon will separate (slip) easily from the stem when fully mature (the separation occurs at the abscission zone located between the fruit and the stem) (Figure 25.20). This stage of fruit maturity is called "full slip" (Figure 25.21) and at this stage muskmelons are at their highest

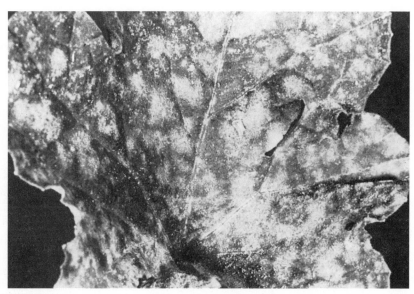

FIGURE 25.18   Muskmelon leaves infected by powdery mildew. Source: Clemson University Extension Service.

FIGURE 25.19   Muskmelon leaves infected by downy mildew. Source: Clemson University Extension Service.

sugar content. If muskmelons are to be kept more than a few days, harvesting at "half slip" is preferred. At half slip only half of the stem separates easily from the melon. Muskmelons for commercial harvest and destined for shipping are generally harvested at half to full slip and will have a sugar content that will range from 8 to 14%. At the peak of harvest, fields are picked every 3 days.

FIGURE 25.20   Muskmelon fruit at full slip stage of maturity will separate easily from the stem.

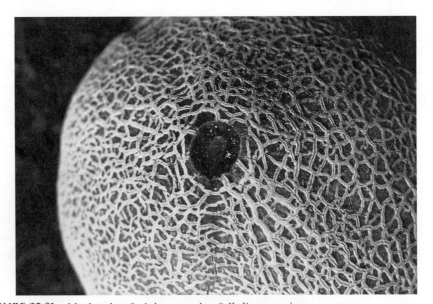

FIGURE 25.21   Muskmelon fruit harvested at full slip maturity.

Honeydew types are green when immature, but turn a creamy white color and begin to produce a sparse netting around the stem. Honeydew melons do not form an abscission zone between the fruit and the stem. Mature honeydew melon fruit will have a sugar content from 10 to 16%. Casaba melons are green when immature and will turn yellow and soften at the blossom end and produce a netting of lines radiating from the stem end when mature.

FIGURE 25.22   Muskmelon must be handled with care after harvest to prevent bruising of the fruit.

Postharvest Handling

The quality of melon fruit peaks with full maturity and deteriorates rapidly afterwards. Cooling is achieved by forced air or by hydrocooling. Cooling rapidly decreases the respiration rate and loss of sugars from the fruit. Muskmelons are stored at 32° to 35°F. Since mature harvested muskmelon fruits are easily sunburned they are typically protected from exposure to direct sunlight.

Market Preparation and Marketing

Muskmelon fruit are handled with care after harvest to prevent bruising and physical damage (Figure 25.22), which could lead to deterioration and loss of marketability. Because muskmelons are often shipped in a firm state to avoid damage they may require storage for a few days at room temperature to enhance fruit softening and for flesh to become juicier.

Water-repellent crates or cartons are used for shipping muskmelons so that the fruit can be top iced for transit. The use of cold forced air moving through the loaded muskmelons also ensures better end-product quality. Muskmelons are typically held in storage at 95% relative humidity.

In grocery stores, muskmelons are often marketed whole or in halves with an overwrapping over the halves. In addition, plastic containers filled with scooped or cut melon and placed in ice are often available.

Uses

Melons are used as a dessert or breakfast fruit or in fruit salads, punches, jellied molds, and similar preparations.

Nutrition Facts

Serving Size 1/4 medium melon (134g)

Amount Per Serving

Calories 50   Calories from Fat 0

% Daily Value*

Total Fat 0g	**0%**
Saturated Fat 0g	**0%**
Cholesterol 0mg	**0%**
Sodium 25mg	**1%**
Total Carbohydrate 12g	**4%**
Dietary Fiber 1g	**4%**
Sugars 11g	
Protein 1g	

Vitamin A 100% • Vitamin C 80%

Calcium 2% • Iron 2%

FIGURE 25.23 Nutritional value of muskmelons. Percent daily values are based on a 2,000 calorie diet. Used with permission from Dole Food Co., Inc.

Nutrition Facts

Serving Size 1/10 medium melon (134g)

Amount Per Serving

Calories 50   Calories from Fat 0

% Daily Value*

Total Fat 0g	**0%**
Saturated Fat 0g	**0%**
Cholesterol 0mg	**0%**
Sodium 35mg	**1%**
Total Carbohydrate 13g	**4%**
Dietary Fiber 1g	**4%**
Sugars 12g	
Protein 1g	

Vitamin A 2% • Vitamin C 45%

Calcium 0% • Iron 2%

FIGURE 25.24 Nutritional value of honeydew melons. Percent daily values are based on a 2,000 calorie diet. Used with permission from Dole Food Co., Inc.

Nutritional Value

Muskmelon is low in calories and is a good source of vitamin A and vitamin C. One medium-sized melon has only 50 calories with no grams of fat and 1 g of dietary fiber (Figure 25.23). This serving of melon provides 100% of the RDA of vitamin A, 80% of the RDA of vitamin C, and 2% of the RDA for both calcium and iron. The aromatic flavor compounds responsible for the characteristic odor of muskmelons are predominantly esters of acetic acid.

Honeydew melons are also low in calories but contain less vitamins than muskmelons. One medium-sized honeydew melon has 50 calories, no grams of fat, and 1 g of dietary fiber (Figure 25.24). This serving of honeydew melon has 45% of the RDA of vitamin C and 2% of the RDA for both vitamin A and iron. Honeydew melons are not a source of calcium.

Squash

Scientific Name

Cucurbita pepo L. (summer squash, courgette, zucchini, marrow)
Cucurbita argyrosperma Huber (pumpkin, cushaw pumpkin, winter squash)
Cucurbita moschata Duch. ex Poir (musky pumpkin, squash or gourd, butternut squash)

Common Names in Different Languages

summer squash, courgette, zucchini, marrow (English); meiguo nan gua (Chinese); courgette, courge pépon (French); Moschuskurbis, Bisakurbis (German); Garten-Kürbis, Zucchini, zuccheta (Italian); pepokabocha (Japanese); tykva tverdokoraja (Russian)

pumpkin, cushaw pumpkin, winter squash (English); mo xi ge gua yü-kwa (Chinese); potiron (French); Riesen-Kurbis (German); zucca gigante (Italian); zucca gigante (Japanese); tykva krupnoplodnaja (Russian); zapallo (Spanish)

musky pumpkin, squash or gourd, butternut squash (English); zhong guo (Chinese); courge muscarde, c. musquec, giraumon (French); Moschuskürbis, Bisakurbis (German); zucca moscada, z. muschiata, z. torta (Italian); nihom kabocha (Japanese); tykva krupnoplodnaja (Russian); calabaza moscada, c. almisclada, c. almiscarada, c. amelonada (Spanish)

Classification, History, and Importance

Most economically important squashes are of three species of *Cucurbita* and belong to the Cucurbitaceae. Other members of the family include cucumber, watermelon, pumpkin, and muskmelon. Summer squashes, *Cucurbita pepo,* are harvested as immature fruit that have a soft skin and are very perishable (1- to 2-week shelf life). Winter squashes, *Cucurbita argyrosperma* and *Cucurbita moschata,* are typically harvested when the fruit is mature and the rind is hard. Winter squashes store well and are usually baked. Winter squashes usually grow on vines, but some are bush types. Included in the winter squash category are hubbard, butternut, gold nugget, acorn, and buttercup.

It is suspected that *Cucurbita pepo* was growing prior to Columbus's voyages to the New World in what is now Mexico City northward into the southwestern United States. *Cucurbita moschata* was more widespread with its center of origin south through the Andes to Peru and north into the southwestern United States.

Plant Characteristics

Summer squash
Height: 30 to 40 in.
Spread: 12 to 16 sq ft
Root Depth: The taproot is usually confined to the upper 2 ft, but can grow 6 ft deep.

Winter squash
Height: 12 to 15 in.
Spread: 12 to 20 sq ft, depending on cultivar

Climatic Requirements

Summer squashes are tender, warm-season vegetable crops requiring 40 to 50 days from planting to harvest. Fruits are harvested only a few days after bloom. Summer squash is grown during the frost-free season and is not tolerant of freezing temperatures. It is adapted to monthly mean temperatures of 60° to 80°F. Summer squash is typically not planted until danger of frost is past and optimum soil temperature for germination is about 85°F with little or no germination occurring below 60°F.

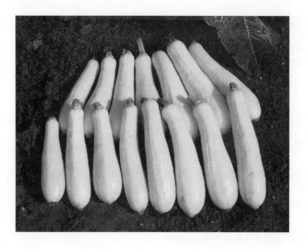

FIGURE 25.25   Straightneck yellow summer squash type. Used with permission from Seminis Vegetable Seed Co., Inc.

The use of plastic mulch, row covers, row orientation, and/or transplants can be utilized to overcome cold-temperature conditions and produce an earlier crop.

Winter squashes are warm-season crops requiring 85 to 120 days from planting to maturity. Acorn squash is one of the more popular winter or hard-shelled types of squash. Acorn squash fruit looks much like an acorn with deep ridges and a shape that is sharply tapered to a point. Acorn squash fruit typically weighs between 1 and 3 pounds.

Field Preparation

Well-drained sandy loam soils with a high organic matter content and a pH of 6.0 to 6.5 are generally the most productive for growing squash. Lighter sandy soils that warm rapidly are often chosen for early spring production. Typically fields are chosen for planting to squash that have not had squash, cucumbers, watermelons, or cantaloupes grown in them for at least 3 years in an effort to reduce the likelihood of soilborne squash diseases.

Fall plowing is often desirable in many areas to eliminate plant debris and residue in which diseases or insects can overwinter. Fields are disked thoroughly prior to planting in the spring. Rows are generally prepared 7 to 10 days in advance and raised beds are formed immediately prior to planting. Bed forming improves the uniformity of the seedbed, thus resulting in more uniform plant stands.

Sixty to 80 lb each of nitrogen, phosphorus, and potassium is generally sufficient on most soils. One-half of the fertilizer is often applied at planting with a subsequent side-dressing of fertilizer when the plants are 8 to 10 in. high (3 to 4 weeks after planting).

Acorn squash grows on a wide variety of soil types, but it will not tolerate poor drainage. Best yields are obtained on sandy soils with a pH between 6.0 and 7.0.

Selected Cultivars

Summer squash types may be divided into the following four groups:

1. Straightneck types are yellow bottle-shaped squash with straight necks (Figure 25.25).

FIGURE 25.26   Crookneck yellow summer squash type. Used with permission from Seminis Vegetable Seed Co., Inc.

FIGURE 25.27   Zucchini type of summer squash. Used with permission from Seminis Vegetable Seed Co., Inc.

2. Crookneck types are yellow bottle-shaped squash with curved necks (Figure 25.26).
3. Scallop types are flattened and ridged around the edge and may be white, green, or striped.
4. Zucchini and cocozelle are vegetables introduced from Italy. Green and yellow types are available, but the green cylindrical zucchinis (Figure 25.27) are the most popular.

FIGURE 25.28   Butternut type of winter squash.

FIGURE 25.29   Acorn type of winter squash. Used with permission from Seminis Vegetable Seed Co., Inc.

Of the summer squashes, the straightneck and zucchini types are most widely grown for commercial markets. Certain yellow squash cultivars contain the precocious yellow gene, which results in a clear yellow stem rather than the traditional green stem. Cultivars with the precocious yellow gene also retain their normal yellow coloring longer when infected with virus(es).

Winter squashes include hubbard, butternut (Figure 25.28), acorn (Figure 25.29), and buttercup types. Acorn squash is available in open-pollinated and hybrid types with vines that are spreading (running types) or compact in growth habit (bush types). The bush types are planted more closely than the running types. The following are some commercially important cultivars of each type.

FIGURE 25.30   Yellow squash grown in the field exhibits a bush-type growth form.

Summer squash

- Straightneck types—Early Prolific, Straightneck, Seneca Prolific
- Crookneck types—Golden Summer, Crookneck, Sundance
- Zucchini types—Black Magic, Aristocrat, Elini
- Scallop types—Patty Pan Hybrid, Peter Pan

Winter squash

- Hubbards—Blue Hubbard, Improved Green Hubbard
- Butternut—Early Butternut, Waltham Butternut
- Gold Nugget—Gold Nugget
- Acorn—Jersey Golden Acorn, Table King, Table Queen, Table Ace, Royal Acorn
- Buttercup—Sweet Mama, Buttercup

Planting

Summer squash seed is generally planted 1 to 1.5 in. deep in rows 36 to 48 in. apart. Final stands are thinned to 18 to 28 in. between plants. If plastic mulch is used, double rows per bed are planted with plants spaced 18 by 18 in. Successive plantings can be made to provide a longer supply of squash during the harvest season. Summer squash fruits grow on bush-type plants that do not spread (Figure 25.30) as is typical of the plants of many of the winter squashes and pumpkins.

Winter squashes such as acorn squash can be planted in 36- to 48-in. rows with 12 to 24 in. between plants. Vining types of squash are planted in 60- to 100-in. rows with 36 to 60 in. between plants in the row. The bush types are frequently higher yielding than the vining types. Summer squash seeds will not germinate until soil temperature is above 68°F; therefore, they are planted to the field well after any potential exposure to frost.

Irrigation

Summer squash plants are irrigated with moderate amounts of water since they root only 3 to 4 ft deep and have many shallow roots. Winter squash plants root deeper and require less frequent irrigation. Moisture stress during fruit development of summer squash can seriously reduce yield. Water is applied early in the day when overhead irrigation is used so that leaves can dry before nightfall.

Culture and Care

The use of black plastic mulch in squash production can enhance earliness of fruit harvest by 7 to 14 days. When plastic is combined with drip irrigation, yields can be doubled or tripled as compared with squash produced on soil without plastic. Row covers increase air temperatures around the plants resulting in earlier fruiting. Earliness of fruiting during the growing season can also be enhanced by using transplants.

Bees are essential for squash flower pollination. If bees are not readily noticeable in the field, growers will typically place one active colony of bees per 1 to 2 acres planted in squash.

Squash plants develop and shade out many weeds. Some weeds such as pigweed, lamb's-quarter, cocklebur, and ragweed can be troublesome if not controlled.

Common Growing Problems and Solutions

Insects   Cucumber beetles attack squash at all stages. Special emphasis on their control is especially important during the seedling stage. The cucumber beetle also carries the disease that causes bacterial wilt. Fall squash is often attacked by pickleworms.

Diseases   Root-knot nematode is a serious pest and soil fumigation is often necessary if nematodes are in the soil. Foliage diseases, mostly powdery and downy mildews, are present in all seasons. These diseases are often worse on a fall crop, requiring fungicide sprays beginning at first sign of disease. Mosaic virus disease can also be serious in the fall. This disease is best delayed by reducing aphid populations that spread the disease and using yellow squash cultivars that have the precocious yellow gene.

Harvesting

Summer squash fruits are harvested while they are tender and have a shiny or glossy appearance (Figure 25.31). Actual desired size of fruit at harvest will depend on the market. Quality of fruit is lost rapidly as the shiny color changes to a dull color and the seeds within the fruit develop hard coats. Summer squash is harvested every other day and sometimes every day under optimum growing conditions.

A short piece of the stem is left attached to the fruit when summer squash is harvested. It is best to cut the squash fruit from the vine rather than pulling it from the

FIGURE 25.31   Yellow squash harvested at proper stage of maturity.

stem. Cotton gloves are sometimes used when harvesting summer squash to avoid scratching and puncturing the fruit. The first summer squash harvested is generally the best quality. A summer squash planting may be harvested for 2 to 3 weeks before the harvesters typically move on to a later sequential planting. This will keep quality of harvested fruit for marketing at its best and will allow a longer marketing season.

Crookneck and straightneck cultivars are harvested when the fruits are 1.25 to 2 in. in diameter. Crookneck cultivars of summer squashes are more difficult to pack than straightneck cultivars and are easily broken at the neck during harvest. Zucchini fruits are harvested when they are 7 to 8 in. long and scallop types are harvested when they are 3 to 4 in. in diameter. Defective and large summer squash fruits are removed and not allowed to remain on the plant since large fruit reduces subsequent fruiting.

Winter squashes are harvested when mature and are normally harvested in one or two pickings. The skin is hard and resists denting by thumbnail pressure when the squash fruits are mature. Winter squash can remain in the field through a light vine-killing frost but should not remain in the field during a hard freeze. The stem of winter squashes is typically cut, leaving a stub attached to the fruit. The harvested fruits are handled carefully to avoid damaging and bruising the skin.

Postharvest Handling

Summer squashes are very tender and are not normally stored except to accommodate holidays, weekends, and so on. They can be held 3 to 4 days at 32° to 40°F with a relative humidity of 90%. Deterioration is rapid after such storage.

Acorn squashes are not washed before packing. Acorn squash can be packed in single- or double-layered boxes and held at 55°F and 70% humidity for up to 5 weeks.

Winter squash can be stored for later marketing. If stored, they should be cured to harden the shell, ensure maturity, and heal any cuts and bruises. Winter squashes

Nutrition Facts

Serving Size 1/2 medium squash (98g)

Amount Per Serving	
Calories 20	Calories from Fat 0
	% Daily Value*
Total Fat 0g	**0%**
Saturated Fat 0g	**0%**
Cholesterol 0mg	**0%**
Sodium 0mg	**0%**
Total Carbohydrate 4g	**1%**
Dietary Fiber 2g	**8%**
Sugars 2g	
Protein 1g	
Vitamin A 6%	• Vitamin C 30%
Calcium 2%	• Iron 2%

FIGURE 25.32 Nutritional value of summer squash. Percent daily values are based on a 2,000 calorie diet. Used with permission from Dole Food Co., Inc.

Nutrition Facts

Serving Size 1/2 cup (78 grams)

Amount Per Serving	
Calories 25	Calories from Fat 0
	% Daily Value*
Total Fat 0g	**0%**
Saturated Fat 0g	**0%**
Cholesterol 0mg	**0%**
Sodium 15mg	**1%**
Total Carbohydrate 4g	**1%**
Dietary Fiber 1g	**4%**
Sugars 2g	
Protein 0g	
Vitamin A 0%	• Vitamin C 2%
Calcium 2%	• Iron 2%

FIGURE 25.33 Nutritional value of spaghetti squash. Percent daily values are based on a 2,000 calorie diet. Used with permission from Dole Food Co., Inc.

are cured for 10 days at 80° to 85°F and 85% relative humidity. Winter squashes are subsequently stored at 50° to 55°F and 50 to 75% relative humidity. Acorn squash can be stored for up to 2 months and butternut squash for 2 to 3 months.

Market Preparation and Marketing

Summer squash is usually precooled prior to shipping and requires a holding temperature at 45° to 50°F at 90 to 95% relative humidity. Summer squash fruits are graded for uniformity and appearance and packed in cardboard containers for marketing.

Winter squash fruit is also handled carefully to avoid damage to the skin or fruit. Winter squash are loaded into large bulk boxes or bulk trucks for marketing. Supermarkets may sell squash sections, prepeeled chunks, or whole fruit.

Uses

Squashes are boiled, baked, stewed, or steamed. They can be served in meat casseroles, soufflés, pancakes, custards, breads, cakes, or pies. Gourds can also be used for decorative purposes.

The flowers of squash can be dipped in a light mixture of flour and water, fried and eaten as a vegetable. Butterblossom is a squash cultivar specifically designed for producing squash blossoms to be eaten. The plant produces a large number of male flowers that should be harvested when they are 8 to 10 in. long. Female flowers can also be used but the center pistil is removed.

Nutritional Value

Summer squash is low in calories and is a good source of vitamin C. One-half of a medium-sized squash has only 20 calories with no grams of fat and 2 g of dietary fiber (Figure 25.32). This serving of summer squash provides 30% of the RDA of vitamin C, 6% of the RDA of vitamin A, and 2% of the RDA for both calcium and iron. Spaghetti squash, which is a winter squash, is also low in calories but has less vitamin C and vitamin A than summer squash (Figure 25.33). Winter squashes are higher in stored carbohydrates than summer squashes and are high in provitamin A.

Pumpkin

Scientific Name

Cucurbita maxima Duch. (pumpkin, giant pumpkin)
Cucurbita argyrosperma Huber (pumpkin, cushaw pumkin, winter squash)
Cucurbita moschata Duch. ex Poir (musky pumpkin, squash or gourd, butternut squash)

Common Names in Different Languages

pumpkin, giant pumpkin (English); yin du nan gua yŭ-kwa (Chinese); potiron, gross courge, giraumon (French); Riesen-Kürbis, Zentnerkürbisse (German); zucca gigante (Italian); seiyou, seiyou-kabocha (Japanese); tykva krupnoplodnaja (Russian); calabaza grande, c. gigante, c. de Castilla, zapallo (Spanish)

pumpkin, cushaw pumpkin, winter squash (English); mo xi ge gua yŭ-kwa (Chinese); potiron (French); Riesen-Kürbis (German); zucca gigante (Italian); zucca gigante (Japanese); tykva krupnoplodnaja (Russian); zapallo (Spanish)

musky pumpkin, squash or gourd, butternut squash (English); zhong guo (Chinese); courge muscarde, c. musquec, giraumon (French); Moschuskürbis, Bisakürbis (German); zucca moscada, z. muschiata, z. torta (Italian); nihom kabocha (Japanese); tykva krupnoplodnaja (Russian); calabaza moscada, c. almisclada, c. almiscarada, c. amelonada (Spanish)

Classification, History, and Importance

Three species of *Cucurbita* (*C. maxima, C. argyrosperma, C. moschata*) make up the majority of the crops that are considered pumpkins. All pumpkin species are members of the Cucurbitaceae family and this group is categorized by plants with sprawling vines and separate male and female yellow flowers produced on the same plant. Other members of the family include cucumber, gourds, watermelon, muskmelon, and squash.

Pumpkins are thought to be of American origin and possibly were grown in Peru as early as 2000 B.C. Pumpkins were used by Native Americans prior to Columbus's voyages to the New World and were part of the first Thanksgiving. Pumpkins were used by early settlers for food and decoration, very similar to how we use them today.

The word pumpkin is derived from a French word *popon*, which originated from the Greek word *pepon*, which means "large melons." By common usage, *Cucurbita* species that have fruits that are round and orange are called pumpkins, while those that have fruits of other shapes and colors are called squash (or fall or winter squash).

Plant Characteristics

Height: 18 to 24 in.
Spread: One plant can spread over 50 to 100 sq ft, depending on the cultivar.
Root Depth: Most of the roots are in the top 12 in. of soil, but the taproot can grow 2 to 3 ft deep.

Climatic Requirements

Pumpkins are warm-season vegetables that require relatively long frost-free growing periods. The optimum soil temperature for germination is 70° to 95°F. Soil temperatures below 56°F suppress germination and optimum growing temperatures are between 64° to 75°F.

Field Preparation

Fields that have no perennial weed problems and have good soil drainage are suitable for growing pumpkins. Long rotations of 3 to 4 years between crops of pumpkins are often practiced. The warmer sandy soils are used for early production. Pumpkins in general require from 75 to 100 lb of total nitrogen. Applications of 40 to 50 lb/acre of nitrogen at planting time with a side-dress of an additional 20 to 40 lb/acre applied along the rows when the vines start to run (6 to 12 in. long) is often done. An ideal soil pH for pumpkins is 6 to 6.5. Since the planting season for pumpkins is late in the season, a final tillage just prior to planting is used to kill any late-germinating weeds.

Selected Cultivars

Many new cultivars of pumpkins are hybrids and are earlier fruiting. Another breeding improvement of pumpkins is the incorporation of an "early coloring" gene so that the fruit turns orange earlier in its development. Cultivars of pumpkins with the earlier coloring characteristic have a somewhat duller finish and a poorer stem than conventional cultivars, but improvements are being made. Common examples of this type are Autumn Gold and Big Autumn.

Growers supplying wholesale markets require pumpkins that have strong, dark-colored handles (stems), a deep or bright color, and heavy weight with a consistent size. Growers for local markets or roadside stands may tolerate more variation in size, finish, and stems. Growers for specialty markets may want pumpkins in unique sizes, shapes, or colors.

One of the major differences in pumpkin cultivars is in the size of the fruit. There are five general categories of pumpkins based on fruit size: giant > 20 lb, jack-o'-lantern 7 to 20 lb, small 5 to 7 lb, baby 2 to 4 lb, and miniature < 1 lb. The predominant market for pumpkins is for jack-o'-lantern types. Smooth-skinned pumpkins are preferred for painting or coloring and white or dark orange-skinned cultivars are novelty types that are increasing in popularity with the consumers.

Planting

Since the pumpkin market begins in late September to early October, the planting season for most cultivars in many areas is in early June. Pumpkins are usually grown in rows 12 to 15 ft apart with plants spaced 2 to 4 ft in the row. Closer in-row spacing encourages faster vine coverage in the row, which can be useful for weed control. Seeds are placed ¾ to 1 in. deep into moist but not wet soil. Watering after planting encourages quick and even emergence in dry weather.

Irrigation

Pumpkins grow with limited amounts of water. However, watering after planting to activate herbicides and encourage rapid, even emergence may be necessary in dry periods. A critical period for water also occurs during blooming and early fruit set. Drip irrigation can be used for pumpkins to reduce foliage wetting and for water use efficiency.

Culture and Care

Weed control is the most critical management practice for growing pumpkins. Weeds can become a serious problem since pumpkins are planted late when many annual broadleaf weeds are rapidly germinating, and the vining habit makes cultivation difficult later in the season. The use of herbicides combined with mechanical cultivation is the best weed-control program. Pumpkins are cultivated just prior to vining (before the vines lay over and start to crawl along the ground) and a disk or field cultivator is used to till row middles until the vines cover the area. Shallow tillage rather than deep cultivation is recommended to conserve moisture and prevent damage to pumpkin roots that may occur with deeper cultivation.

Common Growing Problems and Solutions

Diseases   Powdery mildew causes a white mold on the upper surface of pumpkin leaves. Two powdery mildew fungi, *Erysiphe cichoracearum* and *Sphaerotheca fuliginea*, can infect pumpkins. Warm daytime temperatures, cool nights, and a high relative humidity are environmental conditions that favor the development of powdery mildew on the leaves. The fungus does not require a film of moisture on the leaf surface for infection and can be a problem even during periods of little or no rain. Fungicides are needed to prevent powdery mildew because cultivars currently being grown have little to no resistance.

The black rot fungus (*Mycosphaerella melonis*) attacks all parts of the pumpkin plant. Small, brown to black spots form on the leaves and vines and vine lesions or cankers often crack and form a brown gummy fluid. The fungus produces numerous pinpoint-sized black fruiting structures on decaying vines. The fungus also attacks the fruit and causes multiple light tan to black circular spots.

Several other fungi including *Fusarium* and *Phytophthora* may also cause significant pumpkin fruit damage during wet summers. Conditions for these fungi are generally excessive rain, poorly drained soils, and continuous planting of pumpkins in the same field. A 3-year rotation with non-cucurbit crops is followed and poorly drained soils are avoided for planting to pumpkins to control black rot and other fruit rots. Fungicides are also applied to the plant to suppress injury from the black rot fungus.

Viruses that may affect pumpkin growth include cucumber, squash, watermelon, and zucchini yellow mosaic. Disease symptoms will vary but generally the plants are stunted. Younger leaves may turn yellow or mottled and are often distorted. Older plants develop a distinct mottling of leaves and fruit. Plants severely affected fail to produce fruit. All of the viruses (except squash mosaic virus, which is vectored by cucumber beetles) are transmitted from plant to plant by aphids. Also, certain viruses (squash mosaic and cucumber mosaic) may be seedborne. Control of viruses is difficult because the use of insecticides on pumpkins to suppress aphid populations has not resulted in effective control of the spread of the viruses. However, weed control in and around the pumpkin planting can suppress aphid populations and decrease viral infections. Reflective mulches can also reduce aphid feeding and lower the incidence of virus diseases.

Insects   Cucumber beetles attack pumpkin seedlings, vines, and both immature and mature fruit. Striped cucumber beetles are insects with chewing mouthparts. These beetles overwinter under debris in and around fields. Adult beetles appear early in the season, often feeding on various alternate host plants. Eventually, the beetles are attracted to newly planted fields. They can burrow into the soil and destroy plants before they break the surface and they can also feed on and kill newly emerged seedlings. Cucumber beetles are colored with a black head and antennae, straw yellow thorax, and yellowish wing covers with three distinct parallel and longitudinal black stripes. The use of insecticides is essential for protecting plants against direct feeding. Barriers (floating row covers) can be utilized to exclude beetles from newly emerging seedlings and young plants. Row covers are impractical under large acreages.

Squash bugs utilize their sucking mouthparts to remove plant juices from pumpkin plants. Adult squash bugs move to plants from various adjacent, protected overwintering sites. Adult females deposit brownish red eggs in clusters on a lower leaf surface. High squash bug populations can drain plants causing them to die and wilt. Reduced yields and poor-quality fruit may result from squash bug-feeding activities. Effective squash bug control is contingent upon timely insecticide spraying. It is also essential to reduce overwintering populations by disking and/or removing foliage and fruit immediately after harvest.

Squash vine borers are the larvae of a clear-winged moth whose size, shape, and flying habits are somewhat similar to wasps for which they are often mistaken. Squash vine borers overwinter as larvae or pupae in cocoons buried in the soil. Larvae bore into the stems where they tunnel and feed. Mature larvae exit stems and then burrow into the soil. Due to extensive disruption of conducting tissues, plants often wilt and die. Holes in the stems and an accompanying ooze signal the presence of squash vine borers. Large white worms with brown heads can be seen if the stems are cut open. Insecticidal controls must be implemented before larvae bore into plants.

Aphids are detrimental because of their ability to transmit several virus diseases. Watermelon mosaic, cucumber mosaic, and zucchini mosaic viruses are transmitted by aphids entering fields. Aphids land on plants and utilize their stylets to probe plant tissue. No insecticide kills aphids rapid enough to prevent disease transmission. Rather, aphicides are useful for preventing buildup or reducing already excessive populations.

Physiological Disorders   Poor fruit set is commonly caused by plants with inadequate spacing between plants. Pumpkins require bees for pollination similar to other members of the cucurbit family. Pollinating insects are required to transfer pollen from male to female flowers for fruit set to occur. Excessive nitrogen applications may prevent development of fruit.

FIGURE 25.34   Pumpkins in the field ready for harvest. Courtesy of M. Peet.

Harvesting

Pumpkins are ready for harvest when the rind or skin has toughened and the stems have lost their succulence (Figure 25.34). A long and undamaged stem handle is desirable for marketing of pumpkins. "Dripping" from the stems where the stems are cut from the vine can be minimized if the stems are cut with pruners or clippers instead of being separated by pulling. Pumpkins are usually windrowed into piles or lines in the field so that trailers or trucks can be driven through the field to load the pumpkins. A light freeze before harvest is not damaging; however, temperatures in the mid to low 20s may be injurious to the plant.

Postharvest Handling

Pumpkins can be stored in a well-ventilated, cool (50° to 55°F) location for 1 to 2 months if necessary. Most growers harvest just prior to market delivery or sales. Pumpkins can be cured at 80° to 85°F with high humidity for 7 to 10 days if longer than 1 month of storage is anticipated. Fruit molds and rots can be reduced by surface sterilization with a bleach and water mixture (1 part bleach to 4 parts water).

Pumpkins are usually handled in bulk or loaded into bulk bins directly from the field. The appearance of some miniature pumpkins and gourds may be improved by dipping or spraying them with shellac. After dipping in shellac, the miniature pumpkins and gourds are allowed to dry before handling.

Harvested pumpkins are handled carefully to avoid cuts and bruises on the fruit. Fruits that are not fully mature or that have been injured or subjected to heavy frost will not store well.

Market Preparation and Marketing

Wholesale sales of pumpkins often begin in mid to late September to supply most retail markets (Figure 25.35) since their active sales of pumpkins begin in early

FIGURE 25.35   Pumpkins on display at a retail stand. Courtesy of M. Peet.

October. The traditional sales period begins in mid-October, culminating with the Halloween season in late October.

Uses

In addition to being used as jack-o'-lanterns at Halloween, pumpkins are also used for ornamental and culinary purposes. The seeds of naked-seeded cultivars do not have seed coats and can be roasted in the oven or sautéed for snacks. Small types of pumpkins are grown primarily for cooking or for pies, and the intermediate and large types are for ornamental purposes.

Review Questions

1. What are the differences between fresh market and processing cultivars of cucumbers?
2. When is the most critical water need for cucumbers?
3. What can cause the bitter taste in cucumbers?
4. Why are cucumber fruits often waxed prior to shipping?
5. When is an appropriate time to plant watermelons in the spring?
6. Why is watering avoided when watermelon fruits are near full size?
7. What are some of the harvest indicators used to determine when watermelon fruits are mature and ready for harvest?
8. What are the differences between eastern and western types of muskmelons?
9. Define *half slip* and *full slip* as they pertain to muskmelon harvesting.
10. What are some of the differences between summer squash and winter squash?
11. What are the five general categories of pumpkins and what are the criteria used to separate pumpkins into these categories?
12. Why are weeds often a problem in the growing of pumpkins?

Selected References

Anderson, C. R. 1998a. *Pumpkins.* Univ. of Arkansas Coop. Ext. Serv. Publ.

Anderson, C. R. 1998b. *Summer squash.* Univ. of Arkansas Coop. Ext. Serv. Publ.

Anderson, C. R. 1998c. *Watermelons.* Univ. of Arkansas Coop. Ext. Serv. Publ.

Anonymous. 1984. *Modern cucumber technology.* Kalamazoo, MI: Asgrow Seed Company.

Ball, J. 1988. *Rodale's garden problem solver: Vegetables, fruits, and herbs.* Emmaus, PA: Rodale Press. 550 pp.

Cook, W. P., C. E. Drye, and R. P. Griffin. 1985. *Growing watermelons in South Carolina.* Clemson Univ. Coop. Ext. Serv. Bull. 121.

Cook, W. P., R. P. Griffin, and A. P. Keinath. 1995. *Commercial cucumber production.* Clemson Univ. Coop. Ext. Serv. Hort. Leaflet 34.

Johnson, J. R., J. W. Rushing, R. P. Griffin, and C. E. Drye. 1987. *Commercial squash production.* Clemson Univ. Coop. Ext. Serv. Hort. Leaflet 14.

Kays, S. J., and J. C. Silva Dias. 1995. Common names of commercially cultivated vegetables of the world in 15 languages. *Economic Botany* 49:115–152.

Latin, R. X. 1984. *Reference guide for melon disease control.* Purdue Univ. Coop. Ext. Serv. Publ. BP-7.

Lindstrom, R. K., J. W. Courter, and J. M. Gerber. 1983. *Commercial muskmelon production in Illinois.* Illinois Coop. Ext. Serv. Publ. VC-30-83.

Marr, C. 1994. *Cucumbers and melons.* Kansas State Univ. Hort. Report MF-668.

Marr, C. W., and N. Tisserat. 1994. *Watermelon.* Kansas State Univ. Coop. Ext. Serv. Publ. MF-1107.

Marr, C., T. Schaplowsky, N. Tisserat, and B. Bauernfeind. 1995. *Pumpkins.* Kansas State Univ. Hort. Report MF-2030.

Motes, J., J. Edelson, W. Robberts, J. Duthie, and J. Damicone. 1994. *Squash and pumpkin production.* Oklahoma State Univ. Coop. Ext. Serv. Publ. F-6026.

Nagel, D. H. 1997a. *Commercial production of acorn squash in Mississippi.* Mississippi State Univ. Coop. Ext. Serv. Publ. 1505.

Nagel, D. H. 1997b. *Commercial production of cantaloupes in Mississippi.* Mississippi State Univ. Coop. Ext. Serv. Publ. 1518.

Nagel, D. H. 1997c. *Commercial production of pumpkins in Mississippi.* Mississippi State Univ. Coop. Ext. Serv. Publ. 1515.

Nagel, D. H. 1997d. *Commercial production of summer squash in Mississippi.* Mississippi State Univ. Coop. Ext. Serv. Publ. 1517.

Nagel, D. H., P. Harris, F. Killebrew, and J. Fox. 1997. *Commercial production of watermelons.* Mississippi State Univ. Coop. Ext. Serv. Publ. 1533.

Nonnecke, I. L. 1989. *Vegetable production.* New York: Van Nostrand Reinhold. 656 pp.

Paris, H. S. 1996. Summer squash: History, diversity, and distribution. *HortTechnology* 6:6–13.

Peirce, L. C. 1987. *Vegetables: Characteristics, production, and marketing.* New York: John Wiley & Sons.

Pew, W. D., B. R. Gardner, P. D. Gerhart, and T. E. Russell. 1981. *Growing cantaloupes in Arizona.* Univ. of Arizona Coop. Ext. Serv. Bull. A86-R.

Sackett, C. 1975a. Persians. *Fruit & Vegetable Facts & Pointers.* United Fresh Fruit & Vegetable Association.

Sackett, C. 1975b. Watermelons. *Fruit & Vegetable Facts & Pointers.* United Fresh Fruit & Vegetable Association.

Schultheis, J. R. 1993a. *Cucumbers for fresh market.* North Carolina Coop. Ext. Serv. Leaflet 14.

Schultheis, J. R. 1993b. *Pickling cucumbers.* North Carolina Coop. Ext. Serv. Leaflet 14-A.

Schultheis, J. R. 1993c. *Summer squash production.* North Carolina Coop. Ext. Serv. Leaflet 24-A.

Schultheis, J. R. 1995a. *Commercial watermelon production.* North Carolina Coop. Ext. Serv. Leaflet 30.

Schultheis, J. R. 1995b. *Growing pumpkins and winter squash.* North Carolina Coop. Ext. Serv. Leaflet 24.

Schultheis, J. R. 1995c. *Muskmelons (cantaloupes).* North Carolina Coop. Ext. Serv. Leaflet 8.

Seelig, R. A. 1967. Honey dews. *Fruit & Vegetable Facts & Pointers.* United Fresh Fruit & Vegetable Association.

Seelig, R. A. 1972. Cucumbers. *Fruit & Vegetable Facts & Pointers.* United Fresh Fruit & Vegetable Association.

Seelig, R. A. 1973. Cantaloupes. *Fruit & Vegetable Facts & Pointers.* United Fresh Fruit & Vegetable Association.

Yamaguchi, M. 1983. *World vegetables.* New York: Van Nostrand Reinhold.

Selected Internet Sites

USDA Agricultural Marketing Service Standards for Grades

Fresh
Cantaloupes   www.ams.usda.gov/standards/canatalop.pdf
Cucumbers   www.ams.usda.gov/standards/cucumber.pdf
Cucumbers (greenhouse)   www.ams.usda.gov/standards/cucumgre.pdf
Honeydew melons   www.ams.usda.gov/standards/honeydew.pdf
Squash (fall and winter types
    & pumpkin)   www.ams.usda.gov/standards/squapumk.pdf
Squash (summer)   www.ams.usda.gov/standards/squshsum.pdf
Watermelons   www.ams.usda.gov/standards/watermelo.pdf

Processing
Cucumbers (pickling)   www.ams.usda.gov/standards/vpcucum.pdf

Index